Twin Cities

T0187792

This dynamic international collection provides a comprehensive overview of twin cities on administrative and international borders across the world. Drawing on contemporary and historical examples, it documents constant and changing features of twinned communities over time.

The chapters explore a variety of urban formations including independent cities located side-by-side; cities that have merged over decades or even centuries and those projected to merge; cities partitioned by treaties and cities duplicated in pursuit of better security, intensified trade or both between neighbouring countries. From Europe to Africa, North America to the Middle East, South America to Asia, this book focuses on relationships between cities, citizens and municipal/international borders. A cartographical contents and editorial commentary guides readers through diverse contributions. The authors ask how far cities are changing or remaining constant in the context of conurbanisation, Europeanisation and globalisation. The book provides a glimpse into the variety of roles twin cities can play globally: from laboratories of integration and para-diplomatic actors to economic and cultural brokers.

This is a valuable, engaging resource for researchers in the fields of geography, urban studies, border studies, international relations and global development. It will be of great use to individuals involved in twin-city initiatives and general readers.

John Garrard was Senior Lecturer in Politics and Contemporary History at the University of Salford until 2011. Although primarily a historian, his central teaching and research interests have lain along the borders with political science, with a particular interest in urban history/politics. He has written about the politics of immigration, power in nineteenth-century towns, democratisation, the nature and politics of scandal, and heads of local government.

Ekaterina Mikhailova is a research fellow at the Faculty of Geography, Lomonosov Moscow State University. In addition she currently is working in the project 'Transformation of Soviet Republic Borders to International Borders' at the University of Eastern Finland. Ekaterina is author of over thirty Russian and English articles on twin cities, cross-border communities and cross-border cooperation. Her research interests include sustainable development of border regions and border cities, cross-border integration and governance.

Global Urban Studies
Series Editor: Laura Reese
Michigan State University, USA

Providing cutting edge interdisciplinary research on spatial, political, cultural and economic processes and issues in urban areas across the USA and the world, volumes in this series examine the global processes that impact and unite urban areas. The organising theme of the book series is the reality that behaviour within and between cities and urban regions must be understood in a larger domestic and international context. An explicitly comparative approach to understanding urban issues and problems allows scholars and students to consider and analyse new ways in which urban areas across different societies and within the same society interact with each other and address a common set of challenges or issues. The books in this series cover topics that are common to urban areas globally, yet illustrate the similarities and differences in conditions, approaches and solutions across the world, such as environment/brownfields, sustainability, health, economic development, culture, governance and national security. In short, the Global Urban Studies book series takes an interdisciplinary approach to emergent urban issues using a global or comparative perspective.

Cities at Risk
Planning for and Recovering from Natural Disasters
Edited by Pierre Filion, Gary Sands and Mark Skidmore

Negative Neighbourhood Reputation and Place Attachment
The Production and Contestation of Territorial Stigma
Edited by Paul Kirkness and Andreas Tijé-Dra

The Millennial City
Trends, Implications, and Prospects for Urban Planning and Policy
Edited by Markus Moos, Deirdre Pfeiffer and Tara Vinodrai

Twin Cities
Urban Communities, Borders and Relationships over Time
Edited by John Garrard and Ekaterina Mikhailova

For more information about this series, please visit:
www.routledge.com/Global-Urban-Studies/book-series/ASHSER-1385

Twin Cities

Urban Communities, Borders
and Relationships over Time

**Edited by John Garrard and
Ekaterina Mikhailova**

Routledge
Taylor & Francis Group

NEW YORK AND LONDON

First published 2019
by Routledge
52 Vanderbilt Avenue, New York, NY 10017, USA

and by Routledge

2 Park Square, Milton Park, Abingdon, Oxon OX14 4RN

First issued in paperback 2020

Routledge is an imprint of the Taylor & Francis Group, an informa business

British Library Cataloguing-in-Publication Data
A catalogue record for this book is available from the British Library
Library of Congress Cataloging-in-Publication Data
A catalogue record has been requested for this book

ISBN 13: 978-0-367-58670-6 (pbk)
ISBN 13: 978-1-138-09800-8 (hbk)

Typeset in Times New Roman
by codeMantra

Frontispiece   Twin City, Georgia (USA) Logo by Eileen Dudley.

Contents

Figures

Tables

Contributors

Waheed Ahmed is a PhD candidate at the Institute of Development Studies at Massey University, New Zealand. Previously, he completed a post-graduate diploma (with distinction) in Development Studies from Massey University and MPhil in Social Anthropology from Quaid-i-Azam University, Islamabad. Before starting his PhD, Waheed worked for over five years with various development organisations in Pakistan and New Zealand. The title of his PhD project is 'Transport and women's social exclusion in urban areas of Pakistan'. In this research, he is looking at women's transport-related social exclusion in urban areas of Pakistan and how women negotiate these restrictions on their travel. Waheed's research interests include the interaction of social, gender and mobility (transport) issues in the complex milieu of cities in both the developed and developing countries.

Indrė Balčaitė's academic interests lie at the intersection of political science, social anthropology and geography. Her research involves studies of migration and borders, ethnicity and power, and interpretive methods of research such as oral history and ethnography. Indrė earned her PhD in Politics at the SOAS University of London in 2016. Her interdisciplinary PhD research project built on extended fieldwork in Thailand and Myanmar (Burma) and explored the interaction of human flows with interstate, ethnic, linguistic and religious boundaries. Having completed a post-doctoral fellowship at the Department of International Relations of Central European University in Hungary, Indrė is currently an independent scholar based in London.

Marco Bontje is Assistant Professor of Urban Geography at the University of Amsterdam, Department of Geography, Planning and International Development Studies. In recent years, his research has focused mainly on creative knowledge cities in Europe and China, with Shenzhen as one of the case studies, and urban geographies of growth and decline in Europe. In 2012, he was on sabbatical leave at City University Hong Kong, which included a research project in Shenzhen. He has contributed to the Shenzhen part of the 'New New Towns' project of the International New

Town Institute. Recent publications include the edited volume *Skills and Cities* (2016, co-edited with Sako Musterd and Jan Rouwendal, published in the Routledge-RSA Regions and Cities series).

Sara A. Buentello is an MA candidate in Sociology and a graduate assistant at the Texas A&M International University Office of Research and Sponsored Projects. She is a 2014 graduate of Texas A&M International University with a BA Degree in Sociology, and a recipient of a Lamar Bruni Vergara Graduate Educational Assistance-Assistantship in 2015–16. Before returning to university as a non-traditional student to undertake her BA, she worked for more than 11 years as a program coordinator for the Center for Housing and Urban Development under the College of Architecture, Texas A&M University-College Station, where she worked to improve the daily lives of colonial (rural) residents living in remote areas of Webb, Zapata, Maverick and Val Verde Counties. She is a member of the Sociological Honor Society Alpha Kappa Delta and is the current president of the Laredo Under Seven Flags Rotary Club. At present, she is developing research on the health impact of proximity to non-potable water in the Texas–Mexico border region.

Ilya Chubarov is a PhD candidate in Human Geography, a senior research fellow at the Institute of Far Eastern Studies of the Russian Academy of Sciences and a member of the Russian Geographical Society and Russia–China Friendship Association. His research interest includes the regional development of China and history of Sino-Russian relations.

László Csorba is Associate Professor of Cultural History at Eötvös Loránd University, Budapest. He was Visiting Professor at Indiana University between 1992 and 1993, Director of the Hungarian Academy of Rome between 1998 and 2007, and Director General of Hungarian National Museum between 2010 and 2016. He has published widely on cultural and political history of Modern Hungary and on the Italo–Hungarian relations, as *Garibaldi élete és kora* (Kossuth, 1988, 2009), *Széchenyi István* (Officina, 1991, 2001, 2010), *The Parliament* (Képzőművészeti, 1993) and *Ricordi Ungheresi in Italia* (Benda-Foto, 2003).

Lanre Davies was born and brought up in Lagos, Nigeria, and he attended the Ogun State University, graduating with a BA degree in History. He later attended the University of Lagos, where he bagged an MA and a PhD degree in History from the Department of History and Strategic Studies. He has since 1995 been teaching in the Department of History and Diplomatic Studies, Ogun State University, Ago-Iwoye, Nigeria, now Olabisi Onabanjo University, and is currently a senior lecturer. His research interests are on Abeokuta and Lagos. His publications include: 'The Creoles in Sierra Leone and Abeokuta: African Modernizers?' *African Journal of International Affairs and Development*, 2003; 'Sociopolitical Unrest in Egbaland 1947–1948: A product of Modernisation or

Autocracy?' *The Journal of History and Diplomatic Studies,* 2004; 'The Dilemma of Constitutional Experiments in Abeokuta, 1831–1898' *LASU Journal of Humanities,* 2006; 'The Rise and Fall of Egba Independence: A Review', *Ife Journal of History,* 2013; 'Infrastructure Policies and Problems in Lagos, 1930–1990', 'Urban Renewal and Associated Problems in Lagos, 1924–1990', *Lagos Historical Review,* 2014 and 'Land Tenure, Population Pressure and Urbanisation in Lagos, 1861–1914' *Ilorin Journal of History and International Studies,* 2015.

Sylwia Dołzbłasz is a geographer and economist working in the Department of Spatial Management, Institute of Geography and Regional Development at the University of Wrocław in Poland. She has a PhD degree in Earth Science in the field of geography with specialisation in political geography and socio-economic geography. Her main research/scientific interests are focused on problems of political borders, border areas development, trans-border cooperation, trans-border networks and border cities, as well as regional and local development and spatial planning (suburban areas, border areas). Her most important international publications include: 'Trans-border co-operation and competition among firms in the Polish–German borderland', *Tijdschrift voor Economische en Sociale Geografie* 2016; 'Different Borders-Different Co-operation? Trans-border Co-operation in Poland', *Geographical Review* 2015 and 'Trans-border openness of companies in a divided city. Zgorzelec/Görlitz case study', *Tijdschrift voor economische en sociale geografie* 103(3) 2012.

Heikki Eskelinen is Emeritus Professor at the Karelian Institute, University of Eastern Finland, Joensuu. He received his academic qualifications from the universities of Helsinki, Oulu and Joensuu in Finland, and also completed post-graduate studies at the University of Reading (UK). After a short period as a civil servant in the beginning of his career, he has had several positions at the University of Eastern Finland (until 2010 the University of Joensuu). He has been in charge of research projects funded by various Finnish, Nordic and European organisations (Academy of Finland, European Commission, ESPON programme, Nordic Council, foundations, etc.). In recent years, his research activities have focused on cross-border interaction and co-operation, and development of peripheral regions. His publications include *Curtains of Iron and Gold: Reconstructing Borders and Scales of Interaction* (co-edited with Ilkka Liikanen and Jukka Oksa; Ashgate 1999) and *The EU–Russia Borderland: New Contexts for Regional Co-operation* (co-edited with Ilkka Liikanen and James W. Scott, Routledge 2013).

Matti Fritsch completed his basic education in Germany, a BA Hons in Environment & Planning, University of Liverpool, and an MSSc in Finland (Human Geography, University of Joensuu, Finland). In 2013, he defended his doctoral dissertation on EU–Russian co-operation in spatial development policy and is now Doctor of Social Sciences (Human Geography). He specialises in regional development policy, border studies, spatial planning and European territorial and cross-border co-operation, particularly in the European North and with Russia. Fritsch has worked as a researcher on a number of projects at the Karelian Institute in Joensuu (University of Eastern Finland). This has included an EU-funded ESPON applied research project on the effects and achievements of territorial co-operation, an FP7 project on regional development in border regions and recently an H2020 project on the role of 'the local' in European Union cohesion policy. Key publications with regard to border studies include 'Reconnecting territorialities? Spatial planning co-operation between Finnish and Russian subnational governments', in Eskelinen, Liikanen & Scott (eds), *The EU–Russia Borderland,* (Routledge 2013); and (with Németh, Piipponen & Yarovoy) 'Whose partnership? Regional participatory arrangements in CBC programming on the Finnish–Russian border', *European Planning Studies*, Volume 23, 12, 2015.

John Garrard was Senior Lecturer in Politics and Contemporary History at the University of Salford until 2011. Although primarily a historian, his central teaching and research interests have lain along the borders with political science. His most representative publications include: *The English and Immigration: A Comparative Study of the Jewish Influx 1880–1910* (OUP 1970); 'Parties, Members and Voters after 1867' in *Historical Journal,* XXI, 1977; *Leadership and Power In Victorian Industrial Towns 1830–80* (Manchester University Press 1983); *The Great Salford Gas Scandal of 1887* (British Gas North Western 1989); 'Craft, Professional and Middle-Class Identity: Solicitors and Gas Engineers c1850–1914', in Alan Kidd and David Nicholls (eds), *The Making of the British Middle Class?* (Sutton 1998); *European Democratisation since 1800* (MacMillan 2000), edited with Vera Tolz and Ralph White; *Democratisation in Britain: Elites Civil Society and Reform since 1800* (Palgrave 2002); 'Democratisation: Historical lessons from the British Case' in *History and Policy: Policy Papers:* Online journal 2006; 'The Nineteenth Century European Experience of Democracy' in Stephan Berger (ed.), *A Companion to Nineteenth Century Europe* (Blackwell 2006) and *Scandals in Past and Contemporary Politics* (Manchester University Press 2006), edited with James Newell; *Heads of the Local State: Mayors, Provosts and Burgomasters since 1800* (Ashgate 2007).

Muhammad Imran is Associate Professor of transport and urban planning at Massey University, New Zealand. His research interests broadly focus on understanding the sustainable transport linkages with governance, climate change and poverty in contemporary cities. His current research explores how institutions can promote sustainable transport in cities in developing and developed countries by arguing for a greater recognition of the role of politics and the influence of discourse on transport decision-making. He has received research grants from the Royal Society of NZ Marsden Fund (2013–16) and the NZ Transport Agency (2008) and has acted as a consultant for the World Bank (2010). Major publication: *Institutional barriers to sustainable urban transport in Pakistan,* Oxford University Press, 2010.

Jarosław Jańczak is Assistant Professor, working on cross-border governance and border twin towns in Europe at the Faculty of Political Science and Journalism, Adam Mickiewicz University, Poznan, Poland and Chair of European Studies, European University Viadrina, Frankfurt (Oder), Germany. He is a political scientist researching cross-border relations at the local level in the context of the European integration process. He is author of several publications: among others, *Border Twin Towns in Europe. Cross-border Cooperation at a Local Level,* Berlin: Logos Verlag, 2013.

Pertti Joenniemi is a visiting researcher at the Karelian Institute, University of Eastern Finland and a senior researcher at the Tampere Peace Research Institute (TAPRI), University of Tampere. His background is in the sphere of International Relations and he has previously worked at the Copenhagen Peace Research Institute (COPRI) and the Danish Institute for International Studies (DIIS). He is the author of several publications mainly in the field of International Relations theory and peace research, and has published in a number of periodicals such as the *Review of International Studies, Cooperation and Conflict, Geopolitics* and *Journal of International Relations and Development.*

Alan Kidd is Emeritus Professor of Social and Regional History at Manchester Metropolitan University. He has published widely on modern British social history. His work on the history of Manchester includes *Manchester: A History,* Carnegie, 2006; 'The rise and decline of Manchester' in *Shrinking Cities Vol. I International Research,* ed., P. Oswalt for the German Federal Cultural Foundation, 2006; *The Challenge of Cholera: Proceedings of the Manchester Special Board of Health 1831–1833,* 2010, (ed. with T.J. Wyke); 'Canals, rivers, and the industrial city: Manchester's industrial waterfront, 1790–1850', *Economic History Review,* 2012 (with P. Maw & T.J. Wyke) and 'Rebuilding Manchester: the wartime debate on post-war reconstruction, 1941–45' in M. Dodge & R. Brook eds., *The Making of Post-War Manchester: Plans and Projects,* 2015.

John C. Kilburn is the Associate Dean of Research at Texas A&M International University, where he also holds the position of Professor of Sociology & Criminal Justice. He received his PhD from Louisiana State University and has published on topics related to violence and border issues in numerous journals in the fields of sociology, social work, criminal justice and psychology (such as *Cities, Criminal Justice Review, Psychological Bulletin, Social Forces,* and *Urban Affairs Review*). His work has been funded by grants from the National Institutes of Health and the National

Science Foundation. In addition to his scholarly research, Dr Kilburn has served as a consultant to more than 20 non-profit organisations, as well as having served in leadership positions on several non-profit boards.

Usman Ladan is Senior Lecturer in the Department of History, Ahmadu Bello University, Zaria, Nigeria, where he was Head of Department from 2005 to 2007. He was also Deputy Director, *Arewa House: Centre for Historical Research and Documentation* of the University. He is the author of a forthcoming book titled: *A History of the City of Maiduguri, 1907–1960*. Dr Ladan is also editor, along with Dr Dipo Fashina, of *The British Colonisation of Northern Nigeria, 1897–1914* by Mahmud Modibbo Tukur.

Francisco Lara-Valencia is an associate professor in the School of Transborder Studies at Arizona State University (ASU). He is the founder of the ASU Program for Transborder Communities (PTC) and current President of the Association for Borderlands Studies. He earned a PhD in Urban and Environmental Planning from the University of Michigan and holds degrees in regional development and economics from El Colegio de la Frontera Norte and the Universidad Autónoma de Baja California in Tijuana, Mexico. His current research work focuses on understanding the role of structural and institutional change on cooperation and sustainable urban governance in international border regions. He is also interested in understanding the interaction between internal and external borders in shaping of residential segregation of immigrant communities in US southwest cities. His recent publications include 'Cross-border narratives of development and space: transborder communities in North America', *Journal of Borderland Studies*, 2016; 'Borders and cities: perspectives from North America and Europe' with C. Sohn, *Journal of Borderland Studies* 28(2), 2013.

Thomas Lundén is Emeritus Professor in Human Geography at the Centre for Baltic and East European Studies at Södertörn University, 1970–2009 board member, vice chairman and chairman, and, since 2014, a honorary board member of the Swedish Society for Anthropology and Geography (SSAG). From 2010, he is the chair of the editorial advisory board of the journal *Baltic Worlds*, www.balticworlds.com, Södertörn University. His main research interests include urban, social and political geography, geolinguistics, border relations and history of geopolitics.

His selected publications include *On the Boundary: About humans at the end of territory*, Södertörn Educational Publications, 2004; *Crossings and Crosses. Borders, Educations, and Religions in Northern Europe*, eds Jenny Berglund, Thomas Lundén and Peter Strandbrink. 2015 Boston/ Berlin Walter de Gruyter Inc.; A Turnover in Border Relations: Sweden and its Neighbors in a 100-Year Perspective, 35–49 in *Boundaries Revisited: A Conceptual Turn in European Border Practices*, ed. Tomasz Brańka and Jaroslaw Jańczak. Berlin: Logos Verlag; *Rudolf Kjellén: Geopolitiken och konservatismen* [Rudolf Kjellén. Geopolitics and conservatism], red.

Bert Edström, Ragnar Björk and Thomas Lundén. Stockholm: Hjalmarson & Högberg, 2014 and *Pommern – ett gransfall i tid och rum* [Pomerania – a border case in time and space]. Lund: Slavica Lundensia 27, 2016.

Peter Martyn completed his post-doctoral research (on the role of the tenement house in Warsaw's urban development from the 1860s to the First World War) at the Geography Department, Salford University, UK. Based at the Polish Academy since the mid-1990s, he has published and edited conference proceedings on themes most typically related to European urban and architectural history since the French Revolution.

Ekaterina Mikhailova is a research fellow at Faculty of Geography, Lomonosov Moscow State University. In addition she currently is working in the project 'Transformation of Soviet Republic Borders to International Borders' at the University of Eastern Finland. As an early career border scholar, Ekaterina is an author of over thirty Russian and English articles on twin cities, cross-border communities and cross-border cooperation. Her research interests include sustainable development of border regions and border cities, cross-border integration and governance. Ekaterina has been involved in collecting empirical data and analysing cross-border interactions along Russian borders with Norway, Finland, Estonia, Belarus, Ukraine and China, as well as at the Swedish–Norwegian and Spanish–French borders. From 2012 to 2014, Ekaterina worked in FP7 research project 'EUBORDERREGIONS'. Ekaterina's recent English journal articles are 'Collaborative problem solving in cross-border context: learning from paired local communities along the Russian border' (2017) and 'Are Refugees welcome to the Arctic? Perceptions of Arctic Migrants at the Russian–Norwegian Borderland' (2018).

Jorge Aponte Motta is a founding member of the Group of Transborder Studies (Grupo de Estudios Transfronterizos-GET) at the Amazonic Research Institute (Instituto Amazónico de Investigaciones IMANI). He holds a degree in Political Science by National University of Colombia, Bogotá campus, a Master's degree in Amazonian Studies by National University of Colombia, Amazon campus, and a Master's degree (Diploma de Estudios Avanzados) in Geography by the Autonomous University of Madrid and a PhD in Geography by the same university.

He explores mainly the role of bordering and urbanisation processes in the production of space of the Amazon region. Among his latest publications are:

Zárate, C., Aponte, J., and Victorino, N. (2017) *Perfil de una región transfronteriza en la Amazonia. La posible integración de las políticas de frontera de Brasil, Colombia y Perú*, Leticia: Universidad Nacional de Colombia.

Hurtado Bautista, A., and Aponte Motta, J. (2017) '¿Hacia un gobierno transfronterizo? Explorando la institucionalidad para la "integración" colombo-peruana', *Estudios Fronterizos* 18 (35), 70–89.

The project involved a comparative analysis of place branding in five European Cities: NewcastleGateshead, Leipzig, Manchester, Malmö and Torino. Taking a discursive institutionalist approach, the thesis concludes that the primary contribution that place branding can make to urban development is as a complimentary intervention to enhance the strategic development and management of cities. Prior to this, Rebecca completed her Undergraduate Degree in Combined Social Sciences at Durham University, followed by Master's degrees in Research Methods at Durham University, and Town Planning at Newcastle University, after which she worked as a planning officer for a local authority in North East England, specialising in open space and employment land policy.

Chung-Tong Wu is the Inaugural Chair of the Henry Halloran Trust Advisory Board at the University of Sydney – an endowed fund supporting urban research. In recognition of his distinguished service to other universities where he served in a number of senior academic posts, he has been honoured with the title of Emeritus Professor by both the University of New South Wales and the University of Western Sydney. He has served on the international editorial board of several international journals and a member of numerous review committees and company and foundation boards. His publications are chiefly focused on urban and regional issues in Asia. Professor Wu's recent research projects include cross-border development, urban restructuring and urban shrinkage and urban and regional development in Myanmar.

Carlos Zárate Botía is a titular professor at the National University of Colombia (Amazonia campus) and researcher at the Amazonic Research Institute (Instituto Amazónico de Investigaciones, IMANI) and coordinates the Group of Transborder Studies (Grupo de Estudios Transfronterizos, GET). Having finished his undergraduate studies in sociology at the National University of Colombia, he received his MSc degree from the Latin-American Faculty of Social Sciences, FLACSO (Ecuador), and his PhD degree in History from National University of Colombia. Currently, he teaches postgraduate courses in history of the Amazon border and directs research projects on border integration in the Amazon. His key publications include:

Zárate, C. (2008). *Silvícolas, siringueros y agentes estatales: El surgimiento de una sociedad transfronteriza en la Amazonia de Brasil, Perú y Colombia. 1880–1932.* Bogotá, Colombia: Unibiblos.
Zárate, C. (ed.) (2012) *Espacios Urbanos y Sociedades Transfronterizas en la Amazonia.* Leticia: Universidad Nacional de Colombia Sede Amazonia.
Zárate, C., Aponte, J., and Victorino, N. (2017) *Perfil de una región transfronteriza en la Amazonia. La posible integración de las políticas de frontera de Brasil, Colombia y Perú.* Leticia: Universidad Nacional de Colombia.

Acknowledgements

Our thanks are due to the following people:

For helping organise and run the conference that formed the basis of this book: Melanie Tebbutt, Craig Horner, Fiona Cosson, James Charnock, Adrian Stores, Jeff Evans, Peter Shapely, Derek Antrobus, Afzal Khan and Katie Milestone, the Manchester Centre for Regional History (now the Manchester Centre for Public History and Heritage) at Manchester Metropolitan University (MMU).

For totally invaluable conference funding: Berthold Schoene, Head of Research and Knowledge Exchange in the Humanities, Languages and Social Sciences Faculty, MMU, and also the Friends of the Manchester Centre for Regional History, MMU.

For assisting Ekaterina Mikhailova to attend the conference, thus enabling the editors to meet and decide to produce the book: MMU and the VERA Centre for Russian and Border Studies at the University of Eastern Finland.

For invaluable insight into policing Manchester and Salford: Gareth Parkin and Marcus Noden.

For thoughtful mentoring and encouragement during the first steps in map-making: Natalya Ryabova.

For rich, honest and inspiring academic debates: colleagues from Institute of Geography, Russian Academy of Sciences, and Faculty of Geography, Lomonosov Moscow State University, particularly the Department of Geography of the World Economy.

For invaluable support, patience and sympathy during the long production process: Eve Garrard; Ekaterina Mikhailova's family – Zinaida, Vladimir, Lyudmila and Oksana; Mike and Yoko Inman; Ruth Anderson, our splendid Editorial Assistant at Routledge; Simon Gunn; Michael Goldsmith; Kristina Golodnikova; Maria Goncharenko; Olga Khimchenko; Anya Kondratyeva; all our authors; and Harry who is John Garrard's remarkable whippet-cross.

Introduction and overview

John Garrard and Ekaterina Mikhailova

When the modest urban settlements (combined population of 1,699 in 2016) of Summit and Graymount in Georgia, USA, merged in 1921, the citizens had a decision to make: what to call their new collective self. They came up with 'Twin City', thus semi-consciously testifying to the current and future importance of such pairings across the world. Twin City had a railway running through; several decades after closing in 1952, it provided the town with its crest and current logo (www.twincityga.com/). In 2016, it elected its first woman mayor, Eileen Dudley who designed the logo, and thereby provided the book with its frontispiece.

This book explores the various sorts of twin city,[1] both within individual states and along the borders between them. It is ambitious in scope: while there have been many articles, and some books, covering and sometimes comparing cross-border urban communities in say East Central Europe or between the USA and Mexico,[2] there has been nothing seeking to explore and compare all types of twin cities on all continents (Figure I.1). This volume starts to fill that important gap. For this purpose, it brings together scholars from across the world, drawn from urban geography, economics, sociology, history and politics, including many leading figures in the twin-city field.

Twin cities[3] are important internationally: rapidly trawling open sources reveals well over 100, that is to say over 200 separate urban entities in this sort of closely intertwined relationship, some small, many very large, located on all continents and subcontinents. As will become evident, twin cities also have significance far beyond mere numbers. National twins have been locations where the essential interdependence between urban places and thus the need to co-ordinate action was first and most urgently demonstrated, even if as we shall see they remained conflictual, unaware and unco-ordinated until fairly recently. Meanwhile, some cross-border pairs, being mainly twentieth- and twenty-first-century creations, have pioneered co-ordination via economies of scale in joint, rather than duplicated, service provision. Some international twins have gained attention from national and regional authorities seeking new centres of economic development. They have also been major initial sites, sometimes 'laboratories', for the cross-national

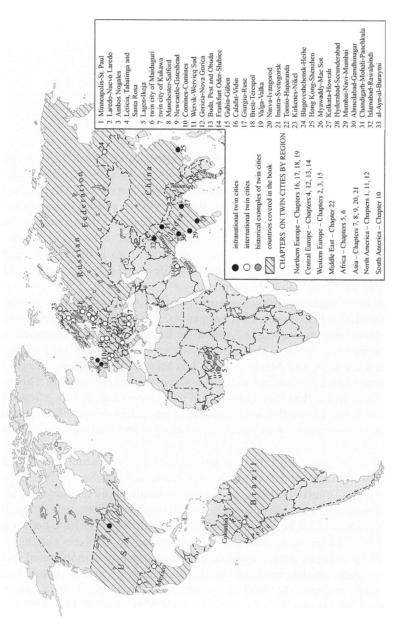

Figure I.1 Towards a world map of twin cities: cases presented in the book by Ekaterina Mikhailova.

mixing and mutual familiarisation of peoples. Partly thereby, they have had and continue having substantial transformative potential to erode international borders. More generally, international twins have been used as special venues, where relations between the adjacent states can be symbolically represented and celebrated as either shields or bridges – via extravagant military rituals like gate-slamming on the Indo-Pakistan border, inter-visiting diplomatic delegations, sport competitions and cultural festivals among borderlanders (particularly within and around Europe). Overall, they can tell us much about the significance of borders, both municipal and international, and borders are important.

* * *

We should start with definitions, simultaneously reviewing some of what has been written about these distinctive communities. Two, perhaps three, ideas are implied. First, the classic version views twin cities historically as nearby urban places that expand/'crash' into each other. This sense of close and converging proximity picks up many twin cities – which mostly have centres 0–5 km apart, common borders, and have often collided. However, this is insufficient especially in the age of conurbations[4] or metropolitan areas: containing multiple towns/cities, a few kilometres apart,[5] often possessing common borders, often merging and becoming indistinguishable in most senses, except possibly the governmental, and (decreasingly) self-identificational, senses. Like twin cities, conurbations have also become increasingly interdependent in terms of their economies, labour markets, transport and communications, social interactions, service-provision, politics, etc. One difference may be that twin cities are mainly geographical and historical accidents, whereas metropolitan areas develop and merge around a dominant industry or set of industries. However, this contention, while working sufficiently for some twin-city species (see below), becomes harder to sustain in face of relatively distant but rapidly merging Indian twins or border-cities generating their own satellites on the other side to jointly fulfil cross-border functions.

   Thus, we must still ask why conurban towns do not qualify for twin-city status when some that do are further apart. This definition of a conurbation from the Cambridge Dictionary Online is unwittingly eloquent here: 'a city area containing a large number of people, formed by various towns growing and joining together'. After all, conurbations have engulfed many twin cities of both the intranational and cross-border sort,[6] including several covered here like Minneapolis–St. Paul (Chapter 1), Manchester and Salford (2), Newcastle–Gateshead (3) and the Danube twins (13). Twin cities can lose their identities in face of conurbations as Pimpri-Chinchwad is apparently doing, engulfed by Pune (India, Chapter 7), and as Maiduguri has increasingly done since Nigerian independence in 1960. This may be the long-term fate of many, unsurprisingly given the engulfing power of conurbanisation.

However, Chapter 5 suggests that, in appropriate political and especially administrative circumstances, twin cities can emerge after conurbanisation: witness Ikeja steadily emerging from metropolitan Lagos to become the Lagos-State capital in 1976, while Lagos was the federal capital between 1914 and 1991.

Our initial definition involving merging proximity is also insufficient because many twin cities probably never were separated except by the river so often marking their common border; rather, they were two administratively separated places growing outwards from adjacent cores. This is true of Manchester and Salford on opposite sides of the Irwell; also 'Newcastle–Gateshead' across the River Tyne. It also characterises many cross-border-cities, including several reviewed here. Furthermore, many border twins were originally one urban entity before an international treaty drew or re-drew the international border and divided them, normally along a river, with their respective national governments trying to push the two halves in increasingly contrasting national directions via propaganda and population-transfer in the decades following border creation. The Mexican–American treaty of 1853, the 1919 Versailles Settlement, and the post-1945 peace settlement were all fertile parents of such twins. Overall, for cross-border, as for intranational, city pairs, simple proximity can never be more than a useful prerequisite for twin-city status (Buursink and Ehlers 1999; Buursink 2001; Joenniemi and Sergunin 2009).

Thus, this classic version must be supplemented by a second idea: these nearby, increasingly indistinguishable urban places must consciously see themselves and/or be seen by others outside and normally governmentally from above – regional, national or international – as 'twin cities'; or at least as especially adjacent, thereby having some special relationship: viewed with regret, pride or some mix of both, depending on the twins and partners concerned. Such twinning or special relationship must have consequences for how they *see* and *act*, and are seen by others; consequences potentially even now distinguishing them from broader conurban areas and their constituent urban centres.

This insistence on perception may restrict twin-city numbers, but they remain highly important, with some regions particularly blessed, especially those containing many nation-states and thus many border crossings. Thus, Mainland Europe has at least fifty cross-border twins (i.e. at least 100 cities), in addition to urban pairs within individual nation-states; the USA has multiple internal twins, while the USA–Mexican border has also proved increasingly prolific. India is an ever-fertile producer of intranational and highly self-perceiving twin cities, driven partly by the electoral cycle.

As this implies, twin cities are still multiplying in our evermore rapidly urbanising and globalising age, both cross-nationally and intranationally. This is so for several interconnected reasons: first, international developments like the Cold War's ending, pressure for European integration, the increasing importance of international trade and migration, and the perceived necessities of international co-operation have all created needs for cross-border twins to enact the consequences. Second, there is the more general advance of transportation systems and technologies – enabling, for example, the 8-km Copenhagen–Malmö bridge; also facilitating, as evident later, the mass-transit lines and motorways enhancing ever-faster connections between quite distant, but now legally and actively twinned, Indian cities.

This raises another point: insistence on close proximity also hits the fact that some conscious twin cities are only somewhat proximate but are nevertheless seen as twins, and this perception brings very significant action in its wake. As evident in Chapter 7, India is producing twins in proliferating numbers, becoming one home of what has been rather variably described as 'the twin-city model'. And this is the product of hyper-rapid urban expansion in area and population – caused, as in the past in Europe and elsewhere, as much by vast inward rural migration as by urban self-generation. Here, somewhat distant (e.g. Ahmedabad–Gandhinagar 22 km) and often vastly populated urban places are perceived by relevant state governments as rapidly converging, certain to merge eventually, and consequently declared twins for planning purposes. Alternatively, one city is duplicated as a satellite outside the borders of another in pursuit of population-relief – as eventually with Mumbai and Navi-Mumbai. Both moves represent attempts to ensure that merging is rational and controlled, rather than chaotic and dysfunctional, for constituent populations and the future of both urban entities. They also seek to ensure that urban partners compliment rather than rival or duplicate each other in economic and other terms. Meanwhile, at least two erstwhile cross-border twins in China have been created to handle the perceived problems of absorbing ex-colonies via notions of 'one country two systems' or 'Special Economic Zones'. Thus, we have Hong Kong–Shenzhen (17 km apart) in this book, and Macau–Zhuhai (9 km).

Such more distant twins are also evident, or potentially evident, elsewhere – even some British observers and decision-makers have begun contemplating this sort of twin city. There has been recent talk about a bizarrely titled place called 'Manpool' or 'Liverchester' in the wake of Manchester and Liverpool's rapid convergent march (Headlam 2014), however improbable this potential creation seems, in name or popular affinities, as a candidate for region-embracing civic pride. We appear, as often in India, to be talking about twin conurbations.

Overall, it seems reasonable to see twin cities as *mental constructs*, but ones with measurable consequences for how their local governments and perhaps citizens act and interact, and/or for what governmental levels above them do, and, as Chapters 12 and 13 argue explicitly, for how each urban partner develops and eventually looks.

For many observers, at least among analysts of cross-border twins along the EU's internal borders or its external boundary with Russia and/or persuaded by notions about 'the twin-city model', there is the implicit or explicit idea that, for urban pairs to be labelled twins, their relationships must not just be special but also functional and positive rather than dysfunctional. Besides mere proximity, there must be active partnership expressed in consciously felt 'common interests and belonging together' (Buursink and Ehlers 1999) or 'cordial relations' (Buursink 2001), or citizens acceptance of twinning initiatives (Bucken-Knapp 2001; Mathieson and Burkner 2001), more generally conscious intentions to do more together than is possible autonomously. These expectations have progressively multiplied, with observers implying that 'real' or 'model' twins should show most of such positive features (e.g. Schulz, Stokłosa, and Jajeśniak-Quast 2002; Joenniemi and Sergunin 2011; Anishenko and Sergunin 2012). Overall, 'proper' twin cities should play broader co-operative and integrative roles: enhancing cross-border interaction; mutual understanding and even dissolving border-relevance in the interests of say European integration. All this should benefit not just each partner but also broader aspirations (Joenniemi and Sergunin 2014), including post-1945 determinations to avoid another European war – the key motive inspiring the EU's foundation. Overall, proper twins should have capacity to reverse or certainly modify often-embittered historical legacies. Consequently, twinning within the EU and around its edges has increasingly emerged from formally negotiated agreements, concluded locally and/or cross-nationally, and specifying forms of co-operation and interaction.

Several qualifying points need making here. First and in recent years, insistence on various positive prerequisites for granting twin-city status particularly to border-cities has tended to yield ground to recognition that twins can exist in various forms and proceed towards positivity at varied speeds (Mikhailova and Wu 2017), thus concentrating on monitoring fluctuations in twin-city development (both regressions and progressions). Chapter 19 explores this theme for Imatra–Svetogorsk. Chapter 16 by Thomas Lundén highlights three levels of relational intensity: between his three sets of Baltic twins: Narva–Ivangorod, Valga–Valka and Haparanda–Tornio. Some observers have added characteristics that twins might expect as they ascend the positivity-scale – money-saving efficiencies (Lundén and Zalamans 2001); mutual economic gains and reciprocal willingness to embrace aspects of each other's culture (Mikhailova and Wu 2017), heritage and identity; and to see the border as something positive – a 'resource' exploitable for trade and tourism (Schulz; Stokłosa, and Jajeśniak-Quast 2002; Herzog and Sohn 2014). The current book reflects some of this in Chapters 16, 18, 20 and 21.

Second, as Chapters 1, 2 and 3 suggest, positivity has not characterised interactions between intranational urban twins like Minneapolis–St. Paul (where, after all, the twin-city term supposedly originated and from which, oddly, many newly founded twin cities – e.g. India and China – still derive inspiration); nor Manchester and Salford, nor Newcastle–Gateshead. In all three, relations historically have been deeply conflictual, even dysfunctional, from the viewpoint of shared problems, even while their very close proximity marked their relationship as special, indeed unavoidable, compared even to other towns nearby. All too often, special twinnedness has involved being 'too near neighbours to be good friends'. This should be unsurprising: after all, human twins, indeed whole extended families, may cordially loathe each other, while remaining twins and families. Quite aside from issues around class, religion or ethnicity, there is clearly nothing automatically unifying about urbanity – indeed the closer the proximity, the less the harmony apparent, particularly where one partner is dominant and the other subordinate. Yet, many leading inhabitants in these internal twins argued in the past, and now more successfully argue, that relations should start becoming more co-operative (even amalgamated). Indeed, for 'Newcastle–Gateshead', amity has been central to their formal partnership since 2000, even if now being eroded by pressures for conurban co-operation. Similar forces (e.g. emerging 'Greater Manchester') help explain the recent harmony emerging between Manchester and Salford, and indeed Minneapolis–St. Paul.

Meanwhile, co-operation or at least co-ordination is not always evident even among cross-border twins outside Europe. Some Mexican–American twins, as evident in Chapter 11, show a mixed picture, unless one allows the drugs underworld inside the world of legitimate functionality. More uncertain still are Blagoveshchensk/Heihe straddling the Russo-Chinese border (Chapter 20); more still, the Omani–UAE border-cities of al-Buraymi and al-Ayn in Chapter 22, these last two being deeply affected by the mutual hostility, lack of trust and latent insecurity of their national governments. Such dysfunctionality has led scholars to characterise their specialness in other terms besides twinnedness – border-crossing, bi-national (Kada and Kij 2004), border-cities (Cañas et al. 2013), companion cities (Sparrow 2001), sister cities (Albuquerque 2007) or even trans-border agglomerations (Buursink 2001; Headlam 2014).

Overall, twin cities arguably come in four varieties: the three implied thus far, plus a new arrival. They share strong 'family resemblances',[7] while revealing significant differences and carrying somewhat different expectations of role and relationship. These types are: first, nearby but administratively separate urban places within one country that are either indistinguishable from the start or rapidly become so. They have built-in interdependence, but their relationships have strongly tended towards conflict and dysfunction, though now may be showing enhanced harmony.

Second, we have mostly indistinguishable urban places that are seen as twin cities and produced by pre-existing or subsequently imposed international borders. These might originally have been one place, later split by an imposed international border (like Valga–Valka on the Estonian–Latvian border generated from the settlement of Walk, reviewed in Chapter 16, the Danube towns in Chapter 13 and other Central European border towns presented in Chapter 14). Alternatively, they were two pre-existing urban settlements facing each other across such a border and discovering reasons to co-operate either on their own initiative like Irún, Fuenterrabía and Hendaye on the Spanish–French border (Garrido 2007) or after encouragement from their central governments (e.g. Chapter 17 Kirkenes–Nikel on the Norwegian–Russian border; Haugseth 2013). Alternatively, cross-border twins can emerge where towns on one side of an international crossing-point find themselves roughly duplicated on the other side due to border-related security concerns, like the fortresses of Narva and Ivangorod on the Estonian–Russian border in Chapters 14 and 16, and/or passage of people and goods across that border, as with Zabaykalsk/Manchouli, railway ports on the China–Russia border (Wang, Cheng, and Mo 2014) and Myawaddy–Mae Sot straddling the Myanmar–Thailand border in Chapter 21. Alternatively again, as with Comines–Wervik in Chapter 15, we find twin towns, both of which have been split by an international border – here between France and Belgium from 1830. Meanwhile, some cross-border twins have deliberately been created by central government to handle some integrative transitional or other process, as potentially with Shenzhen for Hong Kong in Chapter 9.

Whatever its origins, cross-national 'twinning' often results from formal agreement to interact and co-operate culturally, environmentally or economically. This is sometimes, as with Kirkenes–Nikel (Chapter 17), initiated by respective national governments pursuing broader diplomatic aims. Sometimes, perhaps initially, it results from independent, even 'defiant', local initiative. Either way, and wittingly or otherwise, twinned urban localities can become agencies in shaping foreign policy (Chapter 17; also Joenniemi and Sergunin 2014), perhaps increasingly ambitious ones, particularly once they pass from initial 'friendship agreements' to more ambitious twinnings following the Cold War. Acceptance in this second branch of the twin-city family, like the one following, often carries clear expectations that the twins will get on well.

One feature of cross-border twins is that they are sites for cross-national mixing of peoples. This does not necessarily distinguish them from other cities, or conurbations, twinned or not. London, after all, is now frequently described as the most multi-ethnic and multi-national city in the world,[8] with the likes of New York and Los Angeles not far behind. Here, however, cross-national or cross-ethnic contact is part of day-by-day experience. For cross-border twins, the experience is more complex. For various reasons (e.g. uneven economic development), these tend to experience one-way migration and thus uneven ethnic composition: one city mostly sending population, thereby remaining ethnically homogeneous, and the other mostly receiving and thus becoming ethnically mixed. For the latter's citizens, intermixing is indeed day-to-day; for the sending-city's citizens, intermixing must be consciously undertaken by a 'visit' or 'trip' across the border, implying maybe some sense of adventure at least initially, though less so if one's journey is primarily to visit migrated-relatives in the receiving city. This variable complexity is evident across most of the border twins in this book.

Our third twin-city variety comprises somewhat adjacent, very large and rapidly expanding urban places that are expected to physically merge and declared twins to attempt control over the process. As Chapter 7 suggests, India has been prolific here. Admittedly, only two of the three criteria above are initially fulfilled; nevertheless, Indian pairings are emphatically seen as 'twin cities', and such labelling often carries clear consequences for what then happens; and they do eventually merge, albeit in hopefully more planned ways than would otherwise be evident. Furthermore, twin-city status is often confirmed by state or union legislation specifying the co-ordination envisaged – thereby formalising relationships more clearly even than cross-national twinning-agreements.

The new kids on the block are best described as 'engineered twin cities'. So far, mainly emerging across international borders, though facilitated by partial border-dissolution following EU treaties, and only in the past 10–20 years, these primarily result from new developments in fast-transit engineering. The cities/conurbations affected are not rammed against their shared international borders, nor likely to collide territorially, nor particularly close geographically. Long enjoying thriving separate existences, their undoubted and developing special sense of togetherness is recent, resulting from transport engineering's ability to shrink effective distance, thereby producing at least temporal, if not exactly adjacent, twin cities. Thus, Copenhagen and Malmö (44 km distant) until recently were increasingly called, and called themselves, twin cities due to the vast increase in intercity visitation, even commuting, facilitated by the remarkable Øresund Bridge – opened in 2000 (Bucken-Knapp 2001; Löfgren 2008).[9] Tallinn and Helsinki, capitals, respectively, of Estonia and Finland and 800 km apart by road, have steadily begun seeing themselves as twins following inter-mayoral initiative and developing rapid-ferry transport across the Gulf of Finland (8,000,000 people making the 2–3 hour journey annually, Mäki 2016), and, most recently, a projected tunnel. Even Dover and Calais have been awarded 'binational city' status with the building of the Channel Tunnel (Heddebault 2001) – though Britain's pending EU exit pushes this into uncharted territory. Meanwhile (in Chapter 7), some more distant Indian pairings can be seen as becoming effective twins by planning linked to rapid transit.

* * *

Having settled on definitions and types, we can now explore some characteristics of this somewhat strange family. These are interlinked and shared by all four branches, even though their precise expression varies. They are: (1) interdependence; (2) tensions between inwardness and openness;

(3) mostly unequal relationships; (4) ongoing formal or informal negotiation and (5) persistence.

All twin cities are closely interdependent either *from or near the start of their existence* or, as with Indian twins and those of the fourth engineered sort, from whenever twinning is formally declared. This distinguishes them from other units in conurbations or 'multi-centred city regions'.[10] So too does each twin's ability to impact deeply the economic, geographical, social and political character of the other, often intentionally. This interdependence is enhanced by the fact that, unlike run-of-the-mill conurban settlements, one town centre can often be viewed from the other. In our book, this is true of many twins as many of the photographs suggest. This may increase the unique sense of togetherness, comparison and emulation, and facilitate awareness of twin-city rituals.

Second, and resulting from such interdependence, all twin communities are subject to inner tensions and their inhabitants presented with choices – now perhaps common to all conurban entities, but which again have been mostly evident *from the start with twin cities, and experienced with particular intensity by twin cities due to adjacency.* Most important here is the shifting tension between separateness/inwardness and openness. This may pose choices for citizens, though individuals can accommodate both possibilities at different times.

Citizens can turn their eyes inward, feeling their city different and the other strange, even alien. However closely-twinned, urban communities can have a positive sense of place, perceived history, civic pride assisted up to World War I and beyond by highly popular-attention-grabbing rituals. For British industrial cities at least, including the two sets of twins featured here, such rituals often expressed a strong sense of locality and accurate perceptions of forefronting industrial progress, even though industrialism also bred interdependence between places in the short and the much broader long term. Ever-accelerating urban interdependence notwithstanding, locality apparently survives into the twenty-first century, albeit in less traceable, more passive and negative ways: most cities, however twinned, retain their own named, often closely followed, sports teams; most retain their own centres or 'downtowns' – surely one identifier of 'place' – even if one is often far larger, even more prosperous, than the other and even if those centres face each other across a modest river; and most retain their own municipalities, even ornate town halls, albeit carrying far less visibility and autonomy than heretofore, but still helping to mark their centres.

Individual urban identity, or at least difference, can also reflect contrasting social/ethnic structures, or the contrasting ways they have been politically negotiated and expressed in each twin. White working-class Salford viewing multi-ethnic Manchester in Chapter 2 exemplifies the first, while conflicts between Minneapolis and St. Paul in Chapter 1 partly reflect the second. For cross-border-cities, such identity can rest on contrasting national loyalties, particularly immediately following separation as newly created or territory-acquiring older nation-states sought to enhance their viability and coherence by nationalistic propaganda, population-shifts, etc. And, as variously evident in Chapters 13–16, once imposed even on previously unified locations, borders can affect how each now separate twin comes to look and feel.

As this implies, individual urban identity can also rest on more negative factors – a sense of grievance, at least among intranational twins, felt by one twin against a frequently dominant partner, further enhanced by the latter's corresponding indifference. This has been emphatically evident among many twins in the past and, more than its positive counterpart, this probably remains vibrant now – here, most notable whenever Salford (Chapter 2) contemplated Manchester or Gateshead contemplated Newcastle (Chapter 3), or indeed St. Paul contemplated Minneapolis (Chapter 1). Such negative identity-inducing feelings may also be evident among cross-border twins, all the more given that dominance-subordination seems to be the typical twin-city relationship (Schulz, Stokłosa, and Jajeśniak-Quast 2002, 3). It can rest also on fear, sometimes realistic, about what lurks on the other side of some national, even merely municipal border: fear of crime as Laredo contemplates Nuevo Laredo (Chapter 11; Kilburn, San Miguel, and Kwak 2013); increased drug addiction in Ciudad Juarez, a city reportedly hosting substantial drug traffic towards el Paso (Case et al. 2008), alongside growing concerns about HIV transmission risks on the US–Mexican border (Ramos et al. 2009). And of course, there is the likely fear of national/ ethnic 'others' over the border, a fear evident in many of the cross-border twins reviewed in this book (notably in Chapters 20 and 22), even in some intranational twins until quite recently. Most recently, and in Europe at least, such fears can only be reinvigorated by the deeply uncertain impact of the ongoing Middle-Eastern 'Exodus' (*Time* 19 October 2015, 1).

Set against such inwardness, twin cities are subject to more open emotions, due to the many, now-intensifying, forces eroding the significance of municipal and national borders and parochial identity. Cities, even if enclosed in many ways, and twinned or not, cannot be islands by virtue of their very urbanity – certainly not in the way villages and small country towns have sometimes been, or imagined. And a twin city, by its nature, is open to the other twin, often thereby the other nation.

There are at least five factors here. First, people, migrants, settlers and tourists, trade and communications of all sorts cross those civic and often national borders. While generalised to all twin cities, the results are more dramatic for international cross-border inhabitants because they confront greater initial strangeness. This can impact both positively and negatively: visitors, shoppers, traders, etc., to 'the other place' can produce ever-widening familiarity for both sides; alternatively, sheer numbers can induce a sense of being engulfed. Either way, as Chapters 18 and 19 argue, ordinary people are as important in determining what happens as facilitating authorities. A key factor here is shopping. In Imatra–Svetogorsk (Chapter 18), this provides the most important reason for border-crossing, a situation probably generalised across many border-locations. After all, shopping has increasingly shaped urban life since at least the 1850s – economically, socially and geographically, and acceleratingly since around 1990. It is unsurprising if it is now a major force underpinning border dis- solution. Another related factor is quality and price differentials, dramat- ically illustrated in Chapter 11, where Americans cross between Los Dos Laredos primarily to access cheaper health-care, and Chapter 12, where the external orientation of commercial and service establishments is evi- denced for Gubin/Güben and Ambos Nogales city centres. The volatility of cross-border flows is illustrated in Chapter 20, where the number and direction of cross-border tourists closely correlate with the Russian ruble– Chinese yuan exchange rate, and also Chapter 22, where the new UAE visa regulation produced a drop in expatriates residing in the Omani border-city of al-Buraymi, who previously benefited from differentials in rent and sala- ries on the Omani–UAE border.

Second, and more uncertainly, there is the likely presence of substantial businesses and business-people with interests and mental horizons straddling city borders, or indeed running well beyond. Such persons were crucial fig- ures leading the regular attempts at amalgamating Manchester and Salford (Chapter 2) through the nineteenth century and undermining Salford's oth- erwise strident resistance. Chapter 3 reveals them equally visible in promot- ing the Newcastle–Gateshead brand from the late twentieth century; and now pushing for unified efforts promoting the entire Tyneside conurbation – perspectives evident as early as 1850 in the Tyne Improvement Commis- sion's formation, or the Tyneside Geographical Society (initially economic in intent) in 1887. Chapter 1 shows business-people in Minneapolis–St. Paul perceiving their interests purely in city-oriented terms, rendering them de- fensive of city interests, thereby enhancing inter-city conflict, but, in recent decades, becoming powerful forces behind twin-city and indeed regional orientations.

Border-city businesses appear equally ambiguous. Many support cross- border economic co-operation, with successful examples in areas like Russo-Norwegian hydro-electric projects (Fors, Espiritu, and Mikhailova 2015), and most spectacularly (at least until President Trump's advent) Mexico's *maquiladoras* – factories importing duty-free raw materials and re-exporting finished goods to the USA, created under the Mexican gov- ernment's Border-Industrialisation Programme from 1965 and contributing substantially to many border-town economies to the present-day. However, national borders are always more potent than municipal ones. Differential local taxation, labour prices, availability of goods on only one side and pop- ular curiosity about what can be seen and bought there can all generate income for someone – whether tourist operators, entertainment entrepreneurs, traders or smugglers. Thus, businesses rarely desire cross-border municipal amalgamation, even assuming co-operative national governments. This is true even though their interests may well lead them to support cross-border trade and co-operation, and cross-border economies generally.

Third, twin cities, whether intra- or cross-national, also experience early and intensifying needs to co-ordinate action over an increasing range of problems (Del Biaggio 2015) – even if those needs remain un-recognised by one or both twins, and even if the rather frequent result is conflict rather than co-operation. These are particularly intense for twin cities, so often divided by rivers that, particularly in urban, still more in-dustrial, settings, are potent sources of jointly inflicted trouble – flooding, multi-various pollution, barriers to communication requiring bridges and tunnels. More generally, many twins share economies, transport needs, criminality, image-problems, social disturbance, weather, migration flows, risk from emergency situations and the like. They might be general to conurbations, but are particularly intense for twin cities: a fact evident in many chapters here.

Given that a key identifier of these odd communities is consciousness of twinned condition, it is unsurprising to discover signs of joint identity, even conscious promotion of twin cities as distinctive brands – both to persuade citizens and advertise to the world outside. This is clearest for Minneapolis–St. Paul (Chapter 1), naturally enough, this being where the conception originated and the inspiration for many other twins. And a per-suasive brand it is, to judge by the 'Twin-Cities' designation, given the vast output of research emanating from the cities' various universities, and its usage for multiple other advertising purposes. Meanwhile, joint-branding is central to Chapter 3 on 'NewcastleGateshead'. Chapter 7 reveals similar and powerful trends in many Indian twins. Many cross-border twins be-have similarly – both regarding branding for the world outside (Gelbman and Timothy 2011) and persuasive ritual and festivities for twinned citi-zens (Asher 2012), whereby cultural difference in particular is represented as something for celebration rather than fear. One splendid example is the 12 km Barents Friendship Ski Race occurring annually since 1994 along the Russian–Norwegian–Finnish border, starting and ending in Russia, but crossing into Finland and Norway and attracting thousands of participants from all three countries (including Kirkenes–Nickel's twin citizens) and from abroad (Fors, Espiritu, and Mikhailova 2015) (Figure I.2). Another is the binational swim, held annually since 2001 between Blagoveshchensk and Heihe across 750 m of the Amur River marking the Russo-Chinese border (gtrkamur.ru/news/2017/07/17/22274). Several thousand miles west, we find 'hands across the border', ceremonies between Brownsville and Matamoros (Dear 2013, 44), and similar mutualities between other Mexican–American twins from the mid-nineteenth century onwards, including *Los Dos Laredos* in Chapter 11.

Figure I.2 People enjoying themselves in Rayakoski before the Barents Friendship
    Ski Race by Adrian Aleksander Dávila Bolaños (2018).

Judging the persuasive impact is harder. Certainly, we have clear evidence
of cross-border interests being asserted politically with the formation of the
Transfrontier Operational Mission (Mission Opérationnelle Transfrontal-
ière), the City-Twins Association in Europe, and from 2011 the US–Mexican
Border-Mayors Association. Increasingly, ambitious partnership agreements may well help enhance cross-border popular identities, particularly
as more people become involved in enacting them, still more as ordinary
citizens are drawn into crossing to the other city, and thus other nation, for
shopping, visiting and other purposes. As several chapters argue, alongside
governmental action or inaction, much of the reality or otherwise of 'twinning' is created by the responses and behaviour of ordinary citizens in their
day-to-day mutualities. Chapter 19 highlights shopping and other visiting as
crucial to the Imatra–Svetogorsk relationship across the Finnish–Russian
border, with interaction continuing, albeit diminishingly, even after state
and local government impetus has faded. Several chapters raise the possibility of joint identities and cultures. Chapter 12 suggests that some citizens
along the Polish–German border are starting to make choices – being Polish
or German, or simultaneously Polish *and* German, or neither (see also
Balogh 2014). Chapter 17 suggests the same along the Norwegian–Russian
border. Amazon border-cities in Chapter 10 seem to be similarly prone, as
do Myawaddy and Mae Sot in the 'trans-border space' between Myanmar
and Thailand in Chapter 21. So too do Los Dos Laredos in Chapter 11,

where 'a unique population' draws on both US and Mexican culture, apparently reinforcing Michael Dear's (2013) more general claim about a 'third nation' emerging from at least the nineteenth century along that border – again at least partly independent of central direction. This will be particularly tested if President Trump erects his 'wall'.

Fourth, conurbations are potent diluters of local identities and borders between and around single cities and twin cities alike. Thereby regions become potential units of governance, even popular attachment – in Britain, for example, Greater Manchester embracing far more than Manchester and Salford (Chapter 2); Merseyside embracing Liverpool and its surrounding towns encouraging inhabitants to conceive themselves as 'Scousers'; Tyneside, embracing people thinking of themselves as 'Geordies', extending far beyond 'NewcastleGateshead', and persuading local businesses to take this as their preferred brand (Chapter 3) – a long-evident regional tendency.

At the far end of such openness lies the wholesale loss of local urban identity and the dissolution or erosion of borders between twin cities (amalgamation), perhaps even nation-states. Some intranational twins have already gone this way. Hungary's Buda and Pest were united into Budapest in 1873 (Chapter 4). Here, previous identities were radically transformed by vigorous urban planning, even though greatly modified ones continued affecting governance for several subsequent decades. In Britain, Rochester and Chatham disappeared into Medway, while Brighton and Hove co-joined under central government direction in 1974. More voluntarily, Canada's Port Arthur and Fort William merged into Thunder Bay in 1970 after fifty years of deliberation. Thus far, there has been no co-joining among Europe's various border-twins, though some has been imagined by previously united border twins: Kerkrade and Herzogenrath on the Dutch–German border were supposed to merge into a strange border-crossing municipality called 'Eurode' (Ehlers 2007, 13); in June 2009, Slubice and Frankfurt Oder on the Polish–German border briefly produced 'Slubfurt' (Asher 2012, 498), holding virtual elections for a virtual council (www.voxeurop.eu/en/content/article/71141-slubfurt-unreal-city-some).

Fifth, twin cities are subject to the same tensions and identities almost inherent in the urban fabric since at least 1800 – those of class, status, religion and ethnicity – and which of their nature, run beyond both city and twin-city boundaries. Indeed, as Chapter 15 shows, these can greatly complicate relationships even between modestly sized and rurally situated twin towns. With Comines–Wervik, each since 1830 distributed on either side of the river Lys and thus the French–Belgian border, we have a fine example. Each town is subject to a national divide while simultaneously negotiating its way around a localised version of the Flemish–Walloon divide, imposed by broader central government policy.

As this begins suggesting, inhabitants and organisations will experience and react to tensions between openness and closure in different ways at different times. People's perception and experience in twin cities, as in other sorts of urban entity, can vary. As Chapter 19 suggests for Imatra–Svetogorsk, many citizens may be unaware or only partially aware of twinned-relationships. Beyond that, as Chapters 2 and 3 suggest, mental horizons and experience can vary according to the socio-economic level – from the poor and very poor, whether in twin cities or not, whose world is circumscribed by supportive street and neighbourhood up to the business and professional apex of the social spectrum, where horizons and interests (in every sense) extend increasingly beyond cities, even twin cities, to regions, nations and the evermore global world beyond.

Meanwhile, as Chapter 8 on Islamabad–Rawalpindi clearly illustrates, experiences, horizons and choice-range in twin cities can vary according to gender and religion. Furthermore, common sense suggests that horizons differ according to whether testimony comes from inhabitants, internal/external immigrants, visitors and/or tourists.

Civil organisations also have and make choices between localised, civic, twinned or now increasingly regional identities, as we shall see in some chapters. For any given association, those choices will be determined by the individuals involved, its function (e.g. political, social, sociable, social welfare), strength or weakness, ideology, the stage of twin-city evolution, relationships or not with wider metropolitan areas and probably whether the twin city that the organisation inhabits is national or cross-border. For example, Manchester and Salford organisations have over time exhibited parochial, Manchester *or* Salford, Manchester *and* Salford and now increasingly *Greater* Manchester affiliations. The same has been true of criminal gangs, with choices varying according to what they are into and often intense local loyalties – juvenile gangs may focus on neighbourhoods, while drugs cartels may be international.

Civil authorities in twin cities also experience the need for choice, still more, as has been the case for Manchester/Salford, Newcastle/Gateshead, and many Indian twin cities, have it variably imposed upon them from above – federal state or national government.

Alongside interdependence and inner tension, a third frequent characteristic of twin cities is inequality. While some twins are more or less equal and competitive, most are locked into unequal relationships of apparent dominance and subordination. Sometimes one city is just much bigger than the other in population, economy and the pulling-power of its centre to a point where the smaller is seen by outsiders, even some insiders, as in Gateshead's case, 'a dirty little lane running down into Newcastle' (Chapter 3). Sometimes, the smaller twin becomes merely the closest of many satellite towns to a dominant regional capital. This inequality can reach a point where the larger twin is seen as sucking the commercial life from, eroding the viability, even the identity of, the smaller (Antrobus 2009).

Yet, this inequality often breeds resentment about subordination, and 'humiliation', becoming a major source, perhaps the only remaining source, of identity for the lesser twin viewing: 'our over-mighty neighbour'. Underpinning this is the often-realistic sense of one-way dependence, where the subordinate city cannot resolve many of its problems or run its services without its often-indifferent partner's co-operation. Such relations are evident among urban units in any emerging conurbation with a dominant core-city (arguably part of the essence of what conurbations are), but, for twin cities, such inequality and dependence is evident in increasingly intense forms from the start.

The fourth characteristic of twin cities follows from everything suggested thus far. Given interdependence, proximity, consciousness of relationship and needs to co-ordinate, often formalised for border-cities by written agreement, it follows that twin cities are locked into frequent negotiation, dialogue and informal networking. Although some of this may need settling by central government or governments, it is often easier and faster to resolve even cross-border issues face-to-face. This applies not just to municipalities, but perhaps also to all organs of governance including relevant civil organisations like charities, even criminal organisations.

It is sometimes said that twin cities tend to merge. In fact, and this is their fifth and very marked characteristic, at the point of writing, they mostly do not. Indeed, more are being generated at the cross-border level and, for planning purposes, within countries like India. Twin cities are resilient. Why is this: ie why do they remain closely linked but still separate? And why do they multiply? Various interconnected reasons suggest themselves.

One is that they start that way. Two cities are founded in close proximity but in separated jurisdictions with contrasting functions perhaps by different ethnic groups – and then, once started, continue. In other words, historical circumstances explain origins, and then historical trajectory (Eskelinen and Kotilainen 2005), inertia and developing vested interest explains twin-city survival and the specifics of what emerges. History and equally people's sense of, and attachment to, that history are important. Lagos and Ikeja's emergence and continuation in Nigeria point this way in Chapter 5. Furthermore, borders, once imposed, enhance the distinctiveness of what develops on each side; this is possible even where a twin develops from what was originally a single city – as suggested by Chapters 14, 15, 16 and 22.

If history, its perception and the interests, inertia and identities developing around its legacy help explain the survival of national twins, we should also note that separated border twins are also actively useful, evermore so as international trade increases. While the intermixing and people-flows border-cities encourage help to dissolve borders,, thereby serving higher European foreign policy purposes, their existence as recognised crossing-points enables that flow to be controlled, something doubly important in the context of the recent Middle-Eastern exodus. Beyond this, borders represent resources for many folk – to trade across, to attract tourists wishing to safely sample alternative cultures, to access cheap alcohol or more liberal drinking regulations and, as evident in Chapter 11 on US–Mexican cities, to sample health and other services on the other side. Border towns are locations where trade-oriented industries, processing plants, etc., can be located, rendering them major points of employment. Borders, as all the chapters suggest, have been and remain important variables.

Beyond this, and here we return to national twins, the twin city, more properly the imagined twin city as an ideal and 'model', has become a major source of inspiration as a way of exerting control over burgeoning Indian urbanisation and trying to ensure that cities expand and grow towards each other in controlled ways. As we shall see, it has become a symbol of modernity. Almost always, the almost platonic ideal derives from the oldest twin city of all – Minneapolis–St. Paul. Twin cities are not just alive, but also very well and actively breeding across all continents. Hence this book.

Notes

1 The term 'cities' simply indicates urban places rather than legal status or largeness.
2 Special Issue on twin cities in *GeoJournal* in 2001; *Geographica Helvetica*, 2007:1, is devoted to Border towns/city boundaries – boundaryless spaces; Llera and Ángeles 2012 'Cross-border collaboration in border twin cities: Lessons and challenges for the Ciudad Juárez-El Paso and the Frankfurt (Oder)-Slubice'; Special section 'Borders and Cities: Perspectives from North America and Europe' in *Journal of Borderlands Studies* in 2013; Covarrubias 2016 'Bridging the socioeconomic gap: integrating cross-border regions through comparing different worlds – region Laredo, Aquitaine-Euskadi and Øresund' (PhD dissertation); Special Issue on theorization of city-twinning of *Journal of Borderlands Studies* in 2016).
3 We are not concerned with the many 'twinned cities' – that is, distant urban places in different countries that become 'twinned' in order theoretically to erode international barriers via exchanging people, culture, educational experience, etc. These are not included even while clearly having affinities with twin cities as defined here.
4 The term 'conurbation' was first used by Patrick Geddes in 1915 (*Cities in Evolution*) to describe conurban areas in Germany, India, Japan, the Philippines, the UK and the USA. Thomas Freeman made another attempt to conceptualise conurbations and intranational twin cities in *The Conurbations of Great Britain* (1959), where he briefly discussed the urban coexistence of Manchester–Salford, Runcorn–Widnes, Newcastle–Gateshead and London–Southwark.
5 Some countries like Japan, the Netherlands and the UK have been prolific generators of 'adjacent urban nodes'; some, like Tyneside in Britain, began becoming conurbations as early as the 1850s (See Michael Barke and Peter J. Taylor, 'Newcastle's long nineteenth century: a world historical interpretation of making a multi-nodal city region', *Urban History*, 42(1), February 2015, 43–69).
6 For an attempt to see twin cities in a metropolitan context, see Meijers, E., Hoogerbrugge, M., and Hollander, K. (2014) 'Twin cities in the process of metropolisation', *Urban Research and Practice*, 7(1), 35–55; Trillo Santamaría, J. M., Lois González, R. C., and Paül Carril, V. (2015) 'Ciudades que cruzan la frontera', *Cuadernos Geográficos*, 54(1), 160–185.
7 John Garrard is indebted to Eve Garrard for this useful way of thinking about the various sorts of twin cities: originating from Ludwig Wittgenstein *Philosophical Investigations* 1953.
8 In 2011, London hosted an estimated 270 nationalities and 300 languages: www.standard.co.uk/news/270-nationalities-and-300-different-languages-how-a-united-nations-of-workers-is-driving-london-6572417.html (accessed March 20, 2018).
9 The situation has changed somewhat, with the Swedish government introducing passport checks between Copenhagen and Malmö from January 2016 till

May 2017, seriously disrupting the fast rail link by doubling the usual 30-minute commute to one hour and obstructing life for 20,000 commuters. Reportedly, 565 of them sued the Swedish government 'claiming the delays had forced them to buy cars, even change jobs'. Meanwhile, Copenhagen has changed its priorities and insists on including Malmö in the Greater Copenhagen region: www.citylab.com/transportation/2017/05/copenhagen-malmo-oresund-bridge-border-passport-checks/525570/ (accessed March 29, 2018).

10 Even though towns within a nineteenth-century 'multi-centred city region', like the German Ruhr, the UK's Tyneside or South-East Lancashire, were economically interdependent because of sharing one or more dominant staple industries, they were nevertheless socially and politically autonomous in ways that closely adjacent twin cities could never be.

References

Albuquerque, P. H. (2007) 'Shared legacies, disparate outcomes: why American south border cities turned the tables on crime and their Mexican sisters did not', *Crime, Law and Social Change*, *47*(2), 69–88.

Anishenko, A., and Sergunin, A. (2012) 'Twin cities: a new form of cross-border cooperation in the Baltic Sea Region?' *Baltic Region*, *1*, 19–27.

Antrobus, D. (2009) 'Broken hearted cities: cut off, cut up and cut down', Paper at the symposium on peripheral cities, University of Paris, 8 December.

Asher, A. D. (2012). 'Inventing a city: cultural citizenship in 'Słubfurt', *Social Identities*, *18*(5), 497–520.

Balogh, P. (2014) *Perpetual borders: German-Polish cross-border contacts in the Szczecin area* (Doctoral dissertation, Stockholm University).

Bucken-Knapp, G. (2001) 'Just a train-ride away, but still worlds apart: prospects for the Øresund region as a binational city', *GeoJournal*, *54*(1), 51–60.

Buursink, J., and Ehlers, N. (1999) 'The binational city of Eurode', *online depository of ERSA conference papers*.

Buursink, J. (2001) 'The binational reality of border-crossing cities', *GeoJournal*, *54*(1), 7–19.

Cañas, J., Coronado, R., Gilmer, R. W., and Saucedo, E. (2013) 'The impact of the maquiladora industry on US border cities', *Growth and Change*, *44*(3), 415–442.

Case, P. et al. (2008) 'At the borders, on the edge: use of injected methamphetamine in Tijuana and Ciudad Juarez, Mexico', *Journal Immigrant and Minority Health*, *10*, 23–33.

Dear, M. (2013), *Why Walls Won't Work: Repairing the US-Mexico Divide*, New York, Oxford University Press.

Del Biaggio, C. (2015) 'Investigating regional identities within the pan-Alpine governance system: the presence or absence of identification with a "community of problems" among local political actors', *Environmental Science and Policy*, *49*, 45–56.

Ehlers, N. (2007) *The Binational City Eurode: The Social Legitimacy of a Border-Crossing Town*, Aachen, Shaker.

Eskelinen, H., and Kotilainen, J. (2005) 'A vision of a Twin City: exploring the only case of adjacent urban settlements at the Finnish-Russian border', *Journal of Borderlands Studies*, *20*(2), 31–46.

Fors, B. S., Espiritu, A. A., and Mikhaylova, E. (2015), 'Norway-Russia Case Study', *Report* no. 1. Euborderregions.

Garrido, E. N. (2007) 'Inter-municipal cooperation in Spain: dealing with micro-scopic local government', in *Inter-Municipal Cooperation in Europe*, Springer, Dordrecht, 169–192.

Gelbman, A., and Timothy, D. J. (2011) 'Border complexity, tourism and international exclaves: a case study', *Annals of Tourism Research, 38*(1), 110–131.

Haugseth, P. (2013) 'Tvillingbysamarbeid i den norsk-russiske grensesonen', in A. Viken, B. S. Fors (eds.), *Grenseliv*. Tromsø, Orkana Akademisk, 21–38.

Headlam, N. (2014) 'Liverchester and Manpool? the curious case of the lack of Intra-Urban Leadership in the twin cities of the Northwest', in J. Diamond, and J. Liiddle (eds.), *European Public Leadership in Crisis?* Bingley, Emerald Publishing Group, 47–61.

Heddebault, O. (2001) 'The bi-national cities of Dover and Calais', *GeoJournal, 54*(1), 61–71.

Herzog, L., and Sohn, C. (2014) 'The cross-border metropolis in a global age: a conceptual model and empirical evidence from the US–Mexico and European border regions', *Global Society, 28*(4), 441–461.

Joenniemi, P., and Sergunin, A. (2009) 'When two aspire to become one: city-twinning in Northern Europe', *DIIS Working Paper 21*.

Joenniemi, P., and Sergunin, A. (2011) 'City-twinning in Northern Europe: challenges and opportunities', *Research Journal of International Studies, 22*, 120–131.

Joenniemi, P., and Sergunin, A. (2014) 'Paradiplomacy as a capacity-building strategy: the case of Russia's northwestern subnational actors', *Problems of Post-Communism, 61*(6), 18–33.

Kada, N., and Kiy, R. (2004) 'Blurred borders: trans-boundary impacts and solutions in the San Diego-Tijuana border region', *Report of the International Community Foundation*.

Kilburn, J., San Miguel, C., and Kwak, D. H. (2013) 'Is fear of crime splitting the sister cities? the case of Los Dos Laredos', *Cities, 34*, 30–36.

Löfgren, O. (2008) 'Regionauts: the transformation of cross-border regions in Scandinavia', *European Urban and Regional Studies, 15*(3), 195–209.

Lundén, T., and Zalamans, D. (2001) 'Local co-operation, ethnic diversity and state territoriality–the case of Haparanda and Tornio on the Sweden–Finland border', *GeoJournal, 54*(1), 33–42.

Mäki, K. (2016), 'Helsinki urban node: Helsinki-Tallinn case', Port of Helsinki, available at Finnish Transport Agency webpage: www.liikennevirasto.fi (accessed March 27, 2018).

Mathieson, U., and Burkner, H.-J. (2001) 'Antagonistic structures in border areas: local milieu and local politics in the Polish-German twin city Gubin/Guben', *GeoJournal, 54*(1), 43–50.

Mikhailova, E., and Wu, C. T. (2017), 'Ersatz twin city formation? the case of Blagoveshchensk and Heihe', *Journal of Borderlands Studies, 32*(4), 513–533.

Ramos, R. et al. (2009) 'A tale of two cities: social and environmental influences shaping risk factors and protective behaviors in two Mexico–US border cities', *Health & Place 15*(4), 999–1005.

Schulz, H., Stokłosa, K., and Jajeśniak-Quast, D. (2002), 'Twin towns on the border as laboratories of European Integration', FIT Discussion Paper, 4.

Sparrow, G. (2001) 'San Diego–Tijuana: not quite a binational city or region', *GeoJournal, 54*(1), 73–83.

Wang, J., Cheng, Y., and Mo, H. (2014) 'The spatio-temporal distribution and development modes of border ports in China', *Sustainability, 6*(10), 7089–7106.

Part I
Intranational twin cities

1   Minneapolis–St. Paul

The iconic twins?

Mary Lethert Wingerd and John Garrard

Our collection on twin cities around the world starts with Minneapolis and St. Paul, the pioneer example and an inspiration and reference point for many other twin cities covered in this book and beyond it. The chapter shows that, despite their easily recognizable global brand, the historic relationship of 'The Twin Cities' has been traditionally antagonistic: a pattern repeated for other intranational twins. Only fairly recently have things changed – though, as will emerge, peace has recently been breaking out across all internal twin cities. The key factors here seem to be connected to the partial fading of local identities and enhancement of central and/or federal–state government roles in the wake of metropolitization and business delocalization. As in the two following cases, the axis is becoming increasingly regional.

Minneapolis and St. Paul are iconic twin cities, and they have permitted themselves to be so labelled since at least the 1890s. They are twins of the sort that arise side by side along a common border (the Mississippi River) and grow outwards. Like many others of their kind, though still municipally self-governing, they now centre a much broader multi-settlement metropolis, sporting its own Metropolitan Council since 1967. Though intertwined from their origins in the 1860s, their relationship was mutually contemptuous, competitive and conflictual over many decades. However, since the late 1960s, their relations have become increasingly co-operative.

These Minnesota twins have never been identical. Rather, they are 'cities with a common identity that retain individual characteristics that mark them out as separate' (Kane and Ominsky 1983, 1). Indeed, they were and are as different as Los Angeles and San Francisco. Both owed their existence to the Mississippi, but have very different economic origins, historic economic bases and, partly in consequence, economic fortunes and patterns of class relations. In what follows, St. Paul gets more attention due to the main author's central expertise, even though it is arguably the subordinate settlement economically. However, the changing patterns of inter-city relations will repeat in the chapters that follow.

* * *

Steam boats could go no further up-river that the site of St. Paul. Consequently, traders, merchants and speculators congregated there, eventually carving a city from the adjacent bluffs and valleys. Hugging the river, it began as a fur-trading post, making its fortune as a transport and mercantile hub. It became a major railway town, for a while seeing itself as the emerging distribution centre for the rapidly populating west. Minneapolis, the slightly younger settlement, was established around St. Anthony Falls, founded by enterprising Yankee capitalists harnessing waterpower to create an industrial dynamo (Figure 1.1). Feeding on the northern forests and western wheat fields, by the 1880s, Minneapolis was producing more lumber than any other US city and was the world's flour-milling capital. In the same decade, its population overtook St. Paul's, to the latter's fury. The embodiment of industrial capitalism, it was destined for greater success than St. Paul with its mercantile base. For all the latter's ambitions, the closing western frontier, transportation advances and refrigerated rail cars meant that it increasingly became just a regional distribution centre with a shrinking market, struggling mightily to maintain even that modest position (Cronon 1991, 233–9). Doubly so, once the Panama Canal opened in 1914. What saved St. Paul from total overshadowing was its early established role as state capital, with a set of distinguished buildings to match.

The two cities have always looked different. St. Paul grew haphazardly, whereas Minneapolis was planned from birth. The Mississippi's west side was closed to settlement till 1854, but well-heeled, politically connected capitalists were poised to take control of the falls long before the land became

Figure 1.1 St. Anthony Falls, the birthplace of Minneapolis by Andrii Ovsiannikov (2017).

officially available, immediately setting about creating a well-ordered homogeneous New England enclave fixed on industrial development. The long-term result was a city of sky-scrapers. Even today, 'Minneapolis with its wide boulevards, organized grid lay-out and modern downtown, stands in striking contrast to ... St. Paul across the river, with its late Victorian architecture, narrower streets and irregularly shaped neighbourhoods' (The Twin Cities, Community-wealth, 1).

From early on, the cities were ethnically and religiously different. St. Paul was heavily influenced by its German, French and Irish Catholic roots, with German and particularly Irish Catholics most politically prominent. Minneapolis was influenced by its early Scandinavian and Lutheran heritage, but politically and economically dominated by its original New England settlers.

All this partly underpins the one historic thing they had in common: as *Fortune Magazine* April 1936 remarked in an article excoriating both cities, 'they hate each other'. As we shall see, this antipathy embraced a wide spectrum of both urban communities. They competitively deployed vastly magnificent cathedrals: St. Paul completing its Catholic monolith (replicating St. Peters in Rome) in 1915, with Minneapolis following with the equally ornate, but Anglican, Basilica of St. Mary in 1926. Periodically, they did less godly battle. Sports were multiple cases in point: 'No sporting rivalry was more bitterly contested over a longer period than the baseball competition between Minneapolis and St. Paul' (O'Neal 2006, 52). Meanwhile, conflict over the 1890 Census saw the cities arresting/kidnapping each other's census-takers and St. Paul newspaper allegations that Minneapolis, that 'strumpet', that 'draggled wench of the streets', was engaged in 'wholesale padding of the returns' (*St. Paul Pioneer Press*, 22 June 1890, 1). Minneapolis businessmen, meanwhile, accused St. Paul of mean-spirited jealousy of its more 'beautiful, stalwart sister', concluding, 'it only shows how meanly and despicably St. Paul people can act' (Johnson 1890, 14). Not entirely unjustifiably, given what we shall see shortly were St. Paul's policing practices.

Their contrasting fortunes and economies, and their respective elites' responses, meant that the twins also showed very different class relations. Connectedly, while both made shows of civic pride, the localized patriotism of St. Paul was arguably far more deep-rooted than that observable in Minneapolis. Civic identity is always problematic (viz. Dirlik 1999) even in the nineteenth century when US (and British) urban 'communities' were far more separated, in-turned and insulated from the wider world than has been the case for their post-1945 counterparts, still more the global-oriented Internet-fed world of the twenty-first century. In many early class-riven American cities, pride in place, though loudly trumpeted, was little more than business boosterism, drawing in few outside the business elites deploying the idea competitively for commercial purposes.

Minneapolis provides a case in point. There, a small cohort of powerful capitalists, descended from families of the original New England settlers, jealously guarded control: dominating politics, finance and business, and suppressing labour activism with an iron hand. This was so successful that, in April 1936, *Fortune* magazine observed that, though Minneapolis possessed the country's largest Scandinavian population, 'socially and financially [it] is still dominated by the original New-England families'. In 1903, Minneapolis business interests established a vigorous branch of the anti-union Citizen's Alliance, dedicated to making Minneapolis the country's 'Open-Shop Capital', a distinction maintained until the city's historic truckers' strike of 1934 ('Revolt in the North-west', *Fortune* April 1936, 116; Nord 1986). The open-shop initiative generated three major strikes between 1901 and 1903, alongside many smaller disputes. In 1901, an all-out attack on the International Association of Machinists virtually closed Minneapolis machine shops; it was further disrupted by a teamsters' strike in 1902; and in 1903 the Milling Trust drove organized labour from the industry. According to the official biographer of General Mills, the unions (previously representing almost every worker in the industry) 'were practically eliminated and never regained their former strength among the workmen in the mills'. Writing about Minneapolis's labour movement, Elizabeth Faue (1991) has delineated the deep class fissures dividing the city: 'community' was grounded in working-class neighbourhoods, perceiving themselves embattled against a dominant business class engaged in civic control.

By contrast, St. Paul's civic leaders, facing inevitable decline and increasing reliance on local and regional markets, discarded aspirations to become a major metropolis, instead of striving to turn city liabilities into virtue. Smaller was now better: St. Paul, unlike its rival, cared about its citizens. It was not a heartless industrial dynamo, but a community. Community bonds logically demanded St. Paulites to patronize local businesses and support the local Democratic Party (the sole Democratic bastion in Republican-dominated Minnesota).

Minneapolis was an ideal foil against which to construct intense civic parochialism. Its civic leaders unceasingly gloated about its economic predominance, and claimed cultural superiority. The Protestant Yankee's 'progressives' controlling Minneapolis looked disdainfully on St. Paul, where German, and particularly Irish, Catholics shared power with old-stock elite. Typically, one assessment described St. Paul as 'a dive, populated ... by politicians (a very lowly breed), Democrats (even more lowly), Irish (another step down), and drunks' (Salisbury in Anderson 1976). To this disreputable litany the writer might have added equally distasteful 'Catholics'.

However, cultural defensiveness and just disliking Minneapolis were insufficient to foster St. Paul's fierce parochialism. To engender broad-based city-wide loyalty, community claims had to prove themselves, by some sort of pay-back across class. St. Paul's economy of civic identity drew settled workers and small proprietors into negotiated compact with business, a project where cultural and material elements inter-mingled.

Organized labour was making enormous strides nationally. Between 1898 and 1904, the American Federation of Labour (AFL) membership increased from 278,000 to 600,000 (US Commissioner of Labor 1906, 490–1). St. Paul's union membership increased nearly a third: in 1900–2, there were 76 locals with over 7,000 members, fully 31% of city wage-earners. Minneapolis saw similarly explosive union growth, there being 36,576 wage-earners of whom 13,140 (36%) were unionists, a near-24% increase on 1900 (Bureau of Labor 1901–2 50, 97, 452–53). But where Minneapolis businessmen dedicated their considerable resources to combating union power, St. Paul's made their peace, acquiescing almost universally in the closed shop.

The reasons were complex. St. Paul's organized labour was deeply rooted. The Trades and Labour Assembly, established in 1882, was home to both the burgeoning trade-union movement, numbering 1,500 in 1885, and 13 assemblies of the more radical and more open Knights of Labour. The latter's demise in the 1890s simply meant increased AFL membership. St. Paul's organized labour was supported by the local Democratic Party, with its influence heavily dependent on working-class votes. But organized labour also had surprisingly strong Catholic Church support, and at least tacit local-business approval. The two were interlinked.

John Ireland, Archbishop of St. Paul, was both internationally and locally prominent in American Catholicism. His undeniable stature, alongside his inveterate boosterism of St. Paul, made him welcome among city elites, who then generously financially supported his episcopal endeavours. But Catholic power also rested on overwhelmingly working-class communicants. Ireland brilliantly balanced the two constituencies. While many senior Catholics opposed organized labour, he championed the Knights of Labour, enjoying being called 'the socialist bishop' – grossly exaggerated but endearing him to working-class Catholics (O'Connell 1988, 229–38, 394). However, he tempered labour support with insistence on private property's preeminent rights, and condemned strikes and boycotts, thereby reassuring businessmen that, in St. Paul anyway, labour insurgency was containable. To reassure working-class Catholics, Ireland appointed Father John Ryan to prominence at the St. Paul Seminary. Ryan became known as American Catholicism's foremost social-justice champion: between 1902 and 1915, railing from the seminary against capitalism's excesses, supporting a living wage, an eight-hour law, protection for picketing and boycotting, and generally advocating working people's interests: earning great applause from St. Paul's *Union Advocate* (*UA*). While Ireland reassured civic elites that the Church would protect their interests, Ryan reassured the working class. Thereby Catholicism became a central mediator of civic tensions.

Thus, St. Paul business leaders could contemplate the advantages of 'moderate' unionism. A massive Great Northern Railway strike in 1894 had triggered costly repercussions across St. Paul's economy, teaching city businessmen that such disruption could easily destroy its precarious economic equilibrium. They quickly saw the advantages later resulting from rationalized negotiations between the railroads and organized workers, applying the same lesson to more immediate labour relations. Few employers could afford stoppages. Furthermore, they soon discovered that, when strikes disrupted Minneapolis operations, St. Paul businesses invariably experienced upsurging orders and profits (Edgar 1925, 235).

Such businessmen hardly became union enthusiasts; pragmatism simply took precedence over ideology. Not only did they lack the resources to unilaterally control labour negotiation, but also St. Paul's commercial economy depended on city workers as both customers and employees. In 1903, some local businessmen tried establishing the St. Paul Citizens' Association (counterpart to Minneapolis's Citizens' Alliance). It floundered after only a few meetings, with tempted local businessmen proclaiming that 'public opinion' would not stand for aggressive opposition to organized labour (Minutes 1903).

Public opinion was never far from city-business minds. With Minneapolis' attractions only a short streetcar-fare away, manufacturers and retailers necessarily relied on less tangible reasons to keep consumers at home. They called for community loyalty, civic mutuality; not 'quantity, but quality', declaring the city's comparatively small manufacturing output resulted from 'the care used in the manufacture'. Urging consumers to 'get the goods that are made [and sold] at home', they reminded working people that they 'favoured the use of Saint Paul labour' as well as products (Saint Paul Almanack 1915, 60).

St. Paul labour meant organized labour. Indeed, business weakness translated proportionally into union strength and the unions fully exploited the leverage derived from the civic compact. When management and labour negotiated, neither thought their interests to be identical. Nonetheless, negotiational discourse was always about community accountability: mutual rights, responsibilities and citizenly obligations. This required mutual compromise, but numbers told the tale. The wages in St. Paul equalled, often exceeded, those in Minneapolis. In negotiating hours and conditions, claims of community responsibility proved to be highly effective. Thus, labour newspapers had good reason to consistently downplay class divisions, emphasizing negotiations as 'harmonious and pleasant throughout ... the best feeling prevails between employers and employees' (*Minnesota Union Advocate*, 15 June 1917, 1).

When infrequent disputes disrupted industrial 'harmony', the unions used similar language to buttress their cause. Thus, unionized workers boycotted a local department store for betraying 'old-time residents' by hiring 'migratory and often incompetent worker[s]' without 'interest in Saint Paul or the community', thereby betraying the civic compact, forcing legitimate members of the community to leave. This argument was powerful in St. Paul. In 1903, Citizens' Association members were justly alarmed when journeymen plumbers accused local plumbing contractors of 'import[ing] men from other cities ... to fill our places'. As they feared, public opinion sided with the plumbers, as did district-court judge William Kelly, refusing an injunction against the strikers. He urged the 'two classes [to] dwell in harmony and agreement', citing Pope Leo XIII's *Rerum Novarum* encyclical, urging 'all the owners of capital' to 'pay heed to the simple and beautiful lessons of justice taught in the encyclical' (Pamphlet, William Mahoney Papers 1903). Local businessmen could not ignore such admonitions; thus, labour mostly won its point.

The unions used similar tactics to uphold wage-rates and improve working conditions. The *UA* regularly reminded local merchants that generous union wage scales meant more consumer dollars fuelling the local economy. By upholding union scales, it argued that, under what was called St. Paul's 'live and let live' plan, they would make 'a bid for a share of this vast amount of money' (*UA*, 5 January 1917, 3). Concurrently, the paper encouraged union members to

> buy in our home town ... because the community which is good enough for me to live in is good enough to buy in ... every dollar I spend at home stays at home ... the man I buy from helps support my schools, my church, my lodge, my home.
>
> (*UA*, 23 February 1917, 8)

These arguments were largely persuasive among union workers. One retired teamster noted that he and his friends seldom went to Minneapolis: 'really a foreign place ... people really felt ... they should spend their money at home ... because if you go to Minneapolis and spend it, it will be gone. It won't turn over eight times like ... in St. Paul' (Gus Larson, interview, Wingerd, 27 July 1996, St. Paul).

St. Paulites assiduously guarded both real and imagined municipal borders. City businesses seldom cooperated with Minneapolis, nor did its businessmen's association encourage new enterprises, fearing further fragmenting their limited market ('The Men and Products of St. Paul', 60). More important was 'the growth of ... present industries'. Labour solidarity also stopped at city limits (just as it stopped at unskilled workers). Minneapolis locals were not allies but competitors for jobs and members. The *UA* regularly reminded readers about 'the sneering ejaculations' about 'poor old St. Paul ... from our self-esteeming and self-satisfied friends and neighbours in our sister city' (*UA*, 12 October 1917, 4). St. Paul's vision was narrow indeed. When crusading Irish-Catholic attorney James Manahan challenged monopolistic railroad rates on behalf of Minnesota's beleaguered farmers, he encountered 'acrimonious scorn' from fellow St. Paulites, including 'overalled mechanics', stenographers, even his office building's elevator boy, as well as 'white-collar clerical workers' (Manahan 1927, 153–4). Any St. Paulite could have told him that the railroad was the city's lifeblood, not suffering hinterland farmers.

Such intense civic loyalty undoubtedly owed much to the fact that it delivered the goods in terms of wages and secure unionism, but the resulting quality of life could also internalize the 'community' attachment St. Paul posed against the world outside. This is most tellingly illustrated by the peculiar system devised to control vice and maintain law and order within city limits. The sole enterprises wherein St. Paul consistently outpaced Minneapolis were breweries and saloons. Alcohol was a cornerstone of the local economy, alongside attendant gambling and prostitution. The city had a long-standing reputation as a 'wide-open' town, a less touted aspect of civic identity, but still integral to the municipal economy. Its budget derived significant income from saloon-licensing and regular fines on gambling dens and brothels. Moreover, city businessmen considered the 'vice economy' an important attraction for out-of-town buyers (Schmid 1937, 32, 35–6; Best 1987).

Unfortunately, these enticements also attracted less desirable visitors. To keep illicit aspects of city business from mushrooming beyond control and still retain sorely needed municipal income, in 1900 Police Chief John O'Connor, aided by brother Richard, Democratic Party boss, engineered an ingenious crime-control system, spreading word that criminals of any sort were safe from prosecution, provided that they checked in at police headquarters within 12 hours of arrival, provided local police with generous gratuities and 'behaved themselves' within city limits. Local judges consistently refused extradition to other cities, and St. Paul placidly embraced the role of criminal haven for over 30 years, becoming a bolt-hole for some of the era's most notorious criminals, including John Dilinger, Ma Baker and Alvin Karpis (Maccabee 1995, 3, 8–9; Regan 1981, 143).

O'Connor's system was an open secret, which respectable St. Paulites commonly endorsed. The city was virtually crime-free; O'Connor almost universally lionized as 'justly renowned'. As the police department's 1904 official souvenir book boasted, 'never' in St. Paul history 'has human life and the property of citizens been so safe and the virtue of women so assured'. If criminal forays outside the city wreaked mischief elsewhere (like Minneapolis, for example), the events went unreported. Only after the criminals broke the city's cardinal rule by kidnapping two prominent local businessmen did public outcry finally end the system.

To outsiders then, St. Paul presented a united front of stubborn parochialism, but competing internal demands kept civic equilibrium in constant negotiation. Irish Catholics were primary brokers here. Though not demographically dominant, they held influential positions across civic life. A common construction of their shared (and largely imagined) persecution as Irish and Catholics created a surprisingly fluid social space, wherein an Irish brotherhood of business leaders, Catholic clergy, union officials, politicians and even local criminal bosses mingled to negotiate for their various constituencies. If outsiders derisively labelled St. Paul an 'Irish-Catholic town', for city residents of whatever class, ethnicity or religion, the Catholic

Church and its Irish were important and valued players in the complicated negotiations keeping the city afloat.

* * *

The old rivalries had one final kick to deliver. In May 1965, perhaps maintaining its long-held self-image as 'the last city of the east' in reaction to Minneapolis's boast about being 'the first city of the west', St. Paul Council decided to move over to daylight-saving time two weeks ahead of the rest of Minnesota, including Minneapolis, notwithstanding fury from the latter's mayor, the state governor and significant local chaos (Roberts 2014). However, the experiment was not repeated. Since then, public inter-city conflict has increasingly been ritualized, leisure-oriented and even mock-nostalgic. The year 2015 saw the twins tie for first place in the Trust for Public Land's nation-wide contest over who had the finest city parks, with the two mayors ceremonially proclaiming the result at their mutual border (*Star Tribune* 20 May 2015). The year 2017 witnessed a long-prepared inter-city tug of war ('Pulling Together') across the Mississippi, ultimately tied again 20-20, but with Minneapolis predictably winning after raising most charitable money. http://minnesota.cbslocal.com/2017/06/10/minneapolis-st-paul-rivalry-pulling-together/

As this implies, overt conflict has now largely been superseded by co-operation, even increasing collective pride. Even describing the indicators of this medium-term sea change implies the causal factors involved: not just for Minneapolis–St. Paul but for increasingly conurbanized cities across North America, Europe and beyond.

Representative sports teams are no longer points of inter-city conflict. Even when locally based, they now derive their titles from the state: for example, the Minnesota Twins in the American League (baseball) or the Minnesota Vikings in the National Football League. Mostly, they embrace both twins.

The twins remain self-governing, separately presiding over wide-ranging municipal fields and pursuing distinctive policies, sometimes into eccentricity. Thus, while Minneapolis provides a garbage-collection and snow-removal service, St. Paul does not, leaving its householders to negotiate (or not) their own deals with myriad private contractors: raising 'questions of basic civics' for Minneapolis's online newspaper (*MINNPOST*, 14 December 2015).

Nevertheless, the municipalities clearly recognize considerable commonality, and increasing need to co-ordinate. Indeed, alongside mutual hostility, some commonality was also implicitly recognized with the formation of the Twin City Rapid Transit Company (TCRT) in 1891 from two previously separated Minneapolis and St. Paul streetcar companies. The two councils franchised and refranchised this company, just as they had consented to the first inter-city streetcar connection in December 1890. They could hardly do otherwise, given the ever-increasing popularity of this rapidly expanding public transport network (peaking in 1920 at 238,388,782 passengers: Transport Deployment Casebook, 9; Diers and Isaacs 2007). This was so even if, as noted, the medium-term impact of its citizens' ability to travel to the 'other' city was to help reinforce St. Paul's determination to contractually engage with its own citizens.

However, the later twentieth, still more the early twenty-first, century has seen far more explicit recognition of interdependence and common interest. In October 2014, the two mayors met to 'talk taxes, roads, co-operation', a 'love-in' climaxed by a joint radio interview, wherein they mutually declared affection for each other's city (*Pioneer Press*, 20 October 2014). This was far from the first such collaboration, nor was the joint mayoral presentation of the Metropolitan Business Plan, with its five integrated regional economic development strategies, at the highly prestigious and influential Brookings Institution in Washington in April 2011 (https://www.brookings.edu/research/metropolitan-business-plans-a-new-approach-to-economic-growth/)

In recent decades, twinnedness has become something to be celebrated and advantageously exploited rather than fatalistically accepted. Nowhere is this more dramatically evident than in the ever-increasing use of the 'twin-city brand' to enhance tourism, to increase the visibility of regional university research, or for voluntary and trading purposes, both by the twins themselves and by towns within the immediately surrounding towns. At the point of writing, the Google search engine reveals well over a thousand economic, sporting, leisure, tourist, welfare, educational and other organizations and activities utilizing some version of the brand to advertise what they do and where they think they are (Images for Twin Cities). Alongside this are many other organizations identifying themselves as 'Minneapolis–St. Paul'. Civil society, broadly interpreted, has apparently embraced the broad identity with some gusto. On this evidence, these cities are not just the oldest internal twins, but have also become the most self-conscious.

In fact, for many involved in such self-branding, the twin-city title has become a synonym for various versions of the wider region, across which they often necessarily now operate. On 2 May 2017, the *Mpls.St.Paul Magazine* found '52 Reasons to Love the Twin Cities' in 'a wide-ranging, free-wheeling, big, bold bear hug to the greatest metro area on earth'. This highlights the fact that Minneapolis–St. Paul stand at the centre of a substantial conurbation, which is often the level where the real co-operation, co-ordination and indeed planning are seen as needing to occur. Since at least the 1960s, the two cities (with a combined 2016 estimated population of 724,394) have become 'merely' the largest entities in a wider metropolitan area – variously estimated in 2016 at 3,041,526 (the seven-county Metropolitan Council Area) and 3,894,820 (the 16-county Minneapolis–St. Paul Combined Statistical Area, stretching into Wisconsin) (www.citypopulation.de/php/usa-combmetro.php?cid=378). This is so even if the twins have now resumed their position as the region's fastest growing settlements (Reported *Star Tribune*, 17 May 2017).

Very significantly, the need to think regionally has been increasingly evident since at least 1967, and the establishment by Minnesota State (based in St. Paul, the state capital) of the Twin-City Metropolitan Council (TCMC). With powers successively widened in 1974, 1976 and 1994, the Council represented recognition of what had been happening, and became a vehicle for its planned enhancement. Able to supersede municipal decisions and actions, and comprising state-governor-appointed local representatives, it has provided and managed services deemed to be regional. These have now included public transport, sewage treatment, water quality, regional and urban planning, population forecasting, ensuring adequate affordable housing and regional parks (Metropolitan Council www.metrocouncil.org, accessed 22/01/2018). With its impact probably enhanced by Minnesota's willingness to countenance the progressive and urban-advantaging notion of tax equalization (*The Atlantic*, March 2015), the Council has become an influential agency of regionalization. This is most easily evident in public transport and regional planning.

Metro Transit has operated since 1967, superseding the nearly defunct TCRT Company. Initially under its own Commission, it came under TCMC control in 1994. Exclusively running buses at first, it moved into light railways from 2004. Competing increasingly successfully with private cars beset by ever-increasing traffic jams, its bus and light-rail services reached 84.5 million trips in 2014, a total whose impressiveness only pales in face of TCRT's own interwar passenger figures at their peak, before private cars began fully taking off. Like most urban public transport systems, both were far better at conveying passengers into the two city centres than between suburbs. Thus, while enabling increasingly far-flung suburbanization, they also helped ensure that it was centre-dependent, thereby enhancing Minneapolis–St. Paul's role within the metropolitan region, as well as that region's broader interdependence. And just like TCRT from 1891, Metro Transit links the twins increasingly together via rapid intercity routes, this time more amicably.

Taken together, Metro Transit and its planning-oriented owner (the TCMC), have been major forces not just in eroding traditional rivalry, but also in ensuring that inter-twin hostility has been overlaid by enduring hostilities embracing both twins (alongside their inner-ring satellites), and their suburbs, particularly those located beyond TCMC boundaries but within the Twin-City Statistical Area. Some sense of this is available from the hostility generated by TCMC's most recent attempt at long-term regional planning. Adopted in May 2014, and based on detailed regional forecasts, *Thrive 2014* comprised detailed plans for the following 30 years in relation to transport, water resources, parks and housing, particularly affordable housing. Outer suburbans saw this as 'a tool for social planners … to design their perfect society and impose it on the rest of us', conjuring fearful images of 'an army of academics, environmental organisations ... transit advocacy and left-wing religious groups' (Katherine Kirsten, *Star Tribune*, 3 August 2013). Against this, City spokespersons saw a situation wherein 'Minneapolis–St. Paul plays the villain (and pays the bills)', disparagingly noting the twins' collective generation of economic growth and tax resources compared to the unproductive 'burbs' (Mike Mullen, *Star Tribune*, 16 June 2017).

However, regionalization is not just metro-government encouraged. The area's increasingly dominant large-scale businesses, based not just in Minneapolis–St. Paul but also in the large inner-ring satellites, seem equally enthusiastic. Witness, for example, the Joint Regional Business Plan produced in 2011 by an alliance of governmental, voluntary and business or-ganizations (https://donjek.com/wp-content/uploads/MSP-Metro-Business-Plan-Executive-Summary.pdf). Initiated significantly in St. Paul rather than Minneapolis, it draws resources, among others, from the TCMC, the new Regional Council of Mayors ('preferring collaboration to the usual territo-rial politics') and the Minnesota Department of Employment and Economic Development, plus the *Itasca Project* – 'an employer-led civic alliance focused on building a thriving economy and improved quality of life' (www.theitascapro-ject.com/). These are perhaps more ambitious than realistic: 'five integrated economic development strategies desirous of position(ing) the Minneapolis–St. Paul Region as the world's premier business location and strengthen(ing) our diverse economy to compete in the global market place'. Nevertheless, they say something about business orientation in the twin-city region.

Conclusion

In recent pages, we have implied some of the inter-linked forces at work replacing inter-twin hostility with co-operation since the 1960s: the develop-ment of mass transit, suburbanization and conurbanization, economic and political regionalization, and the associated planning emphases. Alongside these have been major changes in the economic/industrial bases of both cit-ies, rendering them less distinctive and, after a period of mutual decline in the 1960s and 1970s, once again more prosperous. This revival from the 1980s has occurred because both twins, indeed the wider twin-city region, have economically shifted to services, high-tech, information technology and finance. One consequence has been that the Minneapolis–St. Paul metropolitan region has become home to some very large businesses indeed: 16 of Minnesota's 17 Fortune 500 companies (estimates vary), plus signifi-cant numbers of major private national and international companies. These bring much employment and income to the twins and/or nearby cities, but their operations and interests extend far beyond even the region.

Amidst this twinned and regional togetherness, there remains just one an-gry fly, aside from centre-city hostility to 'the burbs'. It rises and gets angry in St. Paul when outsiders take 'Minneapolis' as a synonym for the Twin Cities, thereby producing 'the snub of St. Paul' (Dave Beal, *MINNPOST*, 19 February 2015).

References

Anderson, C. G. (ed.) (1976) *Growing up in Minnesota: Ten Writers Remember Their Childhoods*, Minneapolis, University of Minnesota Press.

Best, J. (1987) 'Looking Evil in the Face: An Examination of Vice and Respectability in St. Paul as Seen in the City's Press, 1865–83', *Minnesota History* 50, 241–51.

Bureau of Labor of the State of Minnesota, Eighth Biennial Report, 1901–1902, St. Paul, Pioneer Press 50, 97, 452–3.

Cronon, W. (1991) *Nature's Metropolis: Chicago and the Great West*, New York: W. W. Norton & Company.

Diers, J. W. and Isaacs, A. (2007) *Twin Cities by Trolley: The Streetcar Era in Minneapolis and St. Paul*, Minneapolis, University of Minnesota Press.

Dirlik, A. (1999) 'Place-Based Imagination: Globalism and the Politics of Place', *Review Ferdinand Braudel Centre* 22(2), 151–87.

Faue, E. (1991) *Community of Suffering and Struggle: Women, Men and the Labour Movement 1915–45*, Chapel Hill, University of North Carolina Press.

Edgar, W. C. (1925) *The Medal of Gold: A Story of Industrial Achievement*, Minneapolis, Bellman Company.

Images for Twin Cities, www.google.co.uk/search?q=twin+cities+logos&rlz=1ClEODB_enGB587GB593&tbm=isch&tbo=u&source=univ&sa=X&ved=0ahUK EwjtkO-h0uTZAhXBK8AKHWNlCQ4QsAQIKA&biw=1455&bih=699, accessed 01/02/2018.

Johnson, C. W. (1890) *Another Tale of Two Cities: Minneapolis and St. Paul Compared*, Minneapolis, publisher unrecorded.

Kane, L. M. and Ominsky, A. (1983) *Twin Cities: A Pictorial History of St. Paul and Minneapolis*, St. Paul, Minnesota Historical Press.

Maccabee, P. (1995) *John Dillinger Slept Here: A Crooks' Tour of Crime and Corruption in St. Paul, 1920–1936*, St. Paul, Minnesota Historical Society Press.

Manahan, K. (1927) Unpublished biography of her father, James Manahan, ca. 1927, 153–4. Box 9, *James Manahan Papers*, MHS.

MINNPOST, www.minnpost.com/cityscape/2015/12/st-paul-s-libertarian-alleys-raise-questions-basic-civics, accessed 10/01/2018.

Minutes of the Meeting of the Executive Committee of the Citizens' Association, 3 December 1903. St. Paul Citizens Committee Papers, 1 box, MHS.

Nord, D. P. (1986), 'Hothouse Socialism: Minneapolis, 1910–1925', in *Socialism in the Heartland: The Midwestern Experience*, D. Critchlow ed., Notre Dame, IN, Notre Dame University Press, 133–66.

O'Connell, M. R. (1988) *John Ireland and the American Catholic Church*, St. Paul, Minnesota Historical Society Press.

O'Neal, B. (2006) '125 Glorious Years', *Minnesota Sports Alamanac*, Minnesota Historical Society Press, St. Paul.

Pamphlet charging the Golden Rule Department Store with unfair labor practices, box 2, William Mahoney Papers, Minnesota Historical Society.

Regan, A. (1981), 'The Irish', in *They Chose Minnesota: A Survey of the State's Ethnic Groups*, J. D. Holmquist ed., St. Paul, Minnesota Historical Society Press, 143.

Roberts, E. (2014), 'Two Cities Two Times', https://streets.mn/2014/04/03/two-cities-two-times/, accessed 12/01/2018.

Schmid, C. F. (1937) *Social Saga of Two Cities: An Ecological and Statistical Study of Social Trends in Minneapolis and St. Paul*, Minneapolis, MN, Minneapolis Council of Social Agencies.

'The Men and Products of St. Paul', *Third Annual St. Paul Almanac 1915.*

The Twin Cities, Community-Wealth.org, https://community-wealth.org/content/twin-cities-minneapolis-st-paul-minnesota, accessed 13/01/2018.

Transport Deployment Casebook, https://en.wikibooks.org/wiki/Transportation_Deployment_Casebook/Twin_Cities_Rapid_Transit, accessed 11/01/2018.

U.S. Commissioner of Labor (1906), *Strikes & Lockouts.* 490-I.

2 'Too near neighbours to be good friends'[1]

Manchester and Salford

John Garrard and Alan Kidd

Moving to Europe, we immediately switch to older settlement pairs with longer relational histories, well-etched in public memories and complicating the intercity relationship. Using Manchester and Salford as an example, the current chapter again unveils persistent and often dysfunctional conflict as a feature of intranational twin cities, again only recently resolving into greater harmony – partly, as with Minneapolis–St. Paul, enhanced by conurbanisation.

Manchester and Salford's political relationship needs setting within their socio-economic and geographical relationships. Understanding their interactions as twin cities involves appreciating the historical roots of their present political distinctiveness and how politics has often over-ridden geographical and socio-economic similarities. This chapter takes an extended overview from these cities' foundation down to the present. Manchester and Salford have always been classic, even extreme, examples of twin cities. Though emerging separately and with separated histories going back over a thousand years, they have never been visually distinguishable – indeed, the boundary of one runs right into the other's historic centre. Equally classic has been their common border marked by a river (the Irwell), albeit modest and, in the nineteenth and for most of the twentieth century, hemmed in by industrial buildings and spectacularly filthy. So too has been their long conflicted history only recently superseded by co-operation; conflict enhanced on Salford's side by its sense of being on the overshadowed end of a classically dominant–subordinate relationship, even though Salford was the third largest urban entity in North West England. As evident later, conurban interdependence (arguably dependence) came much earlier to Salford than to Manchester, though it eventually visited both, and was positively embraced by Manchester.

Yet Salford was not always subordinate. Although the area boasts Roman remains, continuous settlement probably dates from around A.D. 1000. Admittedly, in the medieval ecclesiastical hierarchy, Salford was part of Manchester parish. However, politically it was more important. Before the 1066 Norman Conquest, Manchester was merely part of the royal manor of Salford, held directly by the king, Edward the Confessor. This manor was the most important place in Salford 'hundred', sometimes entitled 'Salfordshire', one of six similarly termed local administrative units covering the large area between the Ribble and Mersey Rivers. Post-conquest, the new king, William I, granted Roger de Poitou this territory, for support at the Battle of Hastings. Roger kept Salford manor, but divided other parts of the hundred into several territories granted to lesser Norman knights in return for military service; among these less prestigious territories was the barony of Manchester, which went to Albert de Gresle – thereby creating a division persisting ever since. Throughout medieval times, Salford manor was held directly by the King or granted to a mighty subject. Eventually in 1351, the duchy of Lancaster was created alongside the County Palatine of Lancaster (or Lancashire), meaning that the Duke of Lancaster ruled in the king's name. Finally in 1399, the then Duke of Lancaster was enthroned as Henry IV, retaining his former titles; henceforth, reigning monarchs have been Dukes of Lancaster and lords of the manor of Salford. Salford never owed obedience to a superior beneath the rank of earl. Manchester belonged to mere barons, by the late sixteenth century sinking even further down the social ladder after acquisition by the Mosleys, a merchant-clothier family. This may help explain why Salford gained its borough charter around 1230, while Manchester waited till 1301 (Tait 1904).

However, neither town was exactly important. Only in the sixteenth century did this urban agglomeration start being noticed. By this time, whatever its social placing, Manchester was clearly becoming dominant, not just over Salford, but within the broader area. John Leland, Henry VIII's topographer visiting around 1540, described Manchester as 'the fairest, best builded, quickest and most populous tounne of all Lancastrershire' (quoted in Kidd and Wyke 2016, 42), while dismissing Salford as 'a large suburb' – a descriptor enwrapping Salford ever since. Reliable population estimates were difficult before the 1801 census. However, in 1773, a street-by-street census estimated 22,481 for Manchester and 4,765 for Salford (Percival 1774).

Manchester's rising importance rested upon the cloth trade. In the sixteenth and seventeenth centuries, it became a regionally important commercial centre; its merchants sold the area's wool and linen cloths throughout the provinces as well as on the London cloth-market, with its separate section named 'Manchester Hall'. Manchester's identity was already associated with textiles. This preeminence over Salford began well before the industrial revolution but that momentous process ensured that 'Manchester', not 'Salford', was the name resonating across the globe and becoming synonymous with industrialisation.

By the early nineteenth century, Manchester was not just a major industrial city, but also 'Cottonopolis', a trading centre for the whole Lancashire cotton-manufacturing area. Further, it soon became internationally synonymous with the key ideas associated with industrialism. Thus, its chief public-assembly venue, the Free Trade Hall, built 1853–56, was named not after a religious or political figure, as with many other European cities, but to honour an ideal. What became known as the 'Manchester School' of economics epitomised the town's contribution to liberal ideology. This view of free-market economics gained international currency through the Anti-Corn Law League as a (sometimes hated) expression of selfish economic behaviour, characterised by German critics in the abstract noun *das Manchestertum* to symbolise what they saw as the British ideology of economic individualism. It spawned numerous sister settlements, also industrial and called Manchester – 36 in the USA alone – while many others around the globe forbore the name, but became known as their region's 'Manchester'.

Successive censuses tell their own story. Already by 1801, Manchester recorded 70,409 against Salford's 13,611; by 1831, this had doubled to 142,026 (Salford 50,810); by 1861 there were 335,722 Mancunians against 102,414 Salfordians. These figures were later boosted by territorial gains for both boroughs, but whereas Manchester expanded its territories in 1885, 1890, 1904, 1909, 1913 and 1931 (reaching nearly seven times its original size), Salford grew just once in 1853 (the year Manchester acquired city status). It did not enlarge further until the 1972 Local Government Act. By 1901, Manchester could boast 644,961 people against Salford's 220,957. By 1921, with cotton depression looming, Salford peaked at 234,045, while Manchester, boosted by new industry, was still growing beyond that year's 730,307.

However, these figures highlight a strange contradiction. Salford has long been overshadowed by Manchester, indeed visually indistinguishable. Yet it was and remains a very big urban place – throughout census-recorded time from 1801, the third most populated city in North West England, second only to Manchester and Liverpool, far larger than any of the region's far more distinct industrial towns. Yet, unlike them, it never managed to generate its own distinguishable centre. One marker and underpinning is the fact that, while most of those towns followed Manchester in celebrating municipal self-government and proud industrialism by building magnificent neo-gothic or neo-classical town halls, the Salford Council commandeered a market hall built in 1821, whose elegant but modest front end presided over an equally modest square located just a kilometre from Manchester's border and historic centre – and which eventually housed not just the municipality but also police, magistrates courts and fire brigade (Figure 2.1). This reflected a wider consequence of overshadowing, highlighted by an anonymous 'Ratepayer' writing rhetorically to the *Salford Weekly News* in 1875, 'Where are our markets, our Exchange, our Free Trade Hall, our theatres, but on the other side of the Irwell?' (SWN 15 May 1875, 3). Contemporaries spoke even more ruefully about railways: Salford hosted two thickening swathes of railway line, laying waste extensive areas before terminating magnificently in Manchester. Finally in the 1880s, Salford Council persuaded the London and North Western Railway Company to locate its 'Exchange' terminus on Salford territory. And so it splendidly did, but, to councillors' outrage,

Figure 2.1 Manchester and Salford: dominance and subordination in town halls.
Above: 'Manchester, grade 1 listed, created by the celebrated architect,
Alfred Waterhouse 1868-67, dwarfing its square' by the Victorian Web
(http://www.victorianweb.org/art/architecture/waterhouse/9.html). Below:
'Salford Town Hall, grade 2 listed, an ex-market hall built 1825-7 and now
turned into flats, and yielding its municipal role to a bigger 1930s building,
acquired from one of Salford's new districts', photo by John Garrard (2017).

on Salford's outermost border with an approach – tellingly called Cathedral Approach – running directly down into Manchester's historic centre.

As this starts implying, conflict distinguished Manchester–Salford relations throughout the nineteenth century, running well into the twentieth. The municipalities clashed regularly, most frequently over what ran between and around them – roads, bridges, railways, water supply, postal services and the evermore disgusting river – often enhancing communication but rarely understanding. The 1830s saw battles over connecting river bridges and who has to pay for building them; the 1840s and 1850s witnessed conflicts over the quality and cost of Manchester's water supply to Salford; in 1845, tension arose over Manchester's decision to move its cotton-trading exchange further away from Salford; in 1847, they became embattled over control of their respective fire services during joint operations on the border, Manchester demanding overall control of both; 1847–49 saw poisonous debates within and between the councils about tolls on connecting roads. In the 1850s, the municipalities battled over control of land for cemetery provision, Manchester claiming the right to sell territory originally jointly purchased during 1793–1830 (when the two were theoretically combined under the Manchester and Salford Police Commissioners). The year 1861 witnessed a brief but fruity dispute about Manchester's refusal to allow the British Association for the Advancement of Science to visit Salford during its annual conference in Manchester. In 1868, Manchester Council, with little consultation and arousing fury across the river, made a parliamentary application to amalgamate the two Courts of Record, effectively attempting (albeit failing) to burgle Salford's into Manchester. The 1870s saw a long-running and bitter dispute over the relocation of Manchester Post Office (effectively also Salford's) further away from the joint border when Salfordians complained that they had 'the postal accommodation of most villages' (letter SWN 15 May 1875, 3).

From 1870 to 1914, there were rhythmic disputes over the inter-running of public transport, particularly trams. In the 1890s, anger arose over building the Manchester Ship Canal and Manchester's perceived refusal of Salford's offer to assist in subsidising the temporarily failing Ship Canal Company (SR 2 May 1891, 4). The sea-accessing Canal was eventually highly successful, partly due to its extensive docklands, built on Salford's side but always tellingly called 'Manchester Docks'. Meanwhile, 1901 saw a conflict about abortive plans for a joint smallpox hospital.

These disputes often drove Salford local government representatives into incandescent rage. In 1837, Police Commissioner J. S. Ormerod said bridging the Irwell was 'simply commencing improvements for the benefit of other people', allowing carts and people to travel more easily into Manchester (MG 12 January 1836, 3). For Salford's mayor, Manchester's behaviour over the Courts of Record represented 'one of the most audacious things ... one corporation ever did to another' (SWN 8 February 1868, 4). On Salford's constantly delayed parliamentary application to build tramways during the 1870s (due to Manchester's perceived refusal to co-operate), Alderman McKerrow demanded, 'why should they dance attendance upon Manchester', while Alderman Bromley likened those advocating co-operation to 'little children waiting for the demands of their parents as if they could not stir a step without Manchester', while the *Salford Weekly News* suggested that Manchester's 'very name has only to be mentioned to produce ... fear and trembling throughout nearly the whole ... council' (SWN 4 October 1870, 3). On the periodic issue of amalgamating the two cities (see later), Alderman Davies in 1883, like many colleagues, denounced Manchester as arrogant and, drawing on some hazy history, said that they should not 'allow an ancient borough like Salford, with its history of a thousand years, to be swallowed ... by (some) mushroom creation of the day!' (SWN 9 June 1883, 3), while Alderman McKerrow described another amalgamation scheme in 1889 as more 'like the rape of the Sabines than ... marriage' (SR 2 November 1889, 8). Meanwhile, the *Salford Reporter*, contemplating recent battles over tramways and amalgamation during October–November 1901, amused its readers with an extended 'Peep into the Civic Government of Manchester: Its Scandals and Blunders'.

Yet, as its language implies, this rage carried strong perceptions of impotence. Manchester's 'over-mighty' presence across 'the sludge' did not just infuriate many Salford councillors; it also widely inhibited what they felt they could do without the latter's co-operation. This was evident in several areas – most notably around the long-running issues of tramways and river pollution.

As implied, the municipality delayed building tramways, restrained by uncertainty about whether Manchester would permit running into its city centre, Salford always facing the problem that 'a tramway ... terminat(ing) only in Salford would not pay' (Alderman Bromley, *SWN* 4 October 1870, 3). The two corporations managed agreement in 1895, but resumed embittered hostilities in 1901, eventually producing a Salford tram/bus station, but typically rammed against its border with central Manchester.

The Irwell's embracing presence meant that nascent conurban interdependence posed much earlier problems for Salford than most other towns: not just with Manchester but also with many other urban settlements. By the time it reached Salford, the river carried industrial and untreated human effluent from the many industrial towns situated on its complex tributaries, donating generously but bearing few of the human and other consequences. Upon arrival, it meandered slowly and dirtily through Salford's heart before straightening somewhat to form the border with Manchester. Tarrying in Salford, it formed 'a thick unbearable scum' in dry weather (Hamblett, SWN 9 September 1993, 4), and 'the poor people who live near it say ... when at its worst, the Irwell kills on sight' (Corbett 1907, 6). In very wet weather, and regularly from 1867, it flooded, oozing aromatically across Salford's Lower Broughton area. Councillors badly wanted to act, creating a River Conservancy Committee in 1867, which appointed an inspector to prevent rubbish-tipping, and planned an intercepting sewer to divert flood waters. Unfortunately, for these to have much impact, Salford needed inter-municipal co-operation to curb pollution and share costs. The many local authorities higher up the river, unaffected by what they did, refused, while Manchester, facing the river's loving caress only on its border, prevaricated, failing to appoint an equivalent inspector and doing little about a joint par-liamentary application for powers to create the sewer. Salford eventually built its own more modest but less effective interception facilities. The situa-tion remained paralysed on these and related schemes for 20 years until 1887 when Manchester's River Conservancy Committee chair, noting pending central government intervention, admitted that his municipality had expe-rienced 'no great desire (to) spend the ratepayers money on such a scheme ... and concluded it would be best ... to display ... masterly inactivity' (SWN 16 July 1887, 4).

Underpinning all this was the fact that Manchester and Salford were intertwined communities at many levels, and in ways far more problematic for Salford's governance than Manchester's. Certainly, Salford could display very positive signs of identity – the same mass civic pride evident in other industrial towns before 1914. Annual mayor-making processions, mayoral entertainments, municipal jubilees and coronation festivities were all just as capable of attracting thousands to watch and participate. The local eco-nomic, social and political elite were just as popularly visible in Salford's in-turned local arena: with elite marriages and funerals attracting comparable attention. All received similarly massive coverage in Salford's two to three local weekly newspapers – which also reported local civil society proceed-ings verbatim (Garrard 1983).

But what was arguably Salford's finest funeral provides the clue to the two cities' interrelationship. In 1876, there died Elkanah Armitage, owner of three Salford factories, sometime Salford Boroughreeve (pre-municipal equiva-lent of mayor), Salford Liberal parliamentary candidate in 1859, and also Manchester alderman, and Mayor 1846–48. His two sons were also Salford councillors. Elkanah was conveyed to his grave, heading a kilometre-long, 100-carriage, procession, comprising his workers, local charities, political parties, representatives of both municipalities and many elite figures from both cities. Watched by thousands, the procession perambulated massively through the main streets of both Manchester and Salford, attracting vast newspaper reportage in both cities (SWN 30 November 1876, 5).

This points to the important fact that many of Salford's (and some of Manchester's) top figures before 1914 had cross-city interests. In particu-lar, many lived in the more salubrious Pendleton and Broughton parts of outer Salford, but had all or part of their business interests in Manchester, a trading centre for the Lancashire cotton region. So too did some Man-chester councillors, sometimes intervening in their neighbour's local politics on Manchester's behalf, 'seeking to set the inhabitants of Salford against its own municipal government' (Council statement SWN 05/09/1862, 3).

More importantly, an average third of Salford's council members were figures of this type (Garrard 1970, 79).

Their presence highlights Salford's overshadowed ambivalence as a community. As noted, its only municipal expansion before 1972 occurred in 1853 when it absorbed the neighbouring Pendleton and Broughton townships. For the townships, the marriage was hardly willing, rather enforced by the threat of governance by county magistrates under provisions set by the 1848 Health of Towns Act. Absorption provided the expanded municipality with considerable new territory and taxable values, and also a strange, decentralised federal government. While meeting monthly in Salford town hall to discuss common policy areas and convening in some general committees, the three districts also met separately. Though markedly less populous, Pendleton and Broughton's combined council representation equalled Salford district's, thereby providing veto power whenever desired. They convened in their own purpose-built town halls, appointed their own district committees and had considerable autonomy. This strange, deeply quarrelsome, sometimes paralytic, arrangement lasted until 1892. One reason was that 'the out-townships', especially Broughton, with populations far more wealthy on average, perceived only limited commonality with Salford district, and Salford generally. As noted, Manchester supplied Salford with water, but Pendleton received and paid for its supply in separate pipes. Broughton, the richest township, housing many of Salford's and some of Manchester's elite, saw itself as Salford's 'aristocratic district' (SR 21 January 1888, 8), even as Manchester's 'west end' (MG 18 June 1853, 7).

All this was expressed most vividly in periodic attempts to amalgamate the two cities. Given two highly adjacent towns with just one effective centre, amalgamation seemed and seems rational to many outsiders. Indeed, Parliament combined them under the Manchester and Salford Police Commissioners in 1793. The Commissioners met together just once, adjourned to opposite ends of the building and never met collectively again – Parliament formalising the *de facto* situation in 1830. Amalgamation campaigns emerged in 1883, 1887–88, 1903–4, 1906 and 1910–11, significantly only once (1883) receiving Manchester Council support. They were led by elite figures in both cities – on the Salford side by those with split interests, particularly out-township figures, especially from Broughton. Amalgamation supporters always included significant numbers of Salford councillors, again almost invariably out-township representatives economically active in Manchester. In the eyes of Salford's council majority and its newspapers, they were 'traitors' (SR 14 April 1888, 4), and this majority (generally two thirds) nearly always comprised people with interests just in Salford: living there and generally owning smaller businesses: modest manufacturers and 'shopocracy' (Garrard 1970, 79).

This communal ambivalence, both about Manchester and the linked issues around its three townships, helps explain Salford's apparent timidity in face of the many examples of Mancunian non-co-operation and occasional intervention. In some policy-making areas, Salford simply 'could not live by itself' (SR 22 June 1901, 5). And if Salford often found autonomous governance very difficult, equally was this true of its numerous civil society organisations. In a market-driven world, with many areas largely government-free, these were highly important. Admittedly, there were significant numbers with titles suggesting that they managed to operate just within Salford or Manchester. Many were political – aiming to pressurise the municipalities or intervene in municipal or parliamentary elections on behalf of party or cause (e.g. local taxpayers, temperance). However, very many economic, charitable, behavioural-reform and other organisations found it essential to straddle the border, operating under joint 'Manchester and Salford' designations. By 1887, there were supposedly 'over thirty'. (Ellis Lever, letter, SWN 8 October 1887, 4) The following sample, many persisting well beyond 1945, gives some sense of their importance and range: the Manchester and Salford Bank for Savings, Chamber of Commerce, Trades Council (local trade unions), Provident Society, Wine and Beer Sellers Association, Temperance League and Grocers Association.

In face of Salford's need for co-operation and frequent anger when it was not forthcoming, Manchester council often 'seemed to care very little' (SWN 10 April 1869, 3). Mostly, it seemed indifferent about amalgamation. One reason perhaps was that it had little motivation to acquire a town whose inner areas were just as poverty-stricken and problem-ridden as its own, and whose wealthier outer ones could hardly match the vast taxable resources Manchester was acquiring through other expansions. Another was probably Salford's size: as many municipalities discovered, absorbing even modest settlements was hard and expensive to negotiate (Garrard 1983). Large towns, however adjacent, probably seemed impossible, given even minimal internal opposition – and Salford would provide plenty. It was after all a prestigious 'county borough' from 1885, acquiring the much-desired city status in 1926. A third reason was probably that 'Cottonopolis', a self-conscious trading centre for the entire cotton-manufacturing area, had its mind on higher things, particularly from the late nineteenth century as it contemplated its role in the emergent 'Manchester region'. Indeed, while Salford often fixated upon Manchester, Manchester's gazed far more widely, even upon the world.

By the 1880s, the Manchester City Council was consciously becoming an agent in developing not just the city's economy, but also the region's. It was the Manchester Ship Canal Company's main investor, even, as noted, refusing help from Salford in 1893. Later (1929–30), the council moved Manchester into the aviation age by founding and owning Manchester Airport (Kidd 2006, 199). Such enterprise assumed regional leadership. The idea of a larger 'Manchester' began emerging as population growth and movement blurred many municipal boundaries and became conceptualised in the minds of town planners and government officials.

Nineteenth-century urban growth was producing city regions more widely, the largest being Greater London. In Lancashire's industrial area, urban interdependence was accelerating from at least the 1880s, as municipalities ventured into and beyond their hinterlands: attempting to incorporate rapidly emerging suburbs; searching for water supplies for rapidly growing populations and meeting other municipalities doing the same; or extending public transport to emergent suburbs and started needing co-ordination with neighbouring councils doing similar things.

For Manchester, successive boundary expansions from 1885 spurred developing ideas about 'greater Manchester'. From the early twentieth century, proponents of urbanisation like Patrick Geddes, who coined the term 'conurbation', promoted the idea of the 'city region', of which 'greater Manchester' was a prime example (Geddes 1915). Although the council stopped boundary extension in 1931, no other modern British conurbation is so dominated by its core city.

The need for planning across some notional 'greater' Manchester first arose in the transport field. As noted, transport provision caused abiding contention between the twin cities from 1870. When previously privately run tram services were municipalised and electrified in 1901, this provoked another acrimonious dispute. In May 1903, reciprocal arrangements were reached as the transport needs of an emerging metropolis slowly overlaid earlier parochial divisions.

By the 1920s, regional planning for transport, and to some extent housing, started being formalised, Manchester taking the initiative. The Manchester and District Joint Town Planning Advisory Committee emerged in 1926, proposing 65 projects for regional and district main roads. In 1928, a Manchester and District Regional Planning Committee emerged to prepare a joint-planning scheme across 14 municipalities (Nicholas 1945). The culmination of this regional-planning era was *The City of Manchester Plan 1945*, offering a blueprint for post-war reconstruction, seeking to deal not only with the impact of wartime bombing but also the Victorian inheritance of decaying housing-stock across central Manchester and inner residential areas. In fact, this plan transcended political boundaries, in several respects embracing 'Greater Manchester', with Salford no more significant than any other affected municipality.

Regarding transport, the 1945 Plan proposed a railway and road-transport interchange called Trinity Station, straddling the border and located between the existing Manchester Exchange and Salford stations, sporting separate Salford and Manchester entrances. This would have involved a culvert of the river, thus partially removing the natural and sometimes psychologically significant barrier between the twins. In the event, this venture never took off partly due to post-war financial constraints, and partly due to changing national-planning priorities – removing initiative from the local state and lodging it firmly within London national government orbit. However, it reveals Manchester's assumption of planning competence over its nearest neighbour (along with other nearby authorities). The ground-rules for such local and regional planning were substantially changed by the 1947

Town and Country Planning Act, not only requiring municipalities to plan, but also determining the limits and frameworks through which it could occur. This introduced an era of greater central control over regional planning.

Administrative reform follows economic reality, but tempered by historical precedent. However, in very real senses, boundaries on the map increasingly meant little. The various elements of Manchester conurbation were merging long before temporary unification by local government reform in 1972. Economic reality rarely balked at municipal boundaries. For example, 'industrial Manchester', indeed 'residential Manchester' and most recently 'commercial Manchester', could not be confined within administrative areas. From at least the 1880s, this could increasingly confuse the unwary. This was marked dramatically when the Ship Canal's docks, the 'Port of Manchester', were located in Salford, and the Trafford Park industrial estate emerged on the Canal banks occupying districts unincorporated into Manchester even now.

Administrative relationships between municipal Manchester and its immediate neighbours, including Salford, fluctuated during the twentieth century. In territorial terms, Manchester's 'imperial phase' was over by 1914, although Wythenshawe's later addition explains the municipality's strangely elongated shape. Despite economic interdependence within the region, and planning co-operation, Salford, by now joined by several surrounding towns, jealously guarded its independence and identity. Local government reforms tended to strengthen the position of Salford and the necklace of towns surrounding Manchester, while recognising core-city dominance. Thus, the 1972 Local Government Act created Greater Manchester as one of the six new metropolitan counties. After Greater London and the West Midlands, it was the third largest of the new local governmental giants, and the only provincial-governmental entity named after its core city. Greater Manchester County encompassed almost 500 square miles, reorganising 2.7 million people into a cluster of newly created metropolitan boroughs. These included the city of Manchester at the centre, with the crescent of industrial towns to the north, and the adjacent western and southern boroughs of Salford and Stockport, plus the new administrative inventions of Trafford on Manchester's south side and Tameside to the east.

As a local government unit, Greater Manchester County and its council were abolished in 1985, but the metropolitan boroughs, including Salford, survived and acquired the habit of co-operation through various joint boards and agencies until their continued interdependence was recognised again in the 2011 creation of the Greater Manchester Combined Authority (GMCA). This reflected central government conversion to the need to strengthen the former industrial centres of the north to try redressing the north–south economic divide. The recent political talk about a 'Northern power-house' has focused attention on Manchester. In May 2017, the first step occurred via a newly elected executive mayor to run Greater Manchester. The mayor oversees policies like transport, social care and housing alongside police budgets, leads the GMCA, chairs its meetings and allocates responsibilities to a cabinet comprising leaders of each of the area's ten municipalities, of which Salford is one. 'Greater Manchester' marches on.

All this supplies the broader context for the increasing inter-twin peace emerging over recent decades. In retrospect, though only retrospectively, the first apologetic dove arrived much earlier, courtesy of central government. Enforced action and co-ordination over river pollution began tentatively in 1891, though more certainly from 1948 and 1963 with the creation of River Boards and River Authorities taking power from municipalities. Public-transport authority has been regionalised since 1968, albeit with local authority representation. In both areas, municipal borders as authority lines have far less meaning than heretofore. This has been equally true of policing. Meanwhile, the Manchester–Salford border has become even less visible than before: the now-salubrious Irwell, as it swings into Manchester's historic centre opposite its cathedral, has recently disappeared under a fountain-adorned public square spanning both cities; Manchester-based office buildings have also crossed the border. Apartment buildings for young Manchester-based professionals have similarly migrated.

There were also positive reasons to co-operate. From the mid-1980s, councillors and council officials from across Greater Manchester discovered that working together enabled them to access EU funding for projects that monetarist-inclined central government now frowned upon (Goldsmith 1993).

Yet both cities retain councils with considerable authority – something unlikely to change anytime soon. Salford remains separate and distinct for many purposes and many people. Identity, as always, is two sided: the image projected and perceived by outsiders, and its meaning for those living within. For many outsiders, Salford seems even less visible than before. Yet its fortunes and reasons for pride have improved in recent decades. Under the 1972 Local Government Act, Salford not merely survived, but greatly enlarged in area and population – while the city of Manchester remained as before. Since then, Salford's population has become half rather than a third of Manchester's, and recently has been rising. This is not just because it occupies more territory, but also originates from city council initiative. In the 1980s, while Manchester's left-leaning Labour Council busily constructed its 'nuclear-free city' image, Salford's more pragmatic Labour-dominated council successfully acquired the now-redundant ship-canal docklands and created a new arts, leisure, media and residential complex called Salford Quays. Among other good things, this has given Salford a major gallery wherein to display paintings by its renowned artist, L.S. Lowry; and brought 'Media City' occupied by the BBC's northern headquarters broadcasting nightly 'from our studios in Salford'.

For Salford insiders, identity remains ambiguous, though now probably harder to locate – partly dependent, as always upon socio-economic position, and both negatively and positively based. Young professionals in the new apartment buildings probably see themselves as 'Mancunians'. Yet the key reason why Salford survived the 1972 local government reforms was that the Royal Commissioners came to believe it had 'real identity' for many Salfordians (Jim Sharpe, Royal Commissioner, to Michael Goldsmith, letter 1974). This was partly negative, rooted, as ever, in 'not being Manchester' – 'a foreign place you did not go into' according to now-retired ex-working-class people interviewed in 2015 and remembering their teenage years in the 1970s (Interview 11 November 2014). Random conversations with even middle-class Salfordians suggest some surviving civic pride in Salford's recent developments even today. In November 2015 and at the other end of the social scale, Paul Massey, a Salford drug-gang leader, was assassinated by rivals. His large funeral attracted hundreds to watch a procession led by his hearse decorated with the words 'A Salford Legend' (*Guardian* 28 August 2015). His had been one of many Salford gangs: like their nineteenth-century forbears (Davies 2008), neighbourhood-based and rarely extending operations across the border.

* * *

We can finish with appropriate ambiguity. Since 1989, the Salford City Football Club has emerged from what was originally an amateur club (Salford Central). Minor league and now fully professional, home matches attract average crowds of over 2000 who are serenaded by Salford's iconic anthem 'Dirty Old Town'. Since 2014, the club has been funded by four ex-players from Salford's 'real' team: Manchester United, visible across the Irwell, and itself just beyond even Manchester's municipal boundary at Old Trafford.

Note

1 'Salford Burgess', SWN 22 June 1861, 4.

References

Corbett, J. (1907), *The River Irwell, Pleasant Reminiscences of the 19th Century and Suggestions for Improvement in the 20th*, E.J. Morten, Salford.

Davies, A. (2008), *The Gangs of Manchester: The Story of the Scuttlers Britain's First Youth Cult*, Milo Books, Croydon.

Garrard, J. (1970), *Leaders and Politics in Nineteenth-Century Salford*, Sociological and Political Studies, Salford University, Salford.

Garrard, J. (1983), *Leadership and Power in Nineteenth-Century Industrial Towns*, Manchester University Press, Manchester.

Geddes, P. (1915), *Cities in Evolution*, Williams, London.

Goldsmith, M. (1993), The Europeanisation of Local Government, *Urban Studies*, 30 (4/5), 683–700.

Kidd, A. (2006), *Manchester: A History*, Carnegie, Lancaster.

Kidd, A., and Wyke, T. (eds.) (2016), *Manchester: Making the Modern City*, Liverpool University Press, Liverpool.

Manchester and District Joint Town Planning Advisory Committee, Report on the Regional Scheme (1926), Manchester.

MG: Manchester Guardian.

Nicholas, R. (1945), *The Manchester and District Regional Planning Committee Report on the Tentative Regional Planning Proposals*, Jarrold & Sons, Norwich.

Pidd, H. (2015), Funeral for Salford's 'Mr Big' takes place with armed police on standby. *The Guardian*, August 28. Available at: https://www.theguardian.com/uk-news/2015/aug/28/funeral-salfords-mr-big-paul-massey-armed-police-on-standby (accessed January 3, 2018).

Percival, T. (1774), 'Observations of the State of the Population of Manchester and Other Adjacent Places', *Philosophical Transaction* 64. The survey is kept in Chethams Library, Manchester.

SR: Salford Reporter.

SWN: Salford Weekly News.

Tait, J. (1904), *Mediaeval Manchester and the Beginnings of Lancashire*, Manchester University Press, Manchester.

3   NewcastleGateshead

A dynamic partnership

Rebecca Wilbraham

Like Manchester and Salford, Newcastle and Gateshead have long separated histories, and they face each other across a river. Like both Manchester and Salford and Minneapolis–St. Paul, they now partly belong to broader and highly populated metropolitan areas, complicating their relationship economically, politically and in terms of identity. And like both of the foregoing twins, their relationship has long been conflict-ridden, has now become far more co-operative, if still ambiguous, and is being partly overlaid by region-wide pressures. As with many twin cities, whether intranational or international, a key variable is central government/s. This chapter will focus primarily on this most recent stage of the relationship.

From the air, Newcastle and Gateshead resemble a single city with a river running through. Historically, what would eventually become the Tyneside conurbation boomed as a result of 'carboniferous capitalism' (Lancaster 2005) and shipbuilding. Described as a 'workshop of the world' (Hudson 2005) in the nineteenth century, more recently the 'twin cities' and the wider conurbation have gone some way towards identifying new economic trajectories after deindustrialisation. Newcastle, a city north of the Tyne and Gateshead, a town to the south, share the iconic landscape of the Tyne Gorge. The seven bridges spanning the Tyne between Newcastle and Gateshead characterise this unique cityscape, a standing testament to their interaction.

Using our editors' typology, these two urban places are administratively separate but with their interdependence inbuilt from the start, having grown outwards concentrically from an ancient river crossing. However, the area also belongs to the wider Tyneside conurbation, including the local authority areas of North Tyneside and South Tyneside. The conurbation is part of England's North-East Region. The twins' mutual interactions have long been defined by relationships with this wider conurbation and region, so often thrown into flux by changes in governance and funding set by national governments. The various scales at work, those of the conurbation, the wider North-East Region and the national scale, have significantly impacted inter-twin relations in recent history, indeed from their inception.

In studying three European intranational/domestic twin cities,[1] Meijers et al. (2014) use the typology of the functional, cultural and institutional dimensions of integration. *Functional integration* is defined by flows: for example, involving people and goods. Functionally integrated urban areas work as 'one daily urban system' (ibid., 38). *Cultural integration* relates to people's identification and attachment to wider urban areas, both personally and in firms' readiness to orient their activities to the integrated urban area. *Institutional integration* refers to the development of platforms for co-ordination and co-operation, including, public, private and public–private organisations. The relevance of this typology for Newcastle and Gateshead will emerge as the chapter proceeds.

Although Meijers et al. use this framework to compare cases, they also advocate using it to explore integration over time. When applied to the changing levels and dimensions of integration between Newcastle and Gateshead in recent history, what becomes apparent is the impact of changing governance contexts on interactions between these neighbouring urban areas. The level of integration between these cities is not inconsequential. Meijers et al. suggest that cities showing greater metropolitisation or integration can use their joint critical mass to build a more diverse industrial base via enhanced labour productivity, access to amenities, services and infrastructure, and resilience to economic shocks. Thus, integration is an economic development issue with implications for the performance of Newcastle and Gateshead and the wider urban area wherein they exist.

In recent times, the cities formed a symbolic collaboration based on branding and culture instigated by the ultimately unsuccessful joint bid to become the European Capital of Culture 2008. A new organisation, the *NewcastleGateshead Initiative* (*NGI*), was established to oversee the bid, which was intended to increase the location's ability to participate in a context of inter-urban competition among UK and European cities; largely for tourism audiences. However, historically, relations between the settlements have been more fraught based on their unequal relationship. Gateshead has long striven to create its own separate identity, while seeming 'a suburb of Newcastle' (Lancaster 2005, 57). Meanwhile, Newcastle has appeared an overbearing neighbour. Frustratingly for Gateshead, this unequal relationship has been reflected in outsiders' narratives. J. B. Priestley saw Gateshead as 'planned by an enemy of the human race' (Hetherington and Robinson 1988, 190) and Samuel Johnson as a 'dirty back lane to Newcastle' (Mick Henry Gateshead Council leader 2002–16, paraphrasing the eighteenth-century visitor NGI 2009, 11). Such insults are long remembered, adding insult to injury (Lancaster 2005). Newcastle did not escape visitors' judgments: in 1727, Daniel Defoe thought it to be 'exceedingly unpleasant' (quoted, Ellis 2001, 1). Yet, such descriptions are rarely referenced.

Newcastle and Gateshead in history

Robert Colls (2005) describes Tyneside's modern history as characterised by fluctuating fortunes, and this theme emerged from when Emperor Hadrian built a bridge over the Tyne, roughly at the location of the present Swing Bridge between Newcastle and Gateshead, to help defend the Roman Empire's northern boundary. Although historians keenly cite this as establishing what would become modern Tyneside, we should note that for 600 years after the Roman Empire's demise little happened there (Pollard and Newton 2009). We must await the late eleventh and early twelfth centuries before a 'continuous urban history' of Newcastle and Gateshead emerges (Pollard and Newton 2009). However, the Tyne crossing remained important: at the start of the medieval period, a ferry crossing replaced the bridge, with the Roman Road still recognised as a major north–south route (Britnell 2009).

A replacement bridge was built in wood between 1071 and 1160 and in stone in 1248 (Britnell 2009). It is over this bridge that these two settlements' histories have been 'inextricably linked', always 'a fraught combination of cooperation and conflict' (Pollard and Newton 2009, xiii) with co-operation over the bridge and conflict, where different jurisdictions meet along the river. This book's Introduction sees resilience as a key twin-city characteristic. But why have they stayed 'connected but separate'? This question is highly relevant for Newcastle and Gateshead, with their centuries-long relationship, close yet essentially separate. The answer apparently resides in historic division and rivalry, wherein the dominant partner, Newcastle, 'was for centuries inclined to use its powers selfishly' (Henry Mess 1928, in Hetherington and Robinson 1988, 205) producing resentment in Gateshead. Another factor has been the area's characteristically fluctuating fortunes, this rivalry being greatest at times of slump and crisis (Pollard and Newton 2009).

Tyneside's changing fortunes are a long-recurring theme. Newcastle, Tyneside's dominant settlement, grew from nowhere to become England's fourth most populous town in the 200 years before 1300 (Pollard and Newton 2009, xix). The country's oldest coal-mining area (Robinson 1988), industrialisation on Tyneside, predated the industrial revolution, with coal production starting in the fourteenth century (Vall 2007). Yet the region suffered regular crises, for example, the Anglo-Scottish wars in the late thirteen/early fourteenth century, agrarian recession and plague, and the severe mid-fifteenth-century economic slump (Pollard and Newton 2009). The rights and privileges held by Newcastle arguably limited its ability to react to economic opportunity, and the eighteenth century's second half was a time of 'lost opportunity' (Ellis 2001, 23). The 1801 census showed Newcastle's growth falling behind other provincial cities. Yet, with its key role in developing steam locomotives and electrical supply systems, Newcastle's growth between 1850 and 1900 meant that it ranked as the world's seventh fastest growing city (Barke and Taylor 2015). In more recent history, Colls (2005) charts the twentieth-century upheaval when the region endured ongoing cycles of death and rebirth. Reborn initially in 1919, the region was dead again by 1934 and reborn in 1945 only to die in the 1980s as part of the spectacular industrial decline of the century's last quarter (Hudson 2005).

Rivalry, particularly connected with Newcastle's historical supremacy and cultural imperialism (Lancaster 2005), is also a continuing theme. It affected all Tyneside settlements, but particularly Newcastle and Gateshead. For example, conflicts over Gateshead's right to hold a market, ship coal and maintain a quay started in the thirteenth century, rumbling on for many more (Pollard and Newton 2009). Rushton (2009, 295) notes that between 1550 and 1700 Newcastle repeatedly attempted to take over its neighbour, with Gateshead maintaining independence 'against all the odds'. Gateshead's position as gateway to the south and a 'suburb' beyond Newcastle's control produced particularly fractious relations with its dominant neighbour. With Newcastle controlling access to the River Tyne, it became the controller of a system of 'carboniferous capitalism'. Heavy industry arose outside the city, yet Newcastle 'became fat on the back of the region's muck and toil' (Lancaster 2005, 55). After its own mines were exhausted, Newcastle did not have its own coal. Gateshead did, and the means of accessing the Tyne, rendering it a 'desirable area of coal production' (Rushton 2009, 321). It was also seen as a 'deplorable zone of free trade' (ibid., 312) and 'a place to where Newcastle's villains fled' (ibid., 321) due to its different jurisdiction. These facts meant that Newcastle's elite wished to bring Gateshead under control.

Amidst these periods of rivalry, there was also co-operation, for example, when the 'economic oligarchy' (Barke and Taylor 2015, 51) of Newcastle coal owners and merchants controlling the Tyne were replaced by the *Tyne Improvement Commission* in 1850. Other Tyneside towns were represented and infrastructure improvement and greater interaction ensued as a conurbation began emerging around this time, largely due to transport developments making commuting possible. Again, Gateshead played a greater role in these new transport systems than other Tyneside towns; until the rail bridge was erected in 1850, anyone wishing to access Newcastle from south of the Tyne would travel by rail to Gateshead and then walk over the bridge. Gateshead was also only second to Newcastle in getting a tram system in the 1870s (Barke and Taylor 2015), again suggesting its prominence over other Tyneside towns.

Relationships between Newcastle and Gateshead, on the one hand, and the wider conurbation and the wider region, on the other hand, impacted back the twin cities' relationship. Gateshead's history since the Middle Ages has been governed not only by Newcastle but also by its status as a 'borough of the Palatinate of Durham' linking it to the wider region. Fragmentation, rivalry and peripherality have been key themes in the North-East Region's development (OECD 2006), as it has a sense of regional distinctiveness (Vall 2007), and pride in its 'cultural separateness' (Lancaster 2005). Perhaps as a result, regional devolution was suggested as early as 1919 (Colls 2005) and, although unsuccessful, a devolution referendum was held in 2004. Newcastle lacks rivals in the region and there are tensions around its supremacy (Lancaster 2005), often contributing to the problematic fragmentation and rivalry, and compounded by continuously fluctuating governance arrangements. This makes it hard to mobilise regional identity behind a 'mode of governance' (OECD 2006) capable of enabling economic development.

Recent relations

The Tyne and Wear County Council was credited with bringing some 'political cohesion' to the conurbation, yet was abolished in 1986. This is the starting point for analysing Newcastle and Gateshead's more recent relationship, my central focus. The framework of functional, cultural and institutional interaction (Meijers et al. 2014) introduced earlier helps us understand the dynamic relationship between Newcastle and Gateshead and reveals three distinct phases.

From 1986 to 1998: going it alone

The Tyne and Wear County Council, comprising the local authorities of Newcastle, Gateshead, North Tyneside, South Tyneside and Sunderland, was established in 1974 and abolished in 1986, one of several labour-controlled Metropolitan county councils the Thatcher Conservative Government saw as 'unacceptable outposts of dissent' (Hetherington and Robinson 1988, 205). The council had provided something of a unifying force, particularly through its structure plan, which outlined the conurbation's planning objectives designed to override 'petty parochial rivalries' (Hetherington and Robinson 1988, 206). One of the Council's economic successes lay in attracting Nissan to Sunderland, the factory opening in 1986. After the Council's abolition, the structure plan remained. However, without adequate support, inter-twin rivalry returned: for example, the Gateshead Council's approval of the Metro Centre – 'a thinly disguised onslaught on the dominance of Newcastle and its Eldon Square shopping centre' (Hetherington and Robinson 1988, 206).

Newcastle and Gateshead's functional interaction has been underway for centuries. This is also true of the towns in what is now the wider Tyneside conurbation. Immediately before this phase began, there was clearly an institutional interaction based on the County Council as a platform of cooperation. The cohesion thereby engendered suggests that there may also have been some increase in cultural integration. With the County Council's disappearance, although functional integration continued, it seems that the institutional and cultural integration was undermined and the local authorities started going it alone.

This was a difficult time for Newcastle, with jobs lost right across the Tyneside economy. For example, manufacturing employment fell 20% during

1978–81 and a further 23% during 1981–84. Shipbuilding and mining also suffered (Robinson 1988). Central government added the *Tyne and Wear Urban Development Corporation* (TWUDC) to the area's governance and development, but it was not designed to fill the vacuum left by the Council's removal. Instead, it had 'the effect of sidestepping local democracy' (Hetherington and Robinson 1988, 205) and sought to develop specific sites rather than any coherent conurban strategy. One such site was Newcastle Quayside. In 1980, *Newcastle City Council* had voted to invest £2.5 million in its quayside over a five-year period. This included the law courts, offices and numerous restaurants and bars largely instigated by the TWUDC, which took over the development in 1987. These activities shifted the area's focus from production to consumption (Nayak 2003), thereby expanding service-sector employment and giving Newcastle an increasing reputation as the 'party city' (Chatterton and Hollands 2001). Significantly increasing student numbers also enhanced demand for leisure (NGI 2009).

Times were also tough for Gateshead with the proportion of employment in mining, ship building, steel and engineering falling from 50% to 3% in the 20 years following the early 1980s (NGI 2009). Long-used to doing things its own way (Rushton 2009), the Gateshead Council rejected the TWUDC, instead of focusing on its own programme of quayside land-use (NGI 2009). The municipality 'came to realise there was a world of bidding out there' (NGI 2009, 11). Gateshead won the bid to host the *National Garden Festival* in 1990, rehabilitating 200 acres of toxic industrial land. Gateshead also had a history of community arts programs and a long interest in cultural policy. A lack of traditional arts institutions required innovative approaches to cultural policies (Obrien and Miles 2010). A plan to significantly invest in regional cultural infrastructure, the *Case for Capital*, was launched in 1995 and the National Lottery with its funding for arts started in 1994. These events at the regional and national scales combined with the local authority's bidding experience laid the foundations for Gateshead's success (Obrien and Miles 2010) in culture-led regeneration.

Although the functional integration, long characterising of the Newcastle–Gateshead relationship, continued, the County Council's abolition meant that the institutional interaction inevitably decreased. Newcastle and Gateshead became unitary authorities losing their platform of co-operation. It also seems there was decreasing cultural integration between the twin cities: the separate approaches to regeneration were not just different attempts to improve local economies and urban environments. In terms of cultural integration, they were also different approaches to reinvention and to 'restoring a sense of pride' (NGI 2009, 11) bound up with local civic identities.

From 1998 to 2011: twin cities in partnership

Long typical twin cities due to their functional interaction, Newcastle and Gateshead's cultural and institutional emergence as twin cities was instigated in 1998 and broadly lasted until 2011. The two council leaders were pictured in a local newspaper with the Tyne Bridge behind them 'burying their traditional hatchet' to usher in 'a new era of cooperation' (NGI 2009, 32). Symbolically, the *Angel of the North*, a large sculpture of an angel with outstretched wings, often credited as a catalyst for the ensuing arts-led regeneration (Bailey et al. 2004), was unveiled the same year. Designed by Antony Gormley and poignantly located on a former pithead, the Angel is believed to be the largest angel sculpture in the world and is seen by 90,000 people every day (Gateshead Council 2017). Widely recognised, its image is regularly used in media directed at local, regional and national audiences. This co-operative era took material form in 2000 with the launch of the NewcastleGateshead brand and the NGI, an organisation largely funded by the two Councils to promote the cities as a single destination. Once again, a symbolic development occurred the same year with the opening of the *Millennium Footbridge* enhancing the cityscape and joining the two cities in a key regeneration area, all in the historical tradition of building bridges across the Tyne.

The brand was used in the *NewcastleGateshead Buzzin'* marketing campaign orchestrated by the NGI in 2001: an 'attempt to radically shift the conurbation's identity away from its industrial past' (Chatterton and Hollands 2001, 9). However, the NGI's primary task was to co-ordinate the cities' joint bid for the *European Capital of Culture* 2008. In 1992, Colls suggested that 'it can only be a matter of time before Newcastle too becomes the city of something' (2005 edition, 25), the bid to become the Capital of Culture started in 2002. The cities lost out to Liverpool in 2003; yet the organisation oversaw the implementation of *Culture 10*, a ten-year cultural programme. Despite failing in its formal objective, NewcastleGateshead's arts-led regeneration is considered a success (Obrien and Miles 2010) particularly on Gateshead Quayside, which has been recognised as providing a world-class example (Bailey et al. 2004).

The promise of arts-led regeneration incentivised co-operation between the twins; working together, they could achieve critical mass, capitalising on Newcastle's civic assets and Gateshead's riverside development sites and municipal expertise (Pasquinelli 2014). Once again, the region's development pattern centred on its river (OECD 2006) with the dramatic shared landscape of the Tyne Gorge and its bridges, a key asset in this regeneration approach. Co-operation between the municipalities and other actors became a 'rational choice', with the NewcastleGateshead scale becoming recognised as the 'most appropriate and even "natural" organisational unit for triggering regeneration' (Pasquinelli 2014, 738). With this, the European bid's brand shifted to a destination brand, used to attract tourism audiences in a competitive environment, and the NGI became a destination marketing organisation (Pasquinelli 2014).

This pathway was reinforced throughout the period with the completion of a series of high profile, and symbolic cultural developments. Just as the Angel and Millennium Bridge heralded the new partnership, further developments provided concrete proof of confidence and investment, thereby reinforcing the partnership approach (Figure 3.1). The *Baltic Centre for Contemporary Art* and the *Sage Gateshead* music venue, opening in 2002 and 2004, respectively, provided the most transformative examples. Virtually every cultural institution and venue in the cities received capital investment at this time. Tourism had increased in importance to the region; in 2007, it was worth £3.9 billion and employed 60,000 people. Furthermore, it contributed 4.5% of the north east's gross value added compared to 4% for primary industries including mineral extraction (House of Commons 2010, 3). NewcastleGateshead's tourist industry significantly expanded; visitor numbers increased, with the city hotels experiencing the highest occupancy rates of any major UK city outside London (House of Commons 2010, Ev45). The twin cities' functional integration increased as they became a joint destination for visitors.

Cultural integration also increased as NewcastleGateshead was recognised as a visitor destination in high-profile national media coverage. There was also some local recognition, although not without controversy. The two municipalities were at significant pains to stress that there was no intention of combining the cities in any more significant way such as shared administration. In fact, the approach advocated in contemporary policy documents was much more akin to traditional twin cities, linked but separate:

Figure 3.1 The Millennium Bridge with Newcastle on the right bank and Gateshead on the left with the Baltic Art Gallery and Sage Music Centre in the foreground, by Visit England (2016).

Gateshead and Newcastle are neighbours but they are different places, with their own histories and identities. These traditions are very important to local people ... they are part of what makes NewcastleGateshead special and distinctive. But from an economic, labour market and property market perspective the two cities function as one place within an increasingly integrated city region.

(1NG 2010, 21)

There was clearly an intention to take a pragmatic approach to the partnership, avoiding contentious identity issues, while capitalising on shared assets and scale.

Not only did functional and cultural integration increase during this phase, but also institutional integration expanded. The NGI was the first organisation working to combine Newcastle and Gateshead for tourism and culture audiences. *Bridging NewcastleGateshead*, a housing-market renewal pathfinder, was launched in 2003. Such pathfinders were introduced by the New Labour Government to deal with areas of low housing demand. This organisation was one of the nine area-based projects in the North and Midlands mobilising local stakeholders to bring sustained regeneration to local housing markets (Townsend 2006). A third organisation, *1NG*, aiming at joint economic development for the cities, was established in 2009 and responsible for the 1Plan, a joint economic development strategy. *1NG* emerged in response to a 2006 OECD territorial review, highlighting the region's challenges in this period of optimism.

The OECD report highlighted a context of problematic regional governance, declaring an imperative need to build a 'strategic region of Newcastle'. It saw the main challenges as poor leadership and strategy (OECD 2006) in the context of a fragmented region, where 'strategies, powers and funding streams often sit at the wrong spatial scale' (Seex et al. 2007, 27). It also suggested that existing image campaigns were unsuitable for promoting the area to business audiences. It emphasised the importance of the urban core, this being what *1NG* and the joint Core Strategy were intended to address. The municipalities interpreted the urban core to mean the two city centres: among 'the most dramatic and memorable of any major place in England' (Gateshead Council and Newcastle City Council adopted 2015, 26). This capitalised on the partnership, which had been already underway and which had proven delivery capacity (Pasquinelli 2014). Although unclear in the report, the OECD was probably referring to Tyneside's four local authority areas (Newcastle, Gateshead, North Tyneside and South Tyneside) rather than the two that became part of *1NG*, suggesting that this organisation had the potential to add to the already cluttered governance, which the territorial review highlighted (Seex et al. 2007).

Once again, this illustrates the importance of the surrounding governance context and scales in determining the twin cities' integration. During most of this phase, the regional scale was administered by The *Government*

Office for the North East (1994–2011) and the regional development agency, *One North East* (1999–2012), with the local authorities sitting beneath, enabling some fluidity of informal co-operation between them. However, this did not long continue. The financial crash starting in 2008 was heralded by the run on *Northern Rock* in 2007. This company had highlighted its links to Newcastle and contributed significantly to the cultural investment prevalent at the time through the *Northern Rock Foundation*. The resulting austerity politics of the Conservative/Liberal Democrat coalition government from 2010 further changed the governance context and scales in play. From 2010, it became evident that regional government would be abolished, with the North-East Region no longer an administrative scale from 2012. For the end of the phase, the stability of existing governance scales was cast adrift, with municipalities invited to submit joint bids to national government to become local enterprise partnerships (LEPs). The *North-East LEP* was established in 2011 on a slightly different geographical basis from the previous regional governance arrangements. This flux in governance scale reduced the stability of all governance arrangements including informal ones like those around NewcastleGateshead.

The NewcastleGateshead scale was further undermined by the austerity post-2010. The period up to 2010 had mostly been funding-rich. Now, cultural infrastructure projects like those linked to NewcastleGateshead and regeneration through culture started looking frivolous amidst austerity. Moreover, both organisations previously established to administer the NewcastleGateshead scale (*Bridging NewcastleGateshead* and *ING*) closed in 2011, due respectively to changes in the national policy and local decision-making following funding cuts. At the peak of Newcastle and Gateshead's institutional integration, there were three organisations working at the NewcastleGateshead scale; there was also an economic development strategy and a Core Strategy planning document. By this phase's end, only NGI and the Core Strategy remained. As noted by Pasquinelli (2014), funding reduction and organisational flux produced a context wherein collaboration costs were high; challenges to decision-making also frustrated the potential for collaboration. Similarly, stakeholders began pressing for collaboration at the Tyneside scale, rational choice was no longer the driving force of collaboration which now continued due to norms established earlier in this phase.

From 2011 onwards: changing relationships

In the previous phase, functional, cultural and institutional interactions were clearly increasing before a significant decrease began. In the third phase, this decrease has been maintained, with NewcastleGateshead only utilised for culture and tourism audiences. Attempts to broaden the scope of NewcastleGateshead have been frustrated. Once again, broader funding and governance changes have been crucial. The theme of organisational flux starting in 2010 has continued, most notably with the final removal of regional governance in 2012 with the regional development agency's closure. This same year also produced a possible new purpose for the NewcastleGateshead scale, when the local authorities established a new joint inward-investment team in the organisation, with 'investment' added to NGI's mission statement, prompting Pasquinelli (2014, 736) to note that NewcastleGateshead was considered 'an appropriate space for embedding a pro-business system'. However, due to funding pressures, Gateshead Council withdrew its support for *Invest NewcastleGateshead* in 2014, throwing this into doubt.

Similarly, a branding project orchestrated by NGI during this phase had difficulty gaining traction among twin-city stakeholders (e.g. business) due to uncertainty over the sustainability of the NewcastleGateshead scale (Wilbraham 2017). They were concerned about the loss of momentum after the period of culture-led regeneration, inadequate promotion of the area as a business location and the return of outdated stereotypes in the national media, for example, Newcastle, the 'party city', the area's alleged public-sector dependence and the image of Tyneside as a failing region unprepared to help itself. All this risked erasing the progress made in re-visioning NewcastleGateshead as a cultural destination. Consequently, stakeholders also wanted further work on place-branding; the NGI has taken a collaborative approach to developing the brand via engagement of stakeholders. The brand aimed to broaden the cities' appeal to include business audiences, but business stakeholders did not identify with the NewcastleGateshead scale, preferring a Newcastle brand embracing the whole of Tyneside (Wilbraham 2017).

A city deal had been agreed in 2012, largely as a precursor to establishing a combined authority, a legal body intended to enable councils to co-operate across boundaries. This deal included both Newcastle and Gateshead, but this time just under the banner of Newcastle, evidently believed to have greater appeal to central government than 'NewcastleGateshead'. Further pressure to collaborate at broader scales was instigated by the establishment of the *North-East Combined Authority* in 2014 on the same geography as the *North-East LEP* with plans for devolution to this scale. Here, municipal collaboration is prioritised more broadly than NewcastleGateshead, with Gateshead imagined as part of the Newcastle conurbation rather than twin-city partner.

New governance arrangements are seemingly emerging. This has produced difficulty. There is no reason to conclude that the problems of fragmentation and mobilisation that the OECD noted in 2006 have been resolved; they may even have been compounded by the organisational flux experienced in the interim. This emerged in 2016 when the *Combined Authority's* devolution plans stalled because the relevant municipalities failed to agree. Nevertheless, the NewcastleGateshead brand is still deployed for culture and tourism audiences, and the NGI successfully bid in 2016 to host the 2018 *Great*

Exhibition of the North. However, although Newcastle and Gateshead's functional integration continues, their cultural and institutional relationship has decreased. In the current context, this looks unlikely to reverse; the pressure is now for increased interaction of all kinds at the city-region scale.

Conclusion

Analysing functional, cultural and institutional integration in Newcastle and Gateshead over time suggests an ongoing process, wherein inter-twin relationships are dynamic. In the first phase, from *County Council* abolition to 1998, the cities followed separate regeneration strategies and experienced decreasing cultural and institutional integration in response to changed funding and governance contexts. The 1998–2011 phase saw increased inter-twin integration of all types in a more funding-rich period. The third phase from 2011 saw uncertainty and changing relationships between the cities and their wider conurbation. Once more, a problematic economic context plus changes in governance and funding instigated at the national scale produced decreased integration.

What is evident from NewcastleGateshead's case, as this volume's editors' note, is that cities can never be islands: influences on twin cities' destinies are not confined to the cities themselves. Over time, varying interactions between other scales, like the wider Tyneside conurbation or the North-East Region, usually engendered by economic turbulence or changing governance and funding contexts set by national governments, have significantly impacted the cities' relationship. Functional interaction between Newcastle and Gateshead has been evident since their inception. However, it seems that the institutional and cultural interactions have often fallen victim to uncertainty. This dynamic relationship has underpinned NewcastleGateshead's recent attempt to navigate the context of interurban competition just as it has been throughout Tyneside's turbulent economic history. This has interesting implications. If increased interaction can improve economic development, the most appropriate scale must be identified, and the institutional and cultural interaction increased at that scale, if Newcastle and Gateshead are to achieve more consistent and robust prosperity.

Note

1 Linköping–Nörrkoping in Sweden, Rotterdam–The Hague in the Netherlands and Poland's Tri-City area.

References

Bailey, C., Miles, S. and Stark, P. (2004) 'Culture-Led Urban Regeneration and the Revitalisation of Identities in NewcastleGateshead and the North East of England', *International Journal of Cultural Policy*, 10:1, 47–65.

Barke, M. and Taylor, P. (2015) 'Newcastle's Long Nineteenth Century: A World Historical Interpretation of Making a Multi-Nodal City Region', *Urban History*, 42:1, 43–69.

Britnell, R. (2009) 'Medieval Gateshead', in Newton, D. and Pollard, A. J. (Eds.) *Newcastle and Gateshead Before 1700*, Phillimore: Chichester, 137–7.

Chatterton, P. and Hollands, R. (2001) *Changing Our 'Toon' Youth, Nightlife and Urban Change in Newcastle*, University of Newcastle Upon Tyne: Newcastle.

Colls, R. (2005) 'The North Reborn?' in Lancaster, B. and Colls, R. (Eds.) (2nd Edition), *Geordies: Roots of Regionalism*, Northumbria University Press: Newcastle, 1–28.

Ellis, J. (2001) 'The 'Black Indies': Economic Development of Newcastle c.1700–1840', in Colls, R. and Lancaster, B. (Eds.) *Newcastle Upon Tyne: A Modern History*, Phillimore: Chichester, 1–26.

Gateshead Council, 'The Angel of the North Facts', Accessed Online 29/12/17, <www.gateshead.gov.uk/Leisure%20and%20Culture/attractions/Angel/Facts.aspx>

Gateshead Council and Newcastle City Council (Adopted 2015) *Planning for the Future: Core Strategy and Urban Core Plan for Gateshead and Newcastle Upon Tyne 2010–2030*, Newcastle and Gateshead.

Hetherington, P. and Robinson, F. (1988) 'Tyneside Life', in Robinson, F. (Ed.) *Post-Industrial Tyneside: An Economic and Social Survey of Tyneside in the 1980s*, Newcastle Upon Tyne City Libraries and Arts: Newcastle.

House of Commons North East Regional Committee (2010) *Tourism in the North East: Third Report of Sessions 2009–10*, The Stationary Office: London.

Hudson, R. (2005) 'Rethinking Change in Old Industrial Regions: Reflecting on the Experience of North East England', *Environment and Planning A*, 37, 581–596.

Lancaster, B. (2005) 'Newcastle – Capital of What', in Lancaster, B and Colls, R. (Eds.) (2nd Edition), *Geordies: Roots of Regionalism*, Northumbria University Press: Newcastle, 53–70.

Meijers, E., HoogerBrugger, M. and Hollander, K (2014) 'Twin Cities in the Process of Metropolisation', *Urban Research and Practice*, 7:1, 35–55.

Nayak, A. (2003) 'Last of the "Real Geordies"? White Masculinities and the Sub-cultural Response to Deindustrialisation', *Environment and Planning D: Society and Spaces'*. 21:1, 7–25.

NGI (NewcastleGateshead Initiative) (2009) *NewcastleGateshead: The Making of a Cultural Capital*, NcjMedia, Newcastle.

Obrien, D. and Miles, S. (2010) 'Cultural Policy as Rhetoric and Reality: A Comparative Analysis of Policy Making in the Peripheral North of England', *Cultural Trends*, 19:1, 2–13.

OECD, (2006) *Territorial Reviews: Newcastle in the North East, United Kingdom*, OECD Publishing: Paris.

Pasquinelli, C. (2014) 'Branding as Urban Collective Strategy-Making: The Formation of NewcastleGateshead's Organisational Identity', *Urban Studies*, 51:4, 727–743.

Pollard, A. J. and Newton, D. (2009) 'Introduction', in Newton, D. and Pollard, A. J. (Eds.) *Newcastle and Gateshead Before 1700*, Phillimore: Chichester, xiii–xxxviii.

Robinson, F. (1988) 'Industrial Structure', in Robinson, F. (Ed.) *Post-Industrial Tyneside: An Economic and Social Survey of Tyneside in the 1980s*, Newcastle Upon Tyne City Libraries and Arts: Newcastle, 1–11.

Rushton, P. (2009) 'Gateshead 1550–1700 – Independence Against all the Odds', in Newton, D. and Pollard, A. J. (Eds.) *Newcastle and Gateshead Before 1700*, Phillimore: Chichester, 295–322.

Seex, P., Marshall, A. and Johnson, M. (2007) *OECD Review of Newcastle in the North East: One Year On*, Centre for Cities and IPPR North: London.

Townsend, T. (2006) 'From Inner City to Inner Suburb? Addressing Housing Aspirations in Low Demand Areas in NewcastleGateshead, UK', *Housing Studies*, 21:4, 501–521.

Vall, N. (2007) *Cities in Decline? A Comparative History of Malmö and Newcastle after 1945*, Holmbergs: Malmö.

Wilbraham, R. (2017) 'Place Branding and Urban Development: A Comparative International Study', Unpublished PhD thesis: Newcastle University.

1NG, (2010) *1Plan*, Gateshead.

4 The defence of old interests in a new city

Buda and Pest in Budapest in the late nineteenth century

László Csorba

Here, we turn to an ex-twin city. Chapters thus far have explored the impact and interplay of city, twin-city and metropolitan pressures in three widely separated urban locations. This one looks at the interaction of the first two factors, and at how former Buda, Pest and Óbuda loyalties were dramatically overlaid by broad city identity once the new Budapest emerged as the Hungarian national capital. Central government and resurgent nationalism was crucial here. However, we also see just what forces must be commandeered to achieve this subjugation of locality – especially forceful town planning and the prospect of becoming one capital in the newly created Austro-Hungarian Dual Monarchy, alongside Vienna.

The name Budapest clearly preserves its memory that the new capital city originated from two separate settlements in the nineteenth century. In addition to the administrative centre (Buda) and the economic centre (Pest), the new, unified city has absorbed even medieval Óbuda itself. The most important tool for unification was town planning (roads, bridges, etc.), managed by the Council of Public Works founded in 1867. This organisation stood above the new urban leadership and was chaired by the Hungarian Prime Minister because a crucial goal of Hungarian politics was the creation and development of a real capital city. This development far exceeded and eliminated the interests of the original Twin Cities and created a new unit. However, at the same time, there was a special distribution of labour between the two river banks: Buda became the most popular resort area of rich Pest citizenship. Below I present the most important elements of this process.

Budapest's most significant historian was Károly Vörös (1926–96). He was the first to analyse the history of Budapest's citizenship (notably Karoly Vörös 1979). His studies brought new results to the territory of social and cultural (including mass-cultural) history. The following overview draws heavily on his research undertaken in the 1960–90s.

Budapest belongs to the group of ex-twin cities, where originally independent settlements, as a result of progressive and mutual development, merged into a new entity, thereby creating a new united city that carried the geographical legacy of its separated past but in increasingly modified and decreasingly strident form. This development followed two complementary lines: (1) small semi-feudal settlements together became a capitalist metropolis and (2) the assimilative pressure of the large ethnic Hungarian rural population, moving into the now united settlement during the nineteenth century's final decades. This process dissolved the original German and Slavic majority, thus forming the capital of Hungary, authentically Hungarian in nature and language. The process was supported by the Hungarian state government, which had regained full powers of internal governance by the 1867 compromise with Austria. The transformation was produced by Pest's absolute domination of the process of unifying the three original towns, with the interests of the other two (Buda and Óbuda) being increasingly subsumed into district politics. The metropolis planning was controlled by governmental supervision, preventing the return of the original particular interests in the new urban policy.

During its course through the Carpathian Basin, the Danube River turns southwards. To the south of the long wide Szentendre Island, it reaches the foot of a modest hill, and then the foot of a larger one. There were, 2500 years ago, large groups of people settled at fordable points. The settlement belonging to the Celtic Eravisci was then conquered by the Romans, who established the city of Aquincum on the Danube's right bank in AD 89. From AD 106, Aquincum was the capital of Pannonia Inferior, the more southerly part of Pannonia, a new Roman Imperial province. Aquincum was located in the northern part of today's Budapest, on territory later called Óbuda ('Old Buda'). Its status is indicated by the fact that emperors sometimes appeared at the governor's palace. Because the Empire's boundary ran along the Danube, construction work was also performed on the river's left bank, in the interests of defence. Thereby the fort known as Contra-Aquincum emerged, on a territory later becoming the city of Pest. After the Western Roman Empire collapsed, Huns, Lombards, Avars and Slavs all settled among the imposing ruins, which even during the Hungarian Conquest in the ninth century were in such good condition that the tribe of the Hungarian grand prince heading the conquering tribal alliance settled here; indeed, this territory became the centre for one of his personal domains. During the early Middle Ages, the smaller settlements (Óbuda and Pest) slowly began growing due to the traffic using the river's crossing places. A royal castle arose here, at Óbuda. This slowly began serving as Hungary's mediaeval capital, since monarchs increasingly resided there. This flourishing was rudely interrupted by the Tatars, sweeping across the Carpathian Basin in 1241. Only walled towns could resist a siege laid by the nomadic horsemen. King Béla IV of Hungary drew a lesson: on today's Castle Hill in Buda, south of Óbuda, he built a fortified settlement into which the German and Hungarian citizens of the smaller towns nearby could move when threatened.

The new settlement developed rapidly in the second half of the thirteenth century, continuing in the fourteenth century under the Angevin kings. Subsequently, under the fifteenth-century rulers, Sigismund of Luxemburg and Matthias Hunyadi, Buda, as it was now called, became a city of European fame and rank; it also became the Renaissance cultural centre of Hungary. Following this, Ottoman rule lasting a century and a half was established in the country and the city. Buda's administrative role was retained because the pashas, with full powers over the province, were based there. Nevertheless, since the Muslim conquerors built only bathhouses and mosques, Buda as a city of townsmen decayed utterly, architecturally sinking into insignificance. The Hapsburgs acquired the title 'kings of Hungary'; thus, the old Hungarian state's northern and western areas became part of the enormous conglomerate belonging to this great European dynasty, whose head was also the 'Holy Roman Emperor'. Then, in the late seventeenth century, Ottoman rule weakened, so much that the Habsburgs – aided militarily by Christian countries in Europe – recovered all of Hungary. But, due to sharp conflict with the Hungarian nobility about distributing centralised power, they refrained from restoring Buda to its old rank and splendour as Hungary's capital fearing to encourage a Hungarian break-away from the Habsburg Empire. Only in the late eighteenth century did they support Buda's development into a minor provincial centre, under Holy Roman Emperor Joseph II, who was also king of Hungary (Ágoston 1997).

The twin cities' relationship was unequal, Buda having more weight than Pest. The previous century of stagnation did not count: Buda remained in public consciousness the Kingdom of Hungary's one-time capital, reinforced by the Royal Hungarian University relocating there during the 1770s. In the 1780s, the Royal Hungarian Chamber, responsible for national finances, also moved there from Pozsony (today's Bratislava) near Vienna. By contrast, Pest was a small, quiet town still based around the port, a settlement whose commerce mostly consisted of deals regarding pigs between Serbian and Macedonian-Wallachian traders (Sztamatopulosz 2009, 33–35). Pest and Buda were linked by a single, albeit important, interest: operating the pontoon bridge, which since 1767 provided permanent cross-river connection for most of the year, dismantled each December to avoid impeding the flow of ice, and reassembled each spring. It was 445 metres long and 9 metres wide, resting on 43 pontoons. For navigation purposes, it was opened several times daily allowing vessels to pass through (Rácz 2012).

The situation changed dramatically in the first half of the nineteenth century. The developing market economy greatly boosted the previously insignificant town of Pest. It did so primarily by increasing transportation along the Danube. A new, huge regional market for agricultural produce, particularly wool, grain, cattle and pigs, formed on the Pest side of the ferry link running across from the foot of Gellért Hill. Seeing this, members of the Hungarian national movement recognised that, if the three settlements of Óbuda, Buda and Pest could be united, an immense and flourishing new national capital could emerge. One of the first to articulate and very effectively publicise this idea was Count István Széchenyi. Effective leader of 'The Bridge Society', he realised that the most effective means of promoting the plan was to build a permanent bridge between Pest and Buda (Csorba 1993). The designer was William Tierney Clark, whose Hammersmith Bridge across the Thames prompted Széchenyi to invite this well-known British engineer to undertake work in Hungary (Sandor 2011). The bridge was completed in autumn 1849, doing what Hungarian reformers intended: it linked the two sides of the river, with hills on one side and a plain on the other (each with different economic attributes in industrial and agricultural terms); it also linked an official city of medieval origin, tiny size and great constitutional significance with a dynamically developing, new economic centre. When, in 1848–49, the struggle between the Habsburg court in Vienna and the Hungarian political elite intensified, an independent Hungarian administration emerged, undertaking armed resistance. It also sent a symbolic message declaring the three cities united, in a decree on 24 June 1849 (Spira 1998). However, the Vienna government defeated the Hungarian War of Independence, and nullified this measure, not wishing to create a city rivalling Vienna.

Nevertheless, the socio-economic processes promoting inter-city union intensified during the following decade, especially after 1860, when the general European economic upturn affected the Hungarian territories. Administrative rationalisation reinforced this trend. The Austrian absolutist government from 1854 onwards made the councils running the cities directly subordinate to the Gubernium (the leading organisation of Hungarian state administration 1849–59), subjecting all three to the same tax regulations. From the early Middle Ages, the city councils were chosen by electors comprising those holding royal free city privileges. When, in 1867, the Habsburg court and the Hungarian political elite compromised creating the Austro-Hungarian Monarchy, the agreement naturally contained plans for a new Hungarian capital. This strengthened the administrative trend: the 1868 Public Education Act treated Buda and Pest as a single school district headed by a new social body elected by both city councils. It comprised 34 members drawn from the cities proportionately to their populations. Even more significant was the new Council for Public Works, created by Act X of 1870. With nine members elected by the Buda and Pest city councils and nine government-appointed (Márkus 1896, 125), this had the expressly political task of developing the three cities into a Hungarian capital, modern from both a political and an economic viewpoint, imposing and technically up to date (Siklóssy 1931). The next step came in March 1871: a competition for town-planning and building regulations for all of what later became Budapest. Simultaneously, negotiations commenced to prepare for immediate union (Vörös 1978, 311).

What kind of interests clashed during that first review of problems surrounding union? Mór Szentkirályi, Pest's mayor, objected in principle:

Why should Pest, becoming evermore Hungarian, be burdened with a German Buda, incapable of development (ibid., 577–581), by shouldering its troubled financial affairs? Buda City Council, led by Ferenc Házmán, viewed co-operation with Pest as similarly burdensome.[1] During the year-long and well-publicised discussions, highly important debates developed around three key issues between the interested cities, on the one hand, and different social and political organisations, on the other hand.

The sharpest conflicts occurred around 'virilism',[2] whose validation was urged by the government during the election of the future capital's representative bodies. According to this idea, the wealthiest men, those paying most tax, had the biggest interest in how the emerging city's affairs were managed. Therefore, they should enjoy more direct representation and a greater proportion of it. During the debate, Buda representatives pointed out that, if all big taxpayers of the three cities were lumped together into one category from which they elected representatives of their own sort, then poorer Buda would be disadvantaged compared to far wealthier Pest. The Hungarian Parliament's central committee, however, accepted this proposal. Accordingly, § 26 of Act XXXVI of 1872 stated:

> Electors[3] shall elect ... half of committee members from a list of the 1200 persons paying the most direct state tax along with the necessary number of alternates ... they shall elect the other half and an appropriate number of alternates from a list of all electors.
>
> (Márkus 1899, 84)

However, wealth's raw power was tempered by providing that taxes paid by members of the intelligentsia could, if they wished, be counted twice. This gave rights to vote and stand for election under virilism to well-off lawyers, doctors and more modest homeowners, trained and working as members of the intelligentsia, as well as to rich but possibly uneducated traders and tradesmen. This overall compromise greatly assisted the new city's social amalgamation (Vörös 1978, 311–312).

Compromise was also reached over the second great issue: how to select the state official to lead the new capital. Rural Hungary was divided into counties. Each county's administration was directed by a *főispán* ('lord lieutenant') whom the interior minister proposed and the king appointed. The three towns stubbornly rejected any such arrangement. Even the most compliant proposal, by Pest, suggested the king appoint any *főispán* from among three candidates chosen by the unified city's assembly. The government reversed the logic of this: according to § 68 of the Act, 'A mayor will head the capital city ... elected by the assembly for a period of six years from among three individuals chosen by the king and approved by the interior minister' (Márkus 1899, 92).

The third great debate focused on how many city officials (actually government employees) could be voting members of the city assembly.

Clearly, the government would have preferred increasing the numbers of its dependents. Here, the three cities won: the new law provided assembly places with voting rights to just seven senior officials alongside the mayor and aldermen (whose number the assembly itself determined in establishing the number of departments). However, the three cities agreed to their police being urgently placed under state control, specifically the Interior Ministry. Important here was the need to free themselves from the enormous costs of policing. Another factor perhaps was how the drama of the Paris Commune coincided with the unification law and how the Commune's suppression produced a large workers' demonstration in Pest commemorating the victims. This may have increased demands for a strong police force among the cities' burghers. Such a force was later ensured by Act XXI of 1881, bringing the police under state control (Vörös 1973).

Discussion of the law amalgamating the three cities into Budapest lasted for two weeks in the Hungarian Parliament's House of Representatives and was accepted in a single sitting of the House of Magnates in late 1872. On 22 December, King Francis Joseph granted royal assent, Budapest emerging the next day. Meanwhile, another decisive step further advanced the earlier interests: determining the boundaries of the capital's various administrative districts. The power lines of city politics clearly emerged in the drawing of these boundaries, especially when we realise which social strata typically lived in which areas.

The proposed districts and their boundaries, alongside other regulations, were soon approved by the Ministry of the Interior, and, on 25 October 1873, the new city assembly convened in the Vigadó Concert Hall in Pest (Vörös 1973).

We can understand why the old interests could not be represented in the new city as in the past if we consider the scale of developments occurring in the two decades up to the city's unification: the three constituent cities were no longer the same places. In 1849, when Austrian troops marched into Pest and Buda, only two – not very long – railway lines led there. By 1873, Budapest had become a hub for railways from all directions reachable in hours from any important part of Hungary. Where, in earlier years, tow-boats, mostly animal-powered, plied the Danube, steamships now chugged between the embankments, where huge stone-built wharves already stretched. Above them, modern palaces and hotels looked across towards the Buda Hills. In 1849, there was just one bridge across the Danube, now the Margaret Bridge (opened 1876) was under construction and, in the south, there was also a railway bridge. On the river banks, warehouses and goods piled high stretched endlessly. On the new city's edge, immense chimneys had replaced poor hovels. Alongside huge factories and mills, dozens of new settlements had emerged. The cityscape had changed: paved and asphalted boulevards were lit by gas; buildings had running water; with loud toots, horse-drawn trams transported passengers to and from evermore distant destinations in the expanding city.

The rhythm of life had changed too: more vibrant, restless even. On street-corners, in doorways and in dubious-looking coffee houses, there was an ever-growing money-chasing. This featured actors of a new type: hawkers, speculators, middlemen and adventurers. The Biedermeier Aquarelle had given way to the stiff woodcut of the 1870s or to the black and deep brown hues of the Realist Society oil painting. All this enormous development was necessarily uneven, with many different faces. Trade in produce had surged, the development of handicraft industries had stalled and the city had become a literary centre, although still not a fine-arts centre. Pest's economy had grown many times over, while Buda continued to be poor (at least alongside Pest). Yet, the intensification, indeed creation, of antagonisms had been intrinsic to the market economy developing within individual sectors – between sectors that were merely stagnating and those on the rise (we can think of the obvious consequences of competition). Since Pest–Buda's entire social and economic modernisation remained very new, it is understandable that antagonisms were stronger and rawer than in other European cities, where development occurred over a longer period. For foreign visitors, the city felt like America: with regard to economic life, it was as though they were in a modern world without tradition or antecedent. However, sometimes surprising, sometimes almost archaic, elements of the exotic surroundings still intruded.

Nevertheless, if its contradictions were shrill, the achievements of this quarter century of city development after amalgamation speak loudly. Chief among these was that the city now became the dynamic hub around which the country's market forces evolved, accelerated and spread outwards, thereby rendering Hungary capable of autonomy within Austria-Hungary. This success had great political and economic effects nationally. The city's special position as the commercial and rail-transport hub (all main lines converged on Budapest) meant that urban trends starting in Budapest, sometimes embryonically, eventually spread across Hungary. This was true organisationally (the city providing models for local governance and market organisation). It was also true of the main problems besetting later generations in fields such as large-scale industry, urbanisation and culture.

This rich, dramatic transformation eventually impacted back on Budapest. In particular, the special interests evident when Buda and Óbuda amalgamated with Pest almost dissolved in face of Pest's ever-growing supremacy. Budapest emerged in such a way that it was Pest first and foremost that became Budapest and, by means of a new apportionment of territory and functions, absorbed into itself everything that Buda and Óbuda had earlier possessed. The weight and role of the earlier twin towns rapidly sank to the level of negotiable district interests with little protest. Emotionally, of course, strong district attachment and identity remained and still remains, but this was transformed into much more modest district politics already strongly dependent on the municipal centre, which was concerned with the immediate problems of inhabited spaces and which attempted to solve these.

The main background factor here was urban change. This meant that, once building regulation around the beginning of the twentieth century had modernised the backward, pre-industrial conditions in the Buda and Óbuda

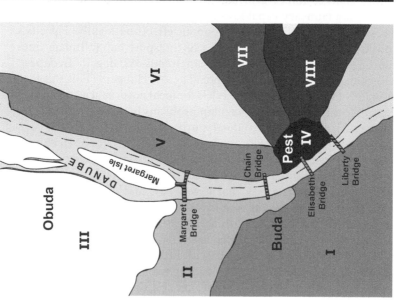

Figure 4.1 Map of bridges and districts of Budapest (based on the historical map of 1928) designed by Ekaterina Mikhailova.

Figure 4.2 Budapest bus in front of the Hungarian Parliament Building. Note the separation of Buda from Pest on the bus logo, by Anna Biriulina (2018).

districts, that side of the river rapidly became a holiday area, then a residential location for better-off strata from those parts of Pest increasingly enveloped in industrial smog. The phenomenon of wealthy Pest citizens purchasing homes in villa areas in the Buda Hills and not just travelling between the two cities at the beginning and end of each summer, but also commuting daily became a crucial motor driving everyday movement of people within Budapest. Accordingly, the new interest representation in the Buda districts was from early-on not just a reformulation of the earlier twin-city position, but also focused on ambition for its immediate living environment by a middle class, which by 1900 already identified with Budapest as a whole. Of course, fitting into this was local patriotic commitment, mainly cultural, and focused on the district's historic past (e.g. Óbuda people always proudly mention the Roman ruins, the amphitheatres, the remains of Aquincum, etc., and their own heritage). However, this implies no challenge to the notion of Greater Budapest, which in historical terms enjoyed total victory (Figure 4.1 and Table 4.1).

The foundations of the new situation were laid primarily by town planning, inspired partly by Haussmann's transformation of Paris in the 1850s. This determined the special allocation of city functions. During 1872, the Council for Public Works produced a town-planning scheme for the Pest

Table 4.1 Districts of Budapest and selected landmarks within their limits

Districts of Budapest			
No. on map	Hungarian name	English equivalent	Toponyms mentioned in the text
I	Krisztinaváros, Vár, Tabán, Kelenföld	Christine district, Castle, Taban, Kelen-field	Gellért Hill, Gellért tér, Castle Hill, Buda's Royal Palace, Lágymányos area, Kelenföld Railway Station, Béla Bartók út, Műegyetem rakpart, Krisztinaváros, Ráczváros, Hegyalja út
II	Víziváros, Országút, Rózsadomb	Water district	Southern Railway Station, Rózsadomb, Rézmál, Szemlőhegy, Vérhalom, Törökvész
III	Óbuda	Old Buda	
IV	Belváros	Downtown	Vigadó Concert Hall, the Church of the Inner City, Kossuth Lajos utca
V	Lipótváros	Leopold district	
VI	Terézváros	Teresa district	Western Railway Station
VII	Erzsébetváros	Elizabeth district	Rákóczi út
VIII	Józsefváros	Joseph district	

Source: Author's elaboration on historical maps of Budapest of 1873 and 1928.

side, followed by another for Buda in 1876 and a plan for Óbuda in 1883. Regarding the Buda side, a basic requirement was that there should be the strongest possible linkage with the Pest side, in the interests of organic urban development. Accordingly, the Danube embankment regulation on the Buda side embraced the whole area between the Kelenföld area and Margaret Bridge, and the wider area around this bridge. A plan was made for the first stretch of a boulevard within Buda from the end of the Margaret Bridge. In continuing this, attention soon turned to planning in Pest, and to additional bridges over the Danube. Accordingly, the most important part of the Buda plan was dominated by the principal structural elements shaping the Kelenföld area, hitherto completely uninhabited. These were tailored partly to the post-regulation line of the Danube embankment, partly to the railway line over the river towards the Kelenföld area, and finally, to the Buda ends of envisioned crossings known today as the Liberty Bridge and Petőfi Bridge. Accordingly, routes laying the foundations for today's *Műegyetem rakpart*, most of today's *Béla Bartók út* and the road leading to today's Petőfi Bridge were drawn.[4]

Meanwhile, one of Pest's main town-planning problems needed solving: designing a new bridge at the busiest point in the downtown area, at a place that had once been a ferry link to the Gellért Hill on the Buda side. However, this was occupied by one of Pest's finest historical buildings: the Church of the Inner City. Thus, when the area was surveyed and regulated, the church was protected and connection ensued from *Rákóczi út* and *Kossuth Lajos utca* to the planned bridgehead via a bold curve missing the church. Construction began in 1897, and naturally stimulated Buda's own town-planning programme, originally drawn up in general terms in 1876. Matters were facilitated by the phylloxera (brown rot) epidemic, which destroyed Buda's vineyards in the 1880s, accelerating the process whereby Buda became a resort area fitted into the urbanising fabric of the city as a whole by enabling buildings on the former vinyards. For a start, the boulevard's line within Buda was decided more exactly. First mooted in 1872, it took account of the fact that traffic travelling from the Western Railway Station and the industrial area of North Pest to the Southern Railway Station could be diverted onto the Margaret Bridge, thus relieving pressure on the overbusy Chain Bridge. Accordingly, Margaret Bridge tolls were abolished in 1885.

The Buda route was built and later continued southwards from the Southern Railway. It ran parallel to Castle Hill and across *Krisztinaváros* to connect with a new bridge at the foot of Gellért Hill. Here was the Buda side's most problematic area – the old *Tabán* district, which obstructed the southern end of the boulevard within Buda. The *Ráczváros* quarter was named after the Serbian grape-growers and merchants settling there in Ottoman times. Perched high on the hillside and stretching along the Danube embankment at the foot of the rocky Gellért Hill, the modest, mostly dilapidated and village-like houses contrasted evermore sharply with the prestigious quarters opposite: Pest's 'Danube row' and Buda's Royal Palace, both imposing sights and both growing.

Here, the city council (and sometimes district councils) initiated regulation. After deciding in 1878 to free Gellért Hill from these shabby hovels, which, along the Danube embankment, were constantly exposed to falling rocks, the council acquired and demolished them. The compulsory-purchase zone gradually extended up the hillside, to the part above today's *Hegyalja út*. However, complete demolition of the Tabán district occurred only in the 1920s, by which time Hungary was an independent state (Schüler 1934).

In the early 1890s, more spectacular changes occurred in the Kelenföld and Lágymányos areas, spurred by constructing the Liberty Bridge at the southern foot of the Gellért Hill. This necessitated demolishing the old Sáros bathhouse, simultaneously outlining what later became *Gellért tér* (a public square). The central government's Council for Public Works planned a residential and factory quarter to the south of this bridge. Also, it intended that thoroughfares should run to the Danube and the Kelenföld Railway Station and wanted these to connect with thoroughfares in Pest via the bridges. But the Council did not neglect the outlying Buda Hills, considering them primarily leisure areas. These included the *Rózsadomb* (near the *Lipótváros*, Pest's wealthiest, most developed district, just across the Margaret Bridge), and the entire *Rézmál, Szemlőhegy, Vérhalom* and *Törökvész* hill country. Regulating these areas too was projected. Streams and drainage ditches in the hills began being separated from roads. Countryside was commandeered for a public park, holiday homes and villas (Gál 1998, 245–253).

These tendencies are clearly shown in the fact that, while average numbers of inhabitants per room in Pest and Buda were pretty much equal between 1880 and 1900, room-occupancy in outer Buda was far lower than that in outer Pest (Vörös 1978, 577–581). In outer Pest, primarily poor workers and day-labourers were crowded into single-storey buildings in conditions far worse than in the flats of the inner areas. However, in outer Buda, many buildings were summer homes or dwellings among hills and forests. If only because of the great distance from other places of work, the latter were inhabited by those with direct links to the land (vineyard workers, market gardeners, etc.).

The re-ordering of the Buda side of the city – which, of course, was approved in Budapest's City Hall located in Pest – induced the wholesale demolition of the urban and geographical foundations for Buda's separate interests as a twin city. Simultaneously, it created a new Buda identity, which was not alien to the earlier heritage: a conscious enjoyment, even celebration, of being the 'more liveable' half of the metropolis. In the late nineteenth century, a newspaper writer highlighted this:

> to find matches for two such different parts of the world as Pest and Buda staring at each other in one and the same city, one would need to travel very far ... The left bank, Pest, is all modern nerves, while the right

bank, Buda, is almost a boarding house. On the left bank, trams race along, the jingling of their bells mixing with people's rapid talk. Over in Buda, acacia bloom falls on the people, who walk calmly and speak softly. But this right bank is not simply calm; also, it is not developing. It is not acquiring huge tenement blocks; it is not overburdened with caprice, tastlessness, and sculpture of every kind. The man of moods flees the desolation of the left bank for Buda, where he can find gateways from Venice, flights of steps from Innsbruck, little fairytale houses, and Baroque palaces with one upper storey.

(From *Budai Hírlap*, 1906)

The poetic words slightly obscure the fact that – as noted – Buda also saw development. It was just different from Pest's, since the division of work in the great city and the geographical separation of its functions traced divergent paths for the different parts stretching either side of the river. The difference perpetuated the old rivalry of the one-time twin cities, but invested it with new content. Buda and Pest rivalry survives today (Figure 4.2). Everyone thinks their particular side is more beautiful, better, more liveable. Pest inhabitants often joke that the only reason for moving to Buda is for a better view of Pest. Nevertheless, the people who bore the changes best were those embracing the notion expressed by the well-known journalist Adolf Ágai in 1872. 'In the future', he wrote, 'we'll be inhabitants neither of Buda, nor of Pest, but of Budapest!' (Ágai 1909, 132).

Notes

1 The overwhelming majority of Buda and Pest's population comprised Germans, Hungarians and Slavs (mostly Serbs) until the mid-nineteenth century. However, large-scale immigration by newcomers in the 1850s altered the ratios. From this point, the Hungarian assimilation of the traditional German population and the Slovaks arriving in several waves accelerated (see Csorba 1997, 92).

2 According to the Act of LXII 1870, the major taxpayers – paying taxes is a manly job, in Latin *officium viriles* – are members of the representative bodies of the municipalities, 50% of whom come from the major taxpayers who are automatically included because of their financial status (see Márkus 1896, 212–213).

3 Determined by property-holding.

4 For details of planning and urban structure, see Preisich (1964, 43–54; 73–82).

References

Ágai, A. (1909), '*Utazás Pestről – Budapestre 1843–1907. Rajzok és emlékek a magyar főváros utolsó 65 esztendejéből*', 2. kiadás [Journey from Pest to Budapest 1843–1907. Drawings and Memories of the Last 65 Years of the Hungarian Capital – 2. edition].

Ágoston, G. (1997), 'History of Budapest from Its Beginnings to 1703', in A. Gerő and J. Poór, eds., *History of Budapest from Its Beginnings to 1703*, Highland Lakes Publishing, NJ, Atlantic Research and Publications, 11–31.

Csorba, L. (1993), 'Budapest-gondolat és városegyesítés' [The Budapest idea and the unification of the city], *Budapesti Negyed* 1, 2, 18–25.

Csorba, L. (1997), 'Transition from Pest-Buda to Budapest, 1815–1873', in A. Gerő and J. Poór, eds., *History of Budapest from Its Beginnings to 1703*, Highland Lakes Publishing, NJ: Atlantic Research and Publications, 69–101.

Gál, É. (1998), 'A rózsadombi villanegyed kezdetei' [Beginnings of the Rose Hill Villa Area], in G. Gyáni, ed., *Az egyesített főváros. Pest, Buda, Óbuda*, Budapest, Városháza Publishing, 245–253.

Márkus, D., ed. (1896), *'Magyar Törvénytár 1869–1871 évi törvényczikkek'* [Hungary Collection of Laws 1869–1871], Budapest, Franklin-Társulat.

Márkus, D., ed. (1899), *'Magyar Törvénytár 1872–1874 évi törvényczikkek'* [Hungary Collection of Laws 1872–1874], Budapest, Franklin-Társulat.

Preisich, G. (1964), *'Budapest városépítésének története II. A Kiegyezéstől a Tanácsköztársaságig'* [The History of the Construction of Budapest II. From the Compromise to the Council Republic], Budapest, Műszaki Könyvkiadó.

Rácz, L. (2012), 'A pest-budai hajóhíd, az éghajlati változások indikátora és áldozata (1767–1849)' [The pontoon bridge of Pest-Buda as an indicator and victim of climate changes (1767–1849)], *Belvedere Meridionale* 24, 2, 4–29.

Sandor, P. V. (2011), 'William Tierney Clark and the Buda-Pesth chain bridge', *Proceedings of the Institution of Civil Engineers-Engineering History and Heritage* 164, 2, 109–122.

Schüler, D. (1934), *'Adalékok a Tabán történetéhez és rendezéséhez'* [Contributions to the History and Settlement of Taban], Statisztikai Közlemények 75/4.

Siklóssy, L. (1931), *'A Fővárosi Közmunkák Tanácsa története. Hogyan épült fel Budapest?* 1870–1930' [The history of the Council for Municipal Public Works. How to build Budapest?], Budapest, Fővárosi Közmunkák Tanácsa.

Spira, G. (1998), 'A pestiek Petöfi es Haynau között' [The Inhabitants of Pest between Petőfi and Haynau], Budapest, Enciklopédia Kiadó.

Sztamatopulosz, A. (2009), 'Görög Vlachok az Osztrák-Magyar Monarchia területén élő görög diaszpórában' [Greek Vlachs living in the Greek Diaspora of Austro–Hungarian Empire], in K. Szabó, ed., *Görög Örökség. A Görög Ortodox Diaszpora Magyarországon a XVII–XIX. Században,* Budapest, Budapesti Történeti Múzeum, 33–42.

Vörös, K. (1973), *'Egy világváros születése'* [Birth of a World City], Budapest, Kossuth Könyvkiadó.

Vörös, K. (1978), 'Pest-Budától Budapestig 1849–1873'. [From Pest-Buda to Budapest], in G. Spira and K. Vörös, eds., *Budapest története a márciusi forradalomtól az őszirózsás forradalomig,* 4, Budapest, Akadémiai Kiadó, 117–320.

Vörös, K. (1979), 'Budapest legnagyobb adófizetői 1873–1917' [Budapest's largest taxpayer], Budapest, Akadémiai Kiadó.

5   Urban Lagos 1927–67

A tale of two cities?

Lanre Davies

With this chapter, we switch continents again and begin examination of African twin cities, all of them arguably ex- rather than currently extant. Both this and the following chapter reveal how city life has been affected by the colonial context. The story of urban Lagos and Ikeja provides insight into the interplay of different historical forces in their establishment, shaping and character – geographical facts on the ground, the calculations of the colonial power, and of post-independence federal and state governments. In the process, these cities have changed from sharing a common border into one city surrounded by an amorphous metropolitan 'other'. Their relationship seems to parallel that of Taipei and New Taipei in Taiwan.

Introduction

In this chapter, I will examine how the political, economic and demographic development of Lagos generated another administrative centre in Ikeja Lagos during 1927–67, as another way twin cities can emerge. The year 1927 saw the separation of Lagos District (which included Ikeja Division) from Municipal Lagos, while 1967 marked the year that Ikeja with other divisions were merged with municipal Lagos to form Lagos State. Lagos' urban history in the studied period is a tale of two cities: Lagos City and Ikeja. What began on Lagos Island, as the seat of the Lagos Oba (the king of Lagos), in the pre-colonial period, became the colonial government's headquarters in 1861. However, the administration and urban development of Lagos colony from 1861 greatly extended its frontier beyond its immediate precinct. This accelerated in 1906 when the colony was merged with the Protectorate of Southern Nigeria, and again in 1914 when both the Southern and Northern Protectorates were merged as Nigeria. Lagos became one big conurbation with its core in Ikeja Division in northern Lagos, which was topographically suitable for such development. I intend exploring the important changes in the administrative development of Lagos 1927–67, and the various inter-governmental relations spanning these changes focusing on the relations among Lagos City Government, Western Regional Government and the Federal government. The chapter seeks to throw light on Ikeja's rise as a distinct administrative and economic centre within Lagos.

Nigerian scholars have explored the twin-city concept by examining towns and cities straddling two territories (Osuntoku 2005) or countries as with Seme and Idi-Iroko in Nigeria and Benin Republic (Adeniran 2007; Asiwaju 1985). Others have explored the effects of population spilling from one city into another (Ibrahim 2008). Internal twin cities are a special case of two adjacent urban centres growing into each other's space and sometimes losing their individual identities with their mutual borders becoming irrelevant.

A prominent feature of urban growth in the period was that it could only proceed northwards. This resulted from Lagos' topography. South of Lagos Island lies the Atlantic; to the east is the lagoon bounded in its southern part by a narrow coastal strip covered by mangrove swamp, forest and creeks. Hence, large-scale settlement was impossible except through tremendous expenditure on drainage and reclamation which the penny-pinching colonial government was unwilling to undertake. Therefore, the only viable area for urban expansion was the mainland to the north in the Ikeja Division. This explains why the two major roads and the only railway linking Lagos with the rest of the country during 1927–67 passed through this area. Right from the beginning, Lagos Island was administratively and physically separated from the Lagos Mainland where Ikeja still predominates as the State capital. Today, Lagos Island remains physically separated from Lagos mainland by the lagoon. Three bridges – Carter Bridge (1901), Eko Bridge (1975) and Third Mainland Bridge (1990) – connect Lagos island to Lagos mainland, attesting not only to physical separation but also to the twinning of Lagos, since the two have since 1967 extended down to their respective waterfronts.

Pre-colonial Lagos

The area called Lagos was known to Europeans by the late fifteenth century: Lagos lagoon had appeared on European maps by 1485 (Agiri and Barnes 1987, 19). A Portuguese man, Duarte Pacheco Pereira, wrote about the trade of Lagos between 1505 and 1508 (Law 1973). Andreas Joshua Ulsheimer, a German surgeon, aboard a Dutch ship, probably visited Lagos in 1603 and described it as a large frontier town inhabited by soldiers (Agiri and Barnes 1987, 18). Lagos is generally believed to have been founded by the Awori, a sub-group of the Yoruba ethnic group. Lagos was conquered by the Benin around the sixteenth century after which Benin soldiers ran its day-to-day affairs for the Oba of Benin. Lagos remained primarily a coastal port until the 1790s, when local entrepreneurs became heavily involved in the slave trade. For several decades, the trade flourished with increasing involvement from European merchants. British attempts to secure their economic interests coupled with the efforts of the abolitionists in Britain ended the trade, causing them to bombard Lagos and impose a consular authority in 1851 (Cole 1975). However, this did not enable them to direct Lagos affairs as they pleased. They could only legitimately intervene on behalf of British subjects in an era of great influx of people into Lagos. From the 1830s to the 1850s, the repatriates, returnees, Europeans and several other peoples from the hinterland flocked into Lagos for trade and other reasons (ibid.). Regulating these varied peoples, while simultaneously trying to abolish the slave trade and promote 'legitimate commerce' instead, was difficult. Hence, the British decided on annexation.

The annexation of Lagos

The 1861 annexation brought Lagos effectively under British colonial rule. Dosunmu ceded the 'port and the Island of Lagos ... with the consent and advice of his council' (Tew 1939). This formalised the British presence. The new colony consisted of a large section under British protection and four small sections (Badagry, Lagos Island, Leckie, and Palma), considered as actual British territory (Barnes 1986, 23). It also included what would eventually be known as metropolitan Lagos. This included the port and Island of Lagos and the nearby villages most of which were in the Ikeja Division. There was also Lagos Municipality consisting of Idumagbo, Alakoro, Iduganran, Marina, Obalende and Race Course (all on Lagos Island) (Odumosu 1999, 10).

According to Barnes (1986, 25), the British from the start intended to administer their newly acquired colony's 62,000 inhabitants as a unit despite their reservations. First, they initially distrusted the chiefs due to their slave-trade involvement; second, residents who were made British subjects had to be governed directly under the laws of the crown; third, the people of the city (Lagos Island) and villages (Ikeja Division) were seen as too fragmented to be subjected to the authority of just one element. British administrators were also uncertain about how to treat them. They saw the city dwellers as relatively urbane, but the villagers in Ikeja Division as very 'backward'. These reservations notwithstanding, a single central administration was introduced for the whole colony.

However, British efforts to develop unified governance were fraught with difficulty from the beginning. Between 1861 and 1886, the administration shifted from Lagos to Sierra-Leone (1866), then the Gold Coast (1874), and back to Lagos (ibid.). In 1886, Lagos received a new charter and separate administration. But strong protest from Lagos leaders (Oba and chiefs) resulted in the original governmental structure (of Resident Governor and Executive and Legislative Councils) being restored for the whole colony until 1914, when the Southern and Northern Protectorates were merged into what became Nigeria, and Lord Lugard became the first Governor General (Sada and Adefolalu 1975).

Importantly, there was no real colonial government outside Lagos colony until 1885. More crucially still, the problem of the physical separation of Lagos Municipality from the Lagos Mainland was yet to be meaningfully addressed. The wide lagoon separating Lagos Municipality from the mainland had not been bridged at all. In pre-1900 Lagos, horses and various horse-drawn carriages seating two, three or four passengers along-side manually operated single-seaters were the means of transport on Lagos Island (Sada and Adefolalu 1975). Communication between Lagos Island (the seat of government) and the mainland was by means of canoes (ibid.). Only in the 1890s did the British make plans for bridges across the lagoon to link the Island with Ido (Carter Bridge) and Ido with Ebute Meta (Denton Bridge).

However, demographic changes in the northern part of Lagos colony were making the British absence from Ikeja Division more problematic. The division's population between 1871 and the 1920s increased by over 300% (Barnes 1986, 26). Additionally, it was becoming ever more heterogeneous and ever harder to govern directly from municipal Lagos. Apart from the in-flux of European merchants consequent upon annexation, many freed slaves from Sierra Leone, Brazil, and Cuba, migrants fleeing from Yorubaland's internecine wars and people from northern Nigeria, among others, flocked to Lagos Island and areas under Ikeja Division.

This large influx of people into Lagos city and the northern part of the colony made the British strengthen Lagos city's administration while still ignoring its urban sprawl northwards. There were attempts at local govern-ment administration like the General Sanitary Board (1899), Lagos Municipal Health Board (1908), and Lagos Township Ordinance of 1917 (Oyesiku 1998, 40). The Health Board gave Lagos city its first separate administration and was thus styled a Municipality. Although, without permanent staff, it was vested with considerable powers and functions and undertook many town-ship improvement schemes formerly undertaken by the central administra-tion. In the colony's western and eastern parts, District Commissioners were asked to take up residence and oversee local affairs. The northern part (Ikeja Division) renamed Lagos District (1927) was still administered from Lagos Island. The Supreme Court and a small staff living and working in Lagos City now ran this part. Thus, the outlying areas' administration, especially the Ikeja Division of Lagos, was not given the required attention until 1927.

From 1917, Lagos local government acquired a new dimension with the passing of the Township Ordinance. This accomplished sub-division classi-fication of towns into first, second and third class townships. This body was given responsibility for tackling the overcrowded and insanitary conditions of the Township and Urban District of Lagos (Oyesiku 1998, 40). Lagos became the only first-class township, and was granted a representative council of be-tween six and twelve members. In practice, it had six members, three elected by the electorate and three appointed by the Governor (Williams 1975).

The evolution and administration of Ikeja district

Since Ikeja was located on an upland site in the north of Lagos colony near the Ogun River estuary, this stretch of land before 1927 was noted for its dominant agricultural land use. The place called Ikeja was founded by one Akeja Onigorun according to oral traditional accounts.

Ikeja acquired prominence in 1927, when Lagos District became administratively separated from Municipal Lagos in order to give it the required administrative attention. This official separation opened a new chapter in the metropolitan area's political, economic and spatial development. Lagos District headquarters were first opened at Agege, but later moved to Ikeja and named Ikeja District. This area in the colonial period comprised places like Ikeja, Agege, Oshodi, Mushin, Akowonjo, Isolo, Itire, Onigbongbo Shomolu, Ogudu, Isheri, Ojota and several other communities (Abegunde 1987). The fact that these areas were rural differentiated Ikeja Division from Lagos Island, which was basically urban. The delay in separating Lagos District's administration from Municipal Lagos has been attributed to the activities of the African Representatives of the Legislative Council who opposed because its provision to include headmen and chiefs of the Ikeja Division in some governmental tasks would allow the metropolitan areas and their patrons develop their own ties to the government (Barnes 1986). The larger a village-patron following, the greater his governmental influence. It was in an effort to curtail them that the colonial government found it desirable to separate Lagos City administration from settlements in the Ikeja Division. Other reasons included outright lawlessness in the form of gangs, armed and violent thefts, and counterfeiting; problems of mutual distrust, suspicion and hostility; the failure of unofficial authorities (village heads) to regulate social interaction. Increased numbers of villages and population also expanded the scale of interaction beyond what could be easily managed by the village system. Moreover, the economic potential of the area in terms of tax-collection had to be tapped (ibid.).

Ikeja District's formal separation from Municipal Lagos entailed establishing a physical colonial presence in this hitherto badly neglected area. There had been no single colonial official assigned to the area, though a few postal workers, doctors and police men worked there. Thus, residents had to travel to Lagos Island for bureaucratic or legal assistance. Also, access to law enforcement and protection required trips to Lagos municipality. As noted, the only official body with direct jurisdiction was the Supreme Court, far removed from the people because of its unfamiliar procedure and legal principles (ibid. 36–37). After separation, Supreme Court Commissioners travelled to hear cases in the District. However, because of residents' alienation from them, native courts were later introduced to Ikeja District in 1928. Furthermore, Ikeja did not enjoy its own police contingent until 1939, when it had 32 (ibid.).

For too long before separation, the area was unvisited by British colonial officials. British authorities in Lagos City were still out of touch with development there even after separation. For instance, I. F. W. Schofield, a Supreme Court Commissioner and the first Resident Officer of the Ikeja District, spent just 16 out of his 315 working days in 1928, touring the area to establish direct contact with local people (ibid.). He ended up relying on village headmen in the division to oversee a taxation system that ironically relied on an indirect collection system. Thus, the application of different governmental standards in Lagos City and the Ikeja District greatly contributed to their separate development. The fact that the two areas were treated differently made them react differently and ultimately define themselves differently within the metropolitan spatial context. Hence, the British goal of uniting them for uniform development was defeated. Ikeja District evolved its own separate system of authority; separate leaders emerged; and a separate political identity also surfaced. This dichotomy arising from serious administrative neglect affected the political outlook of metropolitan Lagos so significantly that, even though Lagos City and Ikeja District belonged to a geographically contiguous and substantially interdependent metropolis, they remained politically divided, each with its own authority system acting independently, and causing the Ikeja area's evolution as another urban authority and ultimately a state capital.

Lagos City Council, Western Region and Federal government's administration of Lagos

Lagos City had been the administrative seat of an expanding British commercial and imperial interest in Nigeria since 1861. It was Nigeria's political capital, and the country's various sections had a sense of both belonging and ownership. It housed the House of Representatives, Nigeria's highest legislative body, and the political elite saw Lagos as a place to advance their political careers. Consequently, its status generated concern among the political class, especially from 1951, when the central government decided to merge the Lagos port and municipality with the Western Region (Adebayo 1987).

The Western Region emerged in 1947 as part of the administrative development of Nigeria when the Western Provinces became the Western Region (Abe/Prof. 2, Regional policy and Cooperation). Ikeja Division had all along been part of the Western Provinces converted to Western Region by 1947. Also, following the constitutional review of 1951, the colony of Lagos was incorporated into the Western Region. The new arrangement made Lagos elect members to the Western Regional House of Assembly, from where a select few would be elected into the Federal House of Representatives (Adebayo 1987). This created conflict between opposition members and the Western Regional government controlled by the Action Group, the political party of the Yoruba of the South West. Leaders of other parties, especially members of the Northern Peoples' Congress and National Council of Nigeria and the Cameroons, felt their political careers at risk if Lagos City continued under Western Regional control (ibid.).

However, merging Lagos City with the Western Region enabled the latter to promulgate the Lagos Local Government Law subjecting Lagos City

Council (LCC) to regional control. The same Act subjected Lagos to supervision by the Western Regional Ministry of Local Government, abolished the office of Mayor, formally enabled traditional chiefs to participate in the city council and made the Oba of Lagos President of LCC. Thus, Lagos City until 1954 was administered by two higher level governments and a local government – the LCC. Its general administration was undertaken by an agency of the government of Western Nigeria, apart from the cluster of administrative functions exercised by the Federal government. In addition, the Federal government was also responsible for maintaining public order, a function delegated to the Nigeria Police (Williams 1975). These layers of administrative responsibility produced jurisdictional conflict, which in turn produced inter-governmental difficulty and serious neglect of urban problems especially outside Lagos City (Adebayo 1987).

The problems then arising between the Federal government and the Western Region government led to Lagos being excised from the Western Region in 1954, and later reconstituted as a federal territory under the 1959 Lagos Local Government Act (Williams 1975). This brought Lagos under the Federal government, which now created a Ministry of Lagos Affairs to supervise the Lagos Town Council, replacing the Western Nigeria Ministry of Local Government. Thereby, the Federal government inherited regional functions from the Western Regional Government. There was a reversion to the pre-federal relations between the central administration and Lagos municipality. Again, normal interaction now prevailed between the local council and the federal authority. However, this did not completely eliminate conflict of jurisdiction since Ikeja Division, which was the other arm of Lagos, remained statutorily under the Western region.

The Federal government's policy responsibility in Lagos was undertaken by several ministries, their constituent departments and subordinate statutory corporations. Most operating responsibilities were in practice delegated to the LCC. The Ministry of Lagos Affairs retained a major role due to exercising the powers of superior government over LCC alongside coordinating and unifying the policies and programmes of all the other ministries and institutions. Its coordinating activities did not eliminate the jurisdictional conflict long plaguing Lagos administration. Divided and overlapping responsibility continued in several spheres like health, transport and water – all critical urban services (Williams 1975).This militated against improving and expanding these services to the Ikeja area because each government agency took an exclusive view of its role rather than a comprehensive view of the demand for these services. The activities of the LCC, which was more directly in touch with the populace, could lead to malpractices, which was why it was supervised by the Ministry of Lagos Affairs. However, the Ministry itself suffered from functional disabilities. Lagos viewed it as an instrument of alien government. The leadership of the Northern Peoples' Congress (now controlling Lagos government and dominated by northerners) was viewed with great suspicion. This affected the performance of the Ministry of Lagos Affairs. It was administratively ill-equipped to achieve its coordinating function.

Therefore, it relied on the ministries it was meant to supervise in order to superintend the LCC, and particularly to approve LCC's estimates (Williams 1975). It forwarded sections of the LCC estimates to appropriate ministries for comments and advice. This however, worked against coordination, since they regarded such responsibility as secondary to their main duties and consequently delayed action. Thus, the LCC budget might not be approved until the approach of another year. Moreover, the ministries did not coordinate their advice, thus defeating the purpose for which it was sought.

Lagos and inter-governmental problems between the Federal and Western Region governments

Serious inter-governmental problems arose between the Federal government and the Western region. Divided responsibility for various functions and the resultant neglect of urban services in the area of planning, extension and improvement that affected the federal territory also affected the whole Ikeja District. The influx of people into Lagos especially more than doubled the population of metropolitan Lagos (Williams 1975). Population explosion created the emerging conurbation disregarding local government jurisdiction between Municipal Lagos and broader metropolitan Lagos, all of which formed an integrated unit for everyday purposes. This situation produced an unprecedented demand for urban services both within Lagos City and the Ikeja suburbs, underscoring the inadequacy of urban infrastructure, financial resources and administrative capacity.

When established, the Western Region created four District Councils (Agege, Ajeromi, Mushin and Ikeja) from the existing Ikeja Division. Its intention was to take governance to the grassroots so as develop the area. However, its role in the metropolitan area paralleled the Federal government's in several respects. Apart from establishing and supervising district councils, its Ministry of Local government occupied a central role in metropolitan administration area as did several other Western Region ministries. They were directly involved in the area's development and administration and supervised activities delegated to the district councils.

The metropolitan dimension of urban services resulted from the lack of a unified governmental structure, the multiplicity of governmental authorities and the absence of urban-wide planning (Williams 1975). This produced unequally distributed resources and services between different sections of the community. Deficiencies in urban infrastructure, financial resources and administrative capability in the Ikeja area where urban services were most urgently needed contrasted badly with what was provided in Lagos City. The political climate of Lagos and the structural arrangement, under which no unit assumed responsibility for area-wide issues and local governments lacked policy-influence, created problems. There was also the problem of geographical diversity of function/authority between Lagos City and Ikeja area. This disparity at the operating level of urban administration raised the cost and often reduced the output to the public of government projects and services. In the area of urban development, land was a major problem. For instance, the Lagos Executive Development Board (LEDB) required plenty of land for cheap housing, which was then in abundance in the Ikeja area outside the Federal territory, where the board had no legal power to act. The result was that public housing and reclamation costs were higher than they should have been had there been synergy between the Federal government and the Western Region. Moreover, the Western Region Housing Corporation's programmes were uncoordinated with those of LEDB (Williams 1975).

Urban sprawl, suburbanisation and the emergence of Ikeja city

We should remember that the area emerging as Lagos State capital was originally part of Lagos Colony's northern district, styled Lagos District, from which Ikeja Division emerged. In tracing the area's suburbanisation, several factors are important. Its geography strongly influenced its evolution as one of Lagos' most developed areas. Thus, it was unnecessary to undertake costly reclamation to provide land for the teeming population that thronged there after 1861, especially in the twentieth century. Lagos City's geographical situation made large settlements almost impossible as this would involve tremendous expenditure on drainage and reclamation works and on bridging the creeks. The area of firm ground suitable for unhindered urban expansion was the mainland towards Ikeja. This explains why apart from the railway, the two major roads linking Lagos with the other parts of Nigeria in the selected period passed through it (Figure 5.1).

Figure 5.1 Urban sprawl in Ikeja-Lagos (the other twin of Lagos). Map by Lanre Davies.

The area's suitability for large settlements affected demography. Those moving to Lagos especially in the early twentieth century preferred living on the mainland towards Lagos District. Several settlements emerged along the railway and motor ways. They also infiltrated the outlying parts in the Mushin, Ajeromi and Surulere axis. New residential areas also decongested central Lagos and resettled people displaced by urban redevelopment like those in Yaba and Surulere. Other settlements filled the space between the city and the pre-existing suburban settlements, for example, the Ilupeju, Obanikoro and Maryland Estates; and the new residential and industrial suburbs like Ojota and Oregun (Sada and Adefolalu 1975).

Indeed, the growth and development of Agege, Ajeromi, Mushin and Ikeja Districts in Lagos District constituted the spatial expression of urban explosion in metropolitan Lagos. As argued earlier, this area was administered by these rural district councils, which were neither empowered nor properly equipped to undertake town planning. Moreover, the absence of planning regulations until 1958 was a major attraction to those individuals migrating to the area to become landlords (Sada 1970). Thus, there was increasing individualisation of property and the break-up of the Awori/Egba communal or village pattern of living.

Another stimulus was Lagos City's economic position as a port and Nigeria's economic nerve centre. Migrants saw Lagos as an *Eldorado*. Most newcomers ended up in the Ikeja area due to the cost of housing, land and rent. The government's economic programme also encouraged people in the area to become farmers growing cash crops for export. Several of these built country houses on their farms, where some resided while others continued living on Lagos Island and commuted to Ikeja at weekends. These farms attracted many labourers, and most lived within the areas with relatives (Davies 2009). The farmsteads later grew into villages, forming part of the Ikeja suburbs. Its soil differentiates it from those of Lagos Island, which formed over low lying sandy plains and marshes near lagoons and creeks. Ikeja's soils were formed on deeply weathered sandstone overlain by the 'coastal plain sands'. The area is upland as opposed to Lagos Island's lowland. Unlike Lagos Island, the soils supported many settlements before and after annexation (Abegunde 1987).

The political neglect of the area until the twentieth century coupled with the differences between the Federal and Western Regional governments enabled Ikeja to be both administratively and physically separated and distinct from LCC. While the neglect lasted, the area was evolving its distinct social and political orientation, like the formation of Community Development Associations ideally suiting it for suburbanisation. Indeed, in the long run, conflicts between the Federal and Western Regional governments produced healthy competition, turning the Ikeja area into Lagos' second most urbanised area by the 1960s, and eventually state capital.

By the end of World War II, suburbanisation was progressing rapidly. It accelerated when the federal government acquired the vast land between Oshodi and Ikeja to construct Ikeja airport to replace the wartime aerodrome at Apapa. It also acquired land between Onigbongbo and Ikeja for the Government Reservation Area, Police College and Army Barracks. As noted earlier, until 1958, Lagos' suburbanisation occurred without town planning authority and building ordinances. Planning regulations were non-existent. However, in 1960, the Western Nigeria Development Corporation, an arm of the Western Regional Government, established the Ikeja Industrial Estate. In 1965, the Ikeja Area Planning Authority emerged. In 1968, it was merged with the LEDB and renamed Lagos State Development and Property Corporation (LSDPC) (Sada and Adefolalu 1975). The idea behind the establishment of the LSDPC was to control hitherto unregulated urban sprawl in Lagos' rural urban fringes.

The federal government's physical presence in the Ikeja area, which was statutorily under the Western Region, challenged the region to establish physical infrastructure facilities for developing the area. Development at Ikeja, Mushin and Ilupeju covered about 2000 acres. The development of Ikeja Industrial Estate started in 1959, when the Western regional government established a 750 acres industrial estate at Ikeja and Ogba villages. Of these, 300 were for industrial and 400 were for residential purposes. The Ikeja industrial complex comprised four units – Oshodi, Gbagada, Matori and Oregun-Ojota (Akinola and Alao 1975). The complex consisted of food, beverage, tobacco, leather, wood, paper, printing, plastic, rubber, metal and publishing among other industries (Akintola-Arikawe 1987). Thus, heavy investment in physical structure by both the Federal and Western Regional governments and the infrastructure facilities put in place in Ikeja made the area emerge as the second most urban division of Lagos. It was just natural that Ikeja would become the state capital after the 12-state federal structure emerged in consequence of the Nigerian military crisis resulting in the 1967–70 civil war. However, given the crisis situation Nigeria found itself in 1967, there was a transition period during which negotiation with the federal and Western State governments for the transfer of functions, institutions and property took place. The federal government and LCC immediately gave some of their buildings for the use of the nascent state, while Lagos also benefitted from rentals of private buildings all within the municipality of Lagos (Olugbemi 1987) until 1976 when the government finally shifted its administrative base to Ikeja.

Conclusion

This chapter has discussed the twinning of Lagos City with Ikeja on the Lagos mainland. It has also shown how the Ikeja area emerged from the big conurbation of Lagos into the capital of Lagos State. The northward growth of Lagos ensured the absorption of neighbouring villages and the transformation of rural areas into suburbs in the Ikeja Division. Relationships of conflict, cooperation and competition were crucial to Ikeja's emergence as the twin of Lagos City. Finally, the struggle to control Lagos for political and economic reasons, which caused the Federal and Western Regional governments to compete for political attention in the Ikeja Division, helped produce the heavy investment and infrastructure facilities in the Ikeja area, especially by the Western Regional government, and the subsequent emergence of the Ikeja area as the second most urban area of Lagos and a state capital. The overall result was two twin cities. By virtue of geographical necessity, migration and transport infrastructure, they were deeply interdependent and interlinked. However, by virtue of physical separation (the lagoon), governmental arrangements and malfunctions, they also became increasingly distinctive.

References

Abegunde, M.A.A. (1987) 'Aspects of the Physical Environment of Lagos' in Adefuye, A. et al., (eds.) *History of the Peoples of Lagos State*. Lagos: Lantern Books, 6–15.

Adebayo, A.G. (1987) 'Lagos: The Choice and Position of a Federal Capital' in Adefuye, A. et al., (eds.) *History of the Peoples of Lagos State*. Lagos: Lantern Books, 306–320.

Adeniran, W. (2007) 'Preliminary Observations on Language Use and Needs in Border Regions: The Nigeria-Benin Experience' in Akinyele, R.T. (ed.) *Academic Disciplines and Border Studies*. Lagos: University of Lagos. 123–142.

Agiri, B.A. and Barnes, S. (1987) 'Lagos Before 1603' in Adefuye, A. et al., (eds.) *History of the Peoples of Lagos State*. Lagos: Lantern Books, 18–32.

Akinola, R.O. and Alao, N.O. (1975) 'Some Geographical Aspects of Industries in Greater Lagos' in Aderibigbe, A.B. (eds.) *Lagos, The Development of an African City*. London: Longman, 108–123.

Akintola-Arikawe, J.O. (1987) 'The Rise of Industrialism in the Lagos Area' in Adefuye, A. et al., (eds.) *History of the Peoples of Lagos State*. Lagos: Lantern Books, 104–127.

Asiwaju, A.I. (1985) *Partitioned Africans: Ethnic Relations across Africa's International Boundaries, 1884–1984*. London: C. Horst & Co Publishings.

Barnes, S. (1986) *Patrons and Power, Creating a Political Community in Metropolitan Lagos*. London: Manchester Publishings.

Cole, P.D. (1975) 'Lagos Society in the Nineteenth Century' in Aderibigbe, A.B. (ed.) *Lagos, The Development of an African City*, London: Longman, 27–58.

Davies, G.O. (2009) 'A Study of the Interconnections between Colonial Land Policies and Urbanisation in Lagos, 1861–1960', University of Lagos, PhD thesis.

Ibrahim, I.B. (2008) 'Legislating for International Border Communities and Development and Transborder Co-operation' in Akinyele, R.T. (ed.) *Borderland and African Integration*. Lagos: Panaf, 21–58.

Law, R.C.C. (1973) 'Contemporary Written Sources' in Biobaku, S.O. (ed.) *Sources of Yoruba History*. Oxford: Oxford University Press, 9–24.

Odumosu, T. (1999) 'Changing Political Boundaries of Lagos State' in Balogun, Y. et al. (eds.) *Lagos State in Maps*. Ibadan: Rex Charles, 10–20.

Olugbemi, S.O. (1987) 'The Administration of Lagos State 1967–1979' in Adefuye, A. et al. (eds.) *History of the Peoples of Lagos State*. Lagos: Lantern Books, 321–340.

Osuntoku, A. (2005) 'Twin Cities across Interstate Boundaries: Case Study of Okemesi and Imesi Ile' in Akinyele, R.T. (ed.) *Contemporary Issues on Boundaries and Governance in Nigeria*. Lagos: Frankard Publishings, 76–90.

Oyesiku, K. (1998) Modern Urban and Regional Planning Law and Administration in Nigeria. Ibadan: Kraft Publishings.

Regional Policy and Cooperation (1947) *National Archives Ibadan, Nigeria*, ABP2329, Abe Prof. 2.

Sada, P.O. (1970) 'The Rural-Urban Fringe of Lagos: Population and Land Use', *The Nigerian Journal of Economic and Social Research*, xxii (2), 225–243.

Sada, P.O. and Adefolalu, A.A. (1975) 'Urbanisation and Problems of Urban Development' in Aderibigbe, A.B. (ed.) *Lagos, The Development of an African City*. London: Longman, 79–107.

Tew, M. (1939) 'Report on Title to Land in Lagos', *National Archives Ibadan, Nigeria* 4, CE/T3.

Williams, B.A. (1975) 'The Federal Capital: A Changing Constitutional Status and Inter-governmental Relations' in Aderibigbe, A.B. (ed.) *Lagos, The Development of an African City*. London: Longman, 59–78.

6 Twin cities in African history

A comparative analysis of the 19th and 20th century capitals of Borno

Usman Ladan

Several twin cities thus far have provided backdrops for the interconnections and interplay between city, twin-city and conurban forces. Both of the historical urban pairs compared here (and this is the first of many chapters to attempt explicit comparison) were functionally twin cities (very adjacent and deeply interlinked). However, like many other twins in this volume, had strong component-city identities rendered the more powerful in these cases because born of class/status and/or ethnic separation and hostilities, in the second case further enhanced by the colonial context. Perhaps unsurprisingly, and strong interdependence notwithstanding, they also share the mutual hostility evident among other internal twins.

The twin-cities phenomenon has been insufficiently studied in Africa. This chapter compares two African twin cities emerging in northeastern Nigeria. African cities have a long and rich history dating back at least a thousand years, especially in the historic empire of Kanem–Borno: the first state system in West and Central Africa. The cities are Kukawa and Maiduguri, Borno's political and administrative capitals in the nineteenth and twentieth centuries, respectively. They developed about eighty miles apart in far northeastern Nigeria. They were intranational twins, but located within a larger border region contiguous to Central Africa, thus exposing them to strong cultural influences from across that border.

The chapter demonstrates that, while the Kukawa and Maiduguri twins differed, the two capitals shared several similarities and dissimilarities, even though serving as separate capitals for two polities under different conditions in different centuries. Because primary sources on Kukawa are relatively sparse, the account of twentieth-century Maiduguri is more detailed.

Kukawa twin city

Foundation

Kukawa was founded in 1814 by a nineteenth-century scholar-warrior known as Sheikh Muhammed El-Amin El-Kanemi, flourishing as the capital of

Kanem–Borno Empire until 1893. El-Kanemi established the new capital while seeking a secure power base from political turmoil in western Borno in the early nineteenth century. Eventually, he seized control in the polity after dislodging the Sayfawa dynasty, thereby founding the El-Kanemi dynasty. Kukawa was named after a baobab tree, known in the Kanuri language as *kuka,* standing near the emergent city (Brennar 1973, 39). The preceding events began with the outbreak of the Sokoto Jihad led by Sheikh Uthman bn Fodio in nearby Hausaland in 1804. Although Kanem–Borno was not part of Hausaland, western Borno's Fulbe population drew inspiration from bn Fodio (a nineteenth-century Islamic scholar and reformer) and launched Jihad in Borno in 1808/09. They attacked and destroyed the Sayfawa capital of Birnin Ngazargamo founded around 1470 AD (Nachtigal 1980, 121). The Sayfawa ruling family's fugitive members appealed for rescue to El-Kanemi, based at the shores of Lake Chad in the east. El-Kanemi mobilized his pupils, advanced upon Ngazargamu, expelled the Jihadist invaders and built Kukawa as his new political base. After a few years of power tussle, El-Kanemi prevailed over the displaced dynasty, usurped extensive state powers and transformed Kukawa into the political and administrative headquarters of the nineteenth-century Kanem–Borno Empire. Unlike twentieth-century Maiduguri, Kukawa was Borno's political capital in a *real* sense. Supreme power lay in the hands of successive Sheikhs, who took key political and administrative decisions over what ultimately happened in Kukawa and in Kanem–Borno Empire, assisted by the resident aristocracy and their subordinates in outlying districts.

However, until the late 1840s, Kukawa was not a twin city. The King of Wadai, who wanted to re-install the vanquished Sayfawa dynasty, raided and destroyed the city in 1846 (Martin 1962, 358). The reigning Sheikh Umar hastily returned to the capital from a trip to save his throne, rebuilding Kukawa as a twin city shortly afterwards. It is unclear why Kukawa was not built as a twin city by El-Kanemi at the beginning, nor exactly why it was rebuilt as a twin city by Sheikh Umar after 1846. Perhaps several factors influenced Umar's decision: a desire for enhanced security (this was a period of significant instability in the polity), class distinction and social status. More significantly, the historic desire for seclusion by rulers in Borno who 'held an ascribed semi-divine status and were accordingly sheltered from commoners' and strangers' eyes until the late-nineteenth century' (Magnavita et al. 2009, 242) was also important. Yet, it is unclear why it was the only known pre-colonial urban centre in Borno built as a twin city. The city grew rapidly: by 1860, its population was around 60,000, remaining thus even in 1892, barely a year before its destruction by a Sudanese invader (Ellison 1959, 323, 328).

When fully developed, Kukawa comprised two towns, half a mile apart, with a combined area of 7–8 square miles. Houses, palaces, public buildings, small markets, wells and even farms were enclosed by a thick mud wall, rising up to twenty feet. The only exceptions, according to a visiting

German traveller, were the main market, large farms and the cemetery, located immediately north of the gate (Barth 1966, 50). The space separating the two towns extended about a mile and was known as 'gemzegenyi', whose meaning remains unclear (Ellison 1959, 323). Since the fifteenth century, major cities of pre-colonial Borno were enclosed by walls due to a high sense of personal and social insecurity. Surprise military attack, theft, robbery and other criminality were constant and pervasive (Cohen 1966, 93). In nineteenth-century Kukawa, intermittent social disorder and political instability became additional factors. Therefore, the walls surrounding each city were built partly from custom, partly from necessity.

Layout

Before twentieth-century British colonization, Borno's cities were laid out according to defined political, administrative, social, religious and cultural requirements, shaped by prevailing custom and external influences; these explain how nineteenth-century Kukawa was built. The only exception was its twin-city character. After 1846, it comprised a *West Town* and an *East Town*, administered by the aristocracy domiciled in East Town. West Town was larger with a perimeter of 5,000 yards, principally occupied by peasants, serfs, traders, visitors and other commoners. It was densely inhabited, possessed a special trading area and housed a well-attended market. Its walls had gates on all four sides, a *Dandal* (promenade) and a lodge for the Sheikh when he visited. East Town, with a perimeter of 3,500 yards, was principally occupied by the Sheikh, palace officials, the nobility, and their servants and dependants (Barth 1966, 49–51). It was sparsely populated, possessing six gates, two each in the east, west and south sides, a large *Dandal*, a Mosque, a small market, institutional buildings, and houses of wealthy merchants, scholars, judges, slaves and the Sheikh's personal bodyguards. The gates of both towns were opened at dawn and closed at sunset, serving as entries and exits. Taxes, caravan tolls and other levies were collected from traders and commoners passing through. Therefore, alongside security, the gates served additional fiscal and administrative functions.

Kukawa's twin city, like twentieth-century Maiduguri, was socially stratified, even though the nature of the social classes differed. Nineteenth-century Borno, unlike Maiduguri, was a feudal society (Cohen 1966, 91), and this shaped Kukawa's social structure. At the social apex stood the Sheikh, followed in descending order by ruling council members, courtiers, commoners and slaves. However, social mobility was higher than later observable in Maiduguri. Slaves, appointed as office-holders, attained better status than commoners, standing higher in the Sheikh's favour (Nachtigal 1879, republished 1980, 245). The hierarchy, as noted, was partly reflected in the difference between East Town and West Town.

Kukawa's settlement patterns were generally determined by status, occupation, territorial origin and prevailing custom. West Town's settlement pattern was significantly shaped by security, supervision along with control of migrants, traders, visitors and the local population. Migrants were settled along territorial and occupational lines in different neighbourhoods: scholars lived in the Mallemti quarter, North African traders in Wasilili, tanners in Mundilmari, and the Shuwa, Tubu, Kanembu and other linguistic groups dwelt in specially designated quarters in West Town's northern part. During El-Kanemi's reign in the early nineteenth century, when the security situation was deemed unstable, North African trading caravans were forbidden from entering Kukawa in his absence (Cohen 1967, 104). Strangers and suspicious elements posing threats, even if mild or indirect, were reported and contained. The settlement pattern may have facilitated such monitoring and the general administration of the city.

The commoner classes in West Town were quartered in a congested, poorly planned and relatively less-guarded location, while the ruling aristocracy and other highly placed figures in East Town were better protected in a spacious, well-planned and highly secure area. This class character of West and East Towns influenced how residents perceived each other. Thus, West-Town dwellers perceived East Town as the residence of the lords, rich, powerful and privileged elements, while theirs was the city of commoners, lacking East Town's facilities, planning, security and tranquillity. Meanwhile, East-Town inhabitants looked down upon West Town as the domicile of their servants, dependent on their patronage and support for survival. Class inequality and social alienation were exacerbated by rampant corruption and favouritism by the ruling aristocracy, which was 'paralysing all … national effort' (Ellison 1959, 326).

We know little about Kukawa's architecture. The aristocracy certainly used red bricks for construction in other cities like Birnin Ngazargamu, Nguru, Machina and Gamboru, but not in Kukawa, where hut and mud buildings predominated. Apparently, in those cities, as in twentieth-century Maiduguri, burnt-brick palaces and intricate clusters of residential compounds represented the political, economic and mystical power of those building them and residing there (Magnavita et al. 2009, 241).

Economy, society and culture

Kukawa's political and administrative functions were sustained by resources derived from pre-colonial Borno's larger economy. The Sheikh controlled land in the polity and distributed huge tracks to his numerous courtiers, titled officials within and outside the palace, and other subordinates, often also based in the city. Kukawa was also a leading trading, manufacturing and administrative centre throughout the century. The urban economy rested on peasant agriculture, handicraft manufacture, trade and commerce. The range and diversity of manufacturing was extensive, including weaving, dyeing, tanning and iron-working (Mahadi 1980, 7).

Manufacturing and trading activity was both locally oriented and directed at much wider markets. Although handicraft manufacture overweighed agriculture, there was noticeable linkage between the city's manufacturing, and agricultural production in the countryside: hoes and axes forged in the city were used in rural domains, while outlying farms supplied agricultural produce for city consumption.

Kukawa's West Town functioned as the principal market, although East Town also had a small market for residents' daily needs. West Town's market occupied a 50-acre space, serving over 10,000 people by around 1870. It was thus the busier but not necessarily the more prosperous of the twins. The nineteenth-century German traveller, Gustav Nachtigal (1831–85), staying in Kukawa from July 1870 to March 1871, wrote that Kukawa was 'always full of strangers, as many merchants as … pilgrims and adventurers' (Nachtigal 1870, 215). Unlike twentieth-century Maiduguri, local, regional and international trade accompanied and stimulated urban manufacture. Regional trade brought Kukawa into contact with large parts of the West African Sahel. Sheikh El-Kanemi successfully established trading and diplomatic relations with the Pasha of Tripoli and the Governor of Fezzan, in the century's first three decades (Boahen, 1962, 351). Thus, Kukawa became part of a wider regional pre-colonial economy, continuing into the early twentieth century. Cross-Saharan trade connected the city with the Maghreb and other places in the Mediterranean littoral, from where it acquired various products (Braudel 1966, 181). In terms of commerce, learning and scholarship, Kukawa belonged more to the Maghreb and Arab world than to sub-Saharan Africa. The Sheikh's fame attracted numerous visitors from Mecca, Medina, Morocco, Tunis, Egypt, Timbuktu and Senegal (Ellison 1959, 325). By the late nineteenth century, Kukawa had become one of pre-colonial Borno's most heterogeneous and cosmopolitan urban centres.

The city also flourished as a Muslim scholarship centre with pronounced Arab cultural influence. Names of individuals, modes of dressing, manners of speech and ways of greeting showed Kukawa's cultural proximity to the Arab world. Terms like 'sheikh', 'qadi', 'sultan' and others, and indeed some Kanuri-language vocabulary, were derived from Arabic. It was the language of teaching and learning in schools, and governmental record-keeping. Unlike twentieth-century Maiduguri, European cultural influence was near absent in Kukawa. European presence in the city was restricted to periodic visits by travellers starting shortly after its foundation.

Some European visitors stayed much longer, but they were few and left no lasting cultural influence. Clearly, European presence in Kukawa, unlike Maiduguri, did not produce colonization or significant settlement. Neither did it spread European cultural influence. As non-Muslims, Europeans were derided as *kerdi* (unbelievers) and misconceived as *keri* (dogs) (Hickey 1985, 154). Consequently, a mixture of Arabic and African cultural influences remained preponderant until the century's end.

Kukawa's West and East Towns never became a single city, though they might have done had it survived. Rabeh bn-Fadlallah, a Sudanese conqueror, invaded Borno and defeated its 30,000-strong army before advancing into Kukawa in May 1893, and devastating it. Kukawa was never rebuilt as a twin city, and it failed to regain its former capital status. In 1931, a European visitor described Kukawa as a desolate city of 'little interest', where 'ruins outnumber the houses', a far cry from Barth's much earlier description: 'a good town with dazzling cavalcades of people' (Ness 1931, 313).

The twin city of Maiduguri

Foundation

Maiduguri's emergence as a twin city directly resulted from British colonization in 1902. For at least a decade beforehand, Borno was thrown into political crisis, paralysing its social, economic and political affairs. Between 1893 and 1902, the displaced El-Kanemi ruling family and much of the population were in flight with the polity thrown into political interregnum. Rabeh expelled the El-Kanemi dynasty from the capital, but failed to substitute himself as an effective ruler. Rabeh finally got into protracted conflict with the French, forcing Britain to intervene militarily to end this double intrusion into its 'sphere of influence'. This culminated in British colonization of Borno in 1902 and Maiduguri's foundation in 1907 (Kirk-Greene 1958, 8). For 53 years, until Nigerian independence in 1960, Maiduguri served as colonial administrative headquarters in Bornu Province, one of the 21 British-created provinces in Nigeria. Unlike Kukawa, Maiduguri was not the Province's capital in any rule-making sense. The Sheikh's powers were taken away, and the preeminent British colonial official based in the city ('the Resident') was subordinate to the Lieutenant Governor in Kaduna, and took key political and administrative decisions implemented locally.

Maiduguri was shaped by the nature of British colonialism. The latter was the by-product of a developed capitalist society, competing with other European powers for foreign territories to colonize as monopolized import–export market outlets. The British selected the area as the site of a new provincial capital in 1906, insisting the Sheikh and his followers to relocate there. Appointed and protected by the British as local potentate, he had to acquiesce. After a few weeks of reluctant grumbling, he persuaded and coerced his followers to migrate to the British-designated capital in January 1907 (Ladan 2002, 137–138). From the outset, the British conceived Maiduguri as a twin city, comprising separate African and European towns. Although euphemistically termed a European Reservation, the latter was a town with all accoutrements. British officials called the African town 'Shehuri', while locals named it 'Yerwa'.

However, Maiduguri's conception and planning as a twin city was not locally determined by British officials in Bornu Province. Rather, it reflected general colonial policy, centrally imposed by Britain's Governor General in Lagos (Lugard 1970, 405–406). After occupying Nigeria, British colonialism did not fashion towns to reflect local situations or prevailing customs. Like their European counterparts, British officials had pre-conceived notions about ideal urban layouts influencing how they planned, established and nurtured colonial cities (Urquhart 1977, 25). They realized they were foreign conquerors establishing hegemony over hostile and restive people needing control and monitoring. Feelings of insecurity and the need to contain potential rebellion were fundamental to Maiduguri's conception and planning as a twin city. This was necessary for the effective administration of the Province because it provided a separate and secure base for the vastly outnumbered British colonial officials and their support staff. Without the twin-city layout, it would have been easier for the masses of the forcefully colonized Africans to rise up in rebellion, and possibly, vanquish British imperial hegemony.

Layout

The British spelt out the spatial segregation of urban residents and the creation of a twin city as the basic principles of its colonial urban policy immediately after colonization. Hence, the twin-city phenomenon was not unique to Maiduguri because it existed in many British colonial urban centres within and outside Nigeria. The policy was first formulated in Nigeria in the 1904 Cantonment Proclamation, which decreed the establishment of a Government Station, a Provincial Office, European hospital, housing quarters and other essential structures around the military garrison. The proclamation explicitly stated that the Government Station shall be separated from a 'native Quarter', and only respectable traders, artisans and other 'natives' engaged in services the cantonment required shall be allowed to reside (Lugard 1970, 406; Urquhart 1977, 26). Hence, Maiduguri began emerging as a twin city as early as 1907: one for the African population (Yerwa or Shehuri) and another for British colonial officials called 'Government Station' or 'European Reservation'. Yerwa's layout reflected broad pre-colonial forms of urban planning, while the European Reservation was built according to the specifications for the British colonial layout.

The principles of twin-city layout were elaborated in the 1917 Township Ordinance (No. 29). The British Governor in Lagos directed that all Europeans in the provincial headquarters must live within their Reservation, with Africans removed into the enclosed town of Yerwa. The European Reservation was created and surrounded by a 440-yard 'Building-Free Zone'. Even clerks, messengers, interpreters and other African employees of the British Provincial Office were removed to the Clerk Quarters outside the Reservation (NAK/Maiprof Acc No.117/1925). These Quarters never constituted a 'town' like the European Reservation since clerks shopped in Yerwa's markets and fraternized somewhat with its African population.

This was possible because the main market was located outside the town wall, nearby the Clerk Quarters, facilitating access to its rich and diverse products.

The European Reservation became the domicile for resident Europeans, whether working for the colonial state, trading companies or as missionaries. Frederick Lugard, British High Commissioner of Northern Nigeria (1900–6) and Governor of Nigeria (1914–19), emphasized that 'a European is as strictly prohibited from living in the native reservation, as a native is from living in the European Quarter' (Dhiliwayo 1986, 164). British officials were directed to live near each other, facilitating an e*spirit*, distinct from the African population. Within a few years, the Ordinance enabled the Reservation to become a city of its own with offices, constabulary and soldiers' quarters, hospital, church, cemetery, club house, recreation grounds, residential plots and market place (Urquhart 1977, 27–28). Thereby resident Europeans met almost all day-to-day requirements locally without need for visiting African Yerwa. They slept and ate in their houses, attended Church and visited the European hospital. Upon death, there was a European cemetery for burial. Inhabitants went to the Reservation's club and shopped in the Trading Layout's canteens. Day-to-day administration was run by the preeminent British colonial official: the Resident, operating from 'the Provincial Office', where many other officials also worked. He directly controlled the provincial police, prison and court, all located within the European town. He also adjudicated all litigation involving Europeans.

The Sheikh headed the Native Administration, presided over separate police, prison and court within Yerwa, and had no jurisdiction over British or European citizens. Legal cases involving Europeans and Africans were tried in the Provincial Court presided over by the Resident, confirming the European town's preeminence (Ladan 2002, 444).

Many of Yerwa's inhabitants had arrived from Kukawa in 1907, and its layout closely followed its predecessor's. The Sheikh's palace was built in the centre, while the *Dandal*, for want of space, was narrower and shorter than Kukawa's (Niven 1982, 155). Like many pre-colonial Bornoan cities, aside from market functions, it also served as the religious, social and cultural venue. The Sheikh's palace soon became surrounded by public buildings and migrant dwellings. The central Mosque was built immediately northwest of the palace, the Native Court to the west of the Mosque, with the central store to the north. The Native Administration Central Office, Divisional Office, Native Authority Workshop and other buildings were located on either side of the Dandal. Houses of slave officials were constructed north of the palace, while free-born aristocrats lived in the south. The slave quarter was established behind the palace to the east, while the occupational quarters followed pre-colonial lines. The non-occupational neighbourhood of Gargar was directly relocated from Kukawa. Finally, Yerwa was enclosed by a wall with gates on all four sides. As in Kukawa, these opened at 6.00 am and closed at 7.00 pm (Seidensticker 1983, 7). Resident Hewby suggested that the wall's purpose was to stop 'armed robbery' and 'stock lifting that has become so common' (NAK SNP 7/6655/1911). The wall enclosed the whole of Yerwa, excluding the European town (which had no surrounding wall), the main market, cemetery and large farms, as previously in Kukawa.

Kukawa's settlement pattern had been determined by class and custom; in twentieth-century Maiduguri, class and race were dominant. Hence, social interaction among the residents of the two cities was necessarily limited, giving spatial expression to class and racial consciousness. The British justified Maiduguri's separated African and European towns on grounds of race, health, security and contrasting European and African ways of life (Lugard 1970, 404–405). Africans saw things differently, perceiving the European town as domicile for foreign colonizers, meticulously planned, spacious, tranquil, amenity rich and out of bounds. They contrasted this with Yerwa, which lacked most of these features. Furthermore, the European town's preeminence was forcefully protected and sustained by the Provincial police and military. Hence, its relationship with Yerwa was one of subordination and super-ordination, apparent inequality, profound resentment, occasional conflict and near-constant tension.

Economy, society and culture

Unlike the nineteenth century, the British created a colonial capitalist economy in twentieth-century Borno. Inheriting a pre-colonial feudal economy, they created a colonial capitalist one. They introduced British currency, demanding tax payment with it to halt the age-old regional and North-Africa-bound trade based on cowry shells, Austrian-minted Maria Theresa dollars and other pre-colonial currencies. In 1917, the Bank of British West Africa also established a branch in Maiduguri, followed by other banks. This coincided with the arrival of numerous European and non-European Trading Companies (NAK/SNP 9/1613/1916).

Maiduguri housed most of Bornu's trading companies, which operated the import–export colonial economy for the remaining colonial period, importing tea, sugar, textile, soap, perfume, bicycles and other consumer items from Europe, and exporting groundnuts, gum Arabic, hides and skin to Britain. By 1939, the colonial economy was dominant, transforming Maiduguri into Bornu Province's economic capital. Instead of trading locally produced food items and manufactured goods with distant and nearby places, Maiduguri increasingly became central to the colonial import–export economy. Yerwa, particularly, became the base for collecting raw materials for export to Europe. In return, it received British-manufactured goods, partly consumed locally, the rest dispatched elsewhere in Bornu, with the European town co-ordinating. By 1960, Maiduguri twin city had become Britain's economic appendage, bound by far more than what had tied Kukawa to North Africa.

Economic Europeanization went side-by-side with spreading European cultural influence. The latter's foundation was laid by establishing the

Provincial School (1915), the Craft School (1920) and the Middle School (1933). Many teachers were foreigners using English as *lingua franca*. Meanwhile, curriculums, uniforms and value-transfer produced graduates who were cultural opposites of the nineteenth-century Koranic schools in Kukawa. The latter had produced Arabic speakers looking up to the Arab world; the former were anglicized elites serving as bridges for diffusing European culture. Cinemas played similar roles from 1950 onwards (Cohen 1967, 49).

Differences and similarities

At a broad level, Kukawa and Maiduguri twin cities shared key similarities: both were centres of trade, administration and defence, flourishing under separate political systems in different centuries. Their twinning in each case was linked to the business of ruling, and to the preferences of the two elites, though also rendered different by Kukawa's feudal setting and Maiduguri's colonial character. However, the twins' component settlements also showed comparisons and contrasts in ways reflecting time and culture. Thus, Maiduguri's twin cities differed in various ways. Until the early 1970s, Yerwa's domestic dwellings largely comprised mud huts, as in nineteenth-century Kukawa. According to an interview with four Liman Amsami compound informants in November 1994, the commonest mud building by the 1930s was a *shoro,* a square mud structure of Sudanese origin, several of which were enclosed by a wall as family compounds. These were, as in Grunebaum's 'Muslim town', inside-out in nature, oriented away from the street, maximizing their Muslim occupants' privacy (Daunton 1992, 204). Their design and pattern were inherited from Kukawa and other urban centres of pre-colonial Borno, and existed across much of West and Central Africa under twentieth-century European colonialism (Hanna and Hanna 1971, 88). The European Town was different. Here, houses were built with mud, later with red bricks and concrete. Unlike those in Yerwa or Kukawa, they were not enclosed by walls but designed to maximize ventilation from the exterior and were European in conception, nature and design.

Furthermore, both Maiduguri twins, unlike Kukawa's, became notable for their red-brick buildings from 1934. However, the use of red bricks by wealthy individuals spread slowly because, as in pre-colonial times, they were declared an exclusive privilege of the Sheikh and nobility (Ladan 2002, 562). The ban ceased around 1945; subsequently, red brick became common in Yerwa's North African quarter, Maiduguri becoming known as the town of red bricks and neem trees (Ness 1931, 308). Burnt-brick technology spread into sub-Saharan Africa from North Africa. As elsewhere in the Muslim world, burnt bricks in tombs, palaces and Mosques indicated the users' social status, economic prosperity and political power (Magnavita et al. 2009, 220). Why they were rare in nineteenth-century Kukawa is unclear.

Concrete buildings emerged in Maiduguri in 1949, becoming common by 1960. The Sheikh's palace and central Mosque in Yerwa were partly rebuilt with concrete in 1960, as were some wealthy notables' houses (BBG 17 January 1995). But concrete houses remained rare in Yerwa even in 1960, only proliferating around 1970 and becoming widespread with Nigeria's oil boom from 1980. Nonetheless, concrete did not radically modify the nature of compounds. A few buildings in the European town showed outward architectural patterns, but the inside-out orientation of compounds within Yerwa remained as they had been in pre-colonial and early colonial days.

Another way Kukawa differed from Maiduguri was the absence of paved roads, tap water, electricity and other urban facilities. However, Maiduguri's European town was ahead of Yerwa in water and electricity supply between 1907 and 1939, and most Yerwa people lacked electricity before 1960. Some used groundnut-oil lamps, as in nineteenth-century Kukawa, kerosene being unknown. Many still resorted to lighting by burning firewood, or simply retired early. Only in the 1970s did electricity start becoming common.

Similarly, all Yerwa's roads before 1940 were identical to those in outlying districts, and indeed Kukawa. Even in the 1910s, they were just enlarged footpaths, cleared of thorns, for pedestrians and ox-carts. Until around 1940, the few motor-fit roads were sand-covered, garbage-clogged and flooded during rains (interview in 1993 with Matawalli Kachallah Borko, a well-known representative of Borno's elite who served in Nigeria's post-independence government). Any significant expansion and improvement had to await independence. However, the most discernible difference between Maiduguri's twins up to 1960 concerned road-naming. Yerwa's roads were named after present or past local notables or Borno towns. Even on the frequent occasions when names were changed, they still drew on Bornoan history, culture or persons. Meanwhile, European-Town streets were christened after nineteenth-century European travellers like Heinrich Barth and James Clapperton, and past or present British colonial officials. (Ladan 2002, 587). Such contrasts were ubiquitous reminders of European Town's foreign character and Yerwa's local character.

Conclusion

Our comparative analysis has shown several things. Overall, nineteenth-century Kukawa showed similarities with twentieth-century Maiduguri but greater differences. Similarly, each town showed similarities with its partner but again greater differences. However, it is also clear that Yerwa in Maiduguri shared greater similarity with its African predecessor in Kukawa than with its European partner. This was because British colonialism focused on domination and trade rather than the wholesale transformation of its colonies. This partially explained its policy of 'indirect rule'. However, these differences and even periodic tensions and conflict did not imply that Kukawa and Maiduguri were other than twin cities. Each had functional and interdependent relationships with its partner, binding it socially, economically, politically, culturally and administratively, such that their respective inhabitants saw them as single entities: they were truly twin cities.

Kukawa twin city lost its capital status after destruction in 1893, with the British bypassing it as they colonized Borno in the early twentieth century. Nonetheless, Kukawa survived and eventually began increasing its population in the 1950s and 1960s. It is currently a single city of 16,000 and the administrative headquarters of Kukawa Local Government Area in Borno State. Meanwhile, Maiduguri's twin-city character began eroding with independence in 1960. It is now a unified city of over 1,000,000, and is Borno State's administrative capital. However, vestiges of the withering social, architectural, morphological and other differences of the former twins are still discernible.

References

Barth, Heinrich (1966), *Travels and Discoveries in North and Central Africa*, Frank Cass, London.

Boahen, Adu (1962), 'The Caravan Trade in the Nineteenth Century', in *The Journal of African History,* 3, 2. 349–59.

Braudel, Fernand (1966), *The Mediterranean and the Mediterranean World in the Age of Phillip II, I*, Fontana, London.

Brennar, Louis (1973), *The Shehus of Kukawa: The Rise and Fall of the Al-Kanemi Dynasty*, Oxford University Press, Clarendon. BBG (Bulama Bukar Goma), 65 years, Farmer and Bulama, Auno Ward (Gwange II), 24th December 1994, 17 January 1995.

Cohen, Ronald (1966), 'Dynamics of Feudalism in Borno,' in Jefrey Butler (ed.), *Boston University Papers on Africa,* II, Boston University Press, Boston.

Cohen, Ronald (1967), *The Structure of Kanuri Society*, PhD, University of Wisconsin, Madison, USA.

Daunton, Martin (1992), 'The Social Meaning of Space: The City in the West and in Islam,' in *Urbanism in Islam: Proceedings of International Conference on Urbanism in Islam (ICUIT),* 1, the Middle Eastern Culture Center, Tokyo, Japan.

Dhiliwayo, Arthur (1986), 'A History of Sabon Garia Zaria, 1911–1950: A Study in Colonial Urban Administration', PhD thesis, Department of History, Ahmadu Bello University, Zaria, Nigeria.

Ellison, Randall Eugene (1959), 'Three Forgotten Explorers of the Latter Half of the 19th Century with Special Reference to Their Journeys to Bornu,' *Journal of the Historical Society of Nigeria,* 1, 4. December.

Hanna, William John, and Judith Lynne, Hanna (1971), *Urban Dynbamics in Black Africa: An Interdisplinary Approach*, Aldine and Atherton, New York.

Hickey, Raymond (1985), 'Filippo da Signi's Journey from Tripoli to Kukawa in 1850', in *The Journal of African Historical Studies,* African Studies Center, Boston University, USA, 18,1, 145–56.

Kirk-Greene, Anthony Hamilton Millard (1958), *Maiduguri and the Capitals of Borno*, NORLA, Zaria, Nigeria.

Ladan, Usman (2002), *'History of the Urbanisation Process in Borno: A Study of the Yerwa (Maiduguri) Area, 1880–1960'*, PhD, thesis, Department of History, Ahmadu Bello University, Zaria, Nigeria.

Lugard, Frederick (1970), *Political Memoranda,* Frank Cass, London.

Magnavita, Carlos et al. (2009), 'Garu Kime: A Late Borno Fired-Brick Site at Monguno, NE Nigeria,' in *The African Archaeological Review,* 26, 3, 219 Springer.

Mahadi, Hauwa (1980), 'A Tentative Reconsideration of the Political Economy of Metropolitan Borno', in *History Research at A.B.U. Vol.V,* Department of History, Ahmadu Bello University, Zaria, Nigeria.

Martin, Bobby Gray (1962), 'Five Letters from the Tripoli Archives', in *Journal of the Historical Society of Nigeria,* 2, 3, 350–72.

Nachtigal, Gustav (1879, republished 1980), *Sudan and Sahara,* III Humphrey J. Fisher, London.

NAK/Maiprof Acc No.117/1925/ Yerwa Town Special Report by Peter Tegetmeir.

NAK SNP 7/6655/1911/ Bornu Province Report for March Quarter 1911.

NAK/SNP 9/1613/1916/ Bornu Province Report for the Year 1915.

Ness, Patrick (1931), 'A Journey to Lake Chad and the Sahara', in *The Geographical Journal,* 77, 4, 305–20.

Niven, Cecil Rex (1982), *Nigeria Kaleidescope,* C. Hurst and Company, London.

Seidensticker, Wilhem (1983), 'Notes on the History of Yerwa', *Annals of Borno,* I, University of Maiduguri Press, Nigeria.

Urquhart, Alvin Willard (1977), *Planned Urban Landscape of Northern Nigeria,* Ahmadu Bello University Press, Zaria, Nigeria.

7 Indian twin cities

John Garrard and Ekaterina Mikhailova

Until now, we have mostly focused on long-established twin cities, whose identities and interactions are partly born of historical trajectory. Here, we turn not just to another continent but mostly to newer, consciously created and managed pairings. Like several following chapters, the approach is heavily comparative, with 20 twins under review. Indian twin cities are generally far larger than those featured thus far, with their relationship directly born of burgeoning urbanisation, indeed conurbanity. Unlike those reviewed until now, but like many of the international border pairs from Chapter 10 onwards, their relationship is expected to become positive and functional from the point of twin-city declaration. Even in Indian twin cities whose relationship was initially customary rather than declared, the expectation of positivity and planning has become increasingly powerful.

Although they are highly, indeed increasingly, important features of its cityscape, India's twin cities have apparently attracted little scholarly attention. We will try to remedy some of that deficiency by drawing on the copious material available from public sources to assess: the context wherein twin cities arise; their characteristics and underpinning intentions; and as far as possible their impact and outcomes.

Context and causes

India's interest in twin cities and the shape it has taken must be set against key features of the background wherein they arise: most notably the burgeoning and apparently chaotic character of urban growth, the propensity for planned responses to that growth, party-political vibrancy at national and state levels, and, most recently, the downward trajectory of political mobilisation.

Most significant has been rapid and accelerating urbanisation, based on inward rural migration at least as much as self-generation, plus perceptions (albeit somewhat inaccurate) that governmental and urban elites have been little more successful at handling the consequential poverty, overcrowding, burgeoning slums, insanitation and ill-health than their European and American predecessors in relation to similarly generated problems in the nineteenth century. This perception holds even though, partly because of that previous experience, problems have been recognised far earlier in the urbanisation cycle, and the understanding and remedies are more potentially effective. Furthermore, India's urbanisation level is lower than many other Asian countries, even if the costs sometimes seem greater.

Thus, although its urban share of population (31% in 2011) is below that of many other developing nations (e.g. China 45%; Brazil 87%), the World Bank Planning Commission forecasts that urban India will be accommodating 600 million people by 2031 (http://planningcommission.nic.in/ hackathon/Urban_Development.pdf). Meanwhile, the 2011 census revealed 65.5 million slum-dwellers, predicted in 2013 to reach 105 million by 2017 (*TOI* 20/08/2013), with a housing shortage estimated in 2012 at 18.78 million units, visiting most harshly upon those at the bottom end. In 2018, India ranked 178th of 180 nations for air pollution (*Economic Times* 24 January 2018). Meanwhile, also highly significant is the fact that urbanisation and urban areas, in India as elsewhere, have increasingly become seen as the only guarantors of economic growth, including the inclusive sort. Commenting in *Urban Development Policy for Karnataka* (ii) in November 2009, the state's Urban Development Department advised taking 'a positive view of urbanisation … it reflects the aspirations of millions of people … (we need) to steer it correctly towards ends that are desirable'.

Painfully acute urban problems are hardly particular to India; being evident across much of industrialising Asia, Africa and Latin America. The commonest cross-continental answer has long been urban (and rural) planning, even if only sometimes enacted. What has distinguished India has been the increasing belief since the 1960s that the highly managed twinning of relatively nearby cities represents a crucial (though obviously not the only) part of that remedy. The twin-city idea has become an important way of thinking about urban planning, a potent symbol of modernity. Partly in consequence pledges about new twin cities ultimately benefiting all have become increasing features of party-political discourse at local, state and federal levels, thereby enhancing twin-city generation. Narendra Modi, both as Gujarat Chief Minister (announcing no fewer than five twinned creations in December 2011) and, more recently, as Indian Prime Minister (twin cities and satellite towns being prominent in the BJP election manifesto of July 2014), has been notably active here: all in his constantly stated belief that cities drive economic growth and provide enduring opportunity to mitigate poverty.

As implied, further fuelling and also complicating this ambiguous but ultimately positive perception of urbanisation, and Indian responses to it, has been vibrant party rivalry – at city, state and federal levels – and the connected fact that the Congress Party's post-Independence hegemony has collapsed completely. This has enhanced the ever-present competitive tendency of democratic politicians to calculate less in terms of amorphous majorities than the perpetual appeasement of small groups (Dahl 1963). In the twin-city context, this is evident in the 'inclusion' of multiple sorts of urban and rural victim groups in political debates on urban planning. This in turn has interacted with the fact that those victims groups in recent years have proved increasingly willing to mobilise on their own behalf.

Characteristics and intentions

One broad consequence of these factors has been a large and increasing crop of Indian twin cities that are both like and dramatically unlike their intranational counterparts elsewhere, indeed having more in common with cross-national twins. Indian twins are of two, arguably three, broad and hazily divided types (Figure 7.1). First, there are pairings whose cities have

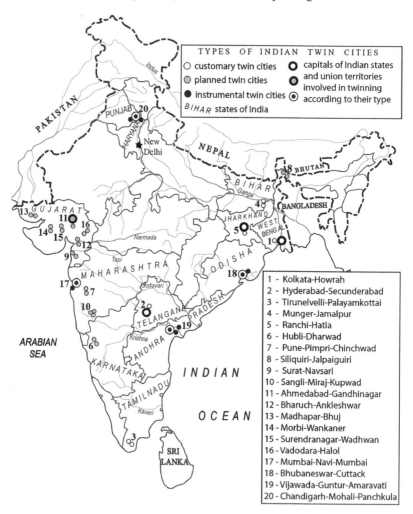

Figure 7.1 Map of Indian twin cities typology by Ekaterina Mikhailova.

either grown into each other in the classic twin-city sense or started that way on opposite river banks, and have come to be commonly seen as twins: all the more for being frequently so labelled on the Internet. In what follows, we draw material from the following contrasting pairs: (1) Hyderabad (currently joint capital of the states of Telangana and Andhra Pradesh) and Secunderabad, separated from each other by the Hussain Sagar Lake; (2) Kolkata (West Bengal's capital) and Howrah, separated by the Hoogli River; (3) Tirunelveli and Palayamkottai in Tamil Nadu state, divided by the Thamirabarani River and (4) Munger and Jamalpur in poverty-stricken Bihar state, separated only by a shrinking strip of land.

Second, there are large and increasing numbers of often-quite-distant cities, whose twinning has been formally declared, often legislatively enabled, and co-ordinated and increasingly planned. This recognition has normally been at state, and sometimes federal (Ministry of Urban Affairs), level. The purpose has been either to take control of two merging cities thought likely to collide in the classic sense, thereby not just avoiding unplanned chaos but also making economically beneficial use of their ever-shrinking hinterland and their supposed complementarity, or to hitch a smaller and poorly performing entity to a larger and more vibrant city nearby to the supposedly mutual benefit of both. We shall draw evidence for planned twins from Hubli–Dharwad in Karnataka, Sangli–Miraj (with Kupwad) and Pimpri-Chinchwad arguably with Pune (in Maharashtra), and Surat–Navsari, Ahmedabad–Gandhinagar, Morbi–Wankaner, Vadodara–Halol and Bharuch–Ankleshwar all in Gujarat state. Our decision to draw heavily on Gujarat is deliberate: it is one of India's most economically vibrant and proudly urbanising states and, partly due to the presence of its combative BJP Chief Minister Narendra Modi 2001–14, it assumed an almost messianistic role in championing both 'the twin-city model', and his more general underpinning view that urbanisation is a positive and socially inclusive economic force.

This type morphs into what is arguably a third *instrumental* twinned category – to facilitate some purpose like creating a new state capital as in Gujarat, Odisha and Andhra Pradesh, or a pressure-relieving satellite, as with Mumbai and highly planned Navi Mumbai across Thane Creek in Maharashtra. There is little room to cover this type beyond noting that examples are marked on the map.

Customary/accidental twins

Customary twins are necessarily very adjacent, their centres separated by a few kilometres and their peripheries either indistinguishable or marked by rivers. They are deeply interlinked by bridges and public transport, and partly in consequence also deeply interdependent, economically and socially. Furthermore, they seem classic examples of dominant–subordinate relationships common among all twin cities, while also demonstrating this model's limits. Thus, Kolkata and Howrah are connected by four bridges, most notably the 'awesome' Howrah Bridge (Figure 7.2), daily carrying 100,000 vehicles and 150,000 pedestrians over the river (https://theculturetrip.com/asia/india/articles/a-brief-history-of-howrah-bridge-one-of-the-worlds-busiest-bridges/). The twins have also recently (November 2017) been linked and greatly boosted by fast transit. Kolkata is the state capital and its 4,496,694 people (2011) outnumber Howrah's 1,072,161 by a factor of 4.3. Kolkata is Eastern and North-Eastern India's main commercial, cultural and educational centre and home to the Kolkata Stock Exchange. However, Howrah retains its own municipality, it is a rapidly growing major industrial centre in its own right and it houses the twin's main rail station, also the main terminus for India's Eastern and South-Eastern Railways. Furthermore, 'subordinate' though it might seem, Howrah's identity seems to be hardly subsumed under Kolkata's, to judge at least by its civil associations: while some exist under the Kolkata–Howrah, or even the Kolkata District, label, considerably more are specific only to Howrah. This includes the Chamber of Commerce, breaking away from Kolkata's in 1991 because business 'forums in Kolkata paid little heed to the problems of Howrah Industries', which were seen as 'unique' (Howrah Chamber of Commerce and Industry (www.howrahchamber.com/about.php).

Hyderabad–Secunderabad seemingly fit the dominant–subordinate model rather better. Hyderabad's 6,809,970 people (2011) outnumbered Secunderabad's 213,698 by a multiple of 31.9. Both have been governed by the Greater Hyderabad Municipal Corporation, alongside the wider metropolis, since 2007 (Secunderabad enjoying separate governance only during 1945–60). Much associational life occurs under a broad Hyderabad

Figure 7.2 Howrah Bridge in Kolkata over the Hugli River by Alex Cheban (2014).

label. However, many associations sport the Twin-City label, indicating, or maybe promoting, twinned identity or at least desire for co-joined activity; aspirations also expressed in yearly 'Miss Twin Cities' celebrations, in 2018 endearingly democratised into 'Mr and Mrs Twin Cities'. Meanwhile, many associations apparently operate only within Secunderabad, a fact perhaps symptomatic of its very separate history as a military cantonment from 1806, initially British but playing a similar role for the Indian military since independence. Though heavily residential, and probably suburban for Hyderabad, Secunderabad is also a major rail centre, the headquarters of the South India Railway. Overall, it is said to 'differ from Hyderabad in its social, cultural, architectural demeanour … life-style and ethos' (www.indiahotels.com/secunderabad/secunderabad-culture.html).

Our two other 'customary' twins fit the same broad, if variable, pattern. Tirunelveli's population in 2011 (474,838) was 5.2 times the size of Palayamkottai (91,171) across the river. They are connected by road and rail bridges, both governed by Tirunelveli Corporation (along with 23 other urban settlements), and most associational life seems Tirunelveli-based. They have economies reliant on both industry and agriculture. However, Palayamkottai's distinctiveness rests partly on its rich educational tradition (with many missionary-founded schools, 16% of the population is Christian), its extreme cleanliness and large pensioner population.

Jamalpur, sometimes described as Munger's suburb, has a population (105,221 in 2011) only one-third its size. Nonetheless, it has been home to a major railway workshop since 1862. This employs 25,000, is reputedly Asia's largest and, since 1888, has been headquarters for the *Indian Railways Institute of Mechanical and Electrical Engineering*, all producing considerable local pride to judge by the city's website (https://thejamalpur.wordpress.com/about/). It is separately governed, albeit by a municipal *council* (for towns up to 100,000) compared to Munger's municipal *corporation* (for towns over 100,000), figures suggesting (alongside its growth forecasts) that Jalampur is due an upgrade.

In fact, as already suggested, the borderline between type 1 and type 2 (customary and planned) twin cities is decidedly hazy, even though the distinction remains valid. Indian predilections for urban management, alongside fears of chaos in face of massive inward migration, have entailed even accidental twins becoming increasingly planned, and planned long term. What is different is that there is no necessary connection between twinning and planning in terms either of date or area planned for. Thus, customary twins become known as such long before any long-term planning occurs or is even considered. Furthermore, while all four customary twins have become subjects of planning instituted by both state and local governments, in three cases the planning has mostly embraced the wider metropolis, while, in the fourth (Munger/Jamalpur), it has embraced the towns separately.

The Kolkata Metropolitan Development Authority has been producing 'perspective plans' since 1966. Of these, only the first iteration was 'bi-nodal', applying just to Kolkata and Howrah. Subsequent 'Master Plans' have been 'multi-nodal', embracing the entire rapidly growing metropolitan region, most recently *Vision 2025* for an area comprising 14,112,536 people in 2011, compared to the twin cities' combined 5,568,855 (*History of Urban Planning Kolkata*: www.kmdaonline.org/home/perspective_plans).

The projected Hyderabad Metropolitan Development Authority (HMDA) *Master Plan 2041* is the most recent of plans dating back to 1981 produced by HMDA and its predecessor (the Hyderabad Urban Development Authority), for the steadily extending metropolitan area (2011 population 7,749,334). It attempts to integrate five previous plans for districts within the region published in the previous few months, all with non-standardised zoning regulations, specifying 24 different land-use types, each with detailed specifications about authorised and non-authorised uses (*TOI* 12/11/2016).

Planned twin cities

Dating when a planned twin city starts is often difficult, suggesting evolutionary trends in perception rather than any sudden appearance on stage. Partly it depends on what one takes as decisive indicators. One might be the emergence of joint municipalities and the consequent explicit recognition of unavoidable interdependence requiring evermore co-ordination. Thus, Pimpri-Chinchwad emerged as twins by virtue of acquiring a joint municipality in 1982; Sangli–Miraj appeared similarly in 1999, likewise Ahmedabad–Gandhinagar in 2011. More surprisingly, the British created a joint corporation for Hubli—Dharwad as early as 1925: 'they were separate municipal entities, but these two cities were so interdependent that a common municipal borough was constituted'. (*Revised City Development Plan for Hubli–Dharwad, 2041*, 63) This was confirmed by Karnataka state in 1962 (this time actually replacing the two separate municipalities), with jointly planning activity being taken further by the Hubli–Dharwad Urban Development Authority in 1987, further reinforced by the Federal Ministry of Urban Development and the World Bank in the most recently updated plan running up to 2041 (ibid.).

In fact, the foregoing may simply be end-products of longer and less certain processes, wherein announcements of pending twin-citydom, while dramatic and often productive of anticipatory economic change in themselves (see later), represent just early steps. The immediate effect is often simply further announcements, at best only slightly more progressed. This is partly because twin-city pledges are often part of party-political electoral battles and thus subject to subsequent political negotiation. Thus, Gujarat's Infrastructure Development Board apparently began preliminary consideration of Surat–Navsari becoming twin cities in March 2008. In April 2009, the 'much-awaited project' received 'a preliminary nod from the state government' via a 'steering committee' to test feasibility (Daily News and Analysis *DNA* 25 March 2009). In January 2010, Surat–Navsari found

its way into Modi's five-twin electoral present for Gujarat (www.navsari.
in/2010/01/15/twin-cities-surat-and-navsari/). By December 2011, there were
supposedly signs that twinning 'was soon to be a reality' with full jointly
planned realisation set for 2025 (*TOI* 29 December 2011). In March 2012, the
state government approved plans for a new four-lane Surat–Navsari highway: 'the first concrete step' (*TOI* 14 March 2012). In May 2013, Modi again
announced twinning, this time to mark Gujarat Foundation Day (*DNA* 3
May 2013). In December 2015, the state Urban Development Department
gave the go-ahead to form a Navsari Urban Development Authority (UDA),
responding to pressure from Surat UDA and embracing 87 villages, thereby
'giving a push' to the idea of the two becoming planned twins (*Master-
plans India*; *TOI* 10 December 2015). In June 2017, there finally appeared a
Twin City Development Strategy for Surat–Navsari Twin Cities drawn up at
Gujerat Infrastructure Development Board's behest; by March 2018 (*TOI*),
there was a Surat–Navsari UDA, and an alluring Facebook logo.

Surat–Navsari now seems firmly embarked upon planned twin-citydom.
The fate of others is less certain. Another of Modi's electorally enticing
five, introduced to the world in February 2010, was Bharuch–Ankleshwar,
being passed to GIDB for preliminary investigation. In January 2012, the
Bharuch–Ankleshwar UDA (BAUDA) emerged apparently ultimately anticipating the 'fusion into one city of the two industrial townships' (www.
bauda.org.in/Organization.aspx), sprouting adverts for building projects.
By May 2013, BAUDA was 'consulting stakeholders', and the 'much-
awaited blueprint' plan emerged in January 2014. However, this apparently
hit increasing trouble, especially from affected farming and builders groups
in December 2015 which by April 2017 had worsened, with paralysis still
apparently prevailing, as we write in March 2018.

If the twinning process is variable in speed and difficulty, this may
partly relate to underpinning intentions, which as noted are similarly varied. Thus, 'bridging the gap', making more mutually beneficial use of the
'thinly populated' and 'underdeveloped' hinterland, seems to be a continuing motivation in Hubli–Dharwad's partnership (*Revised Development Plan
2041*, 50, 57). However, all five of Modi's twins projected in December 2011
were at least partly fuelled by his idea of 'the twin-city model': the desire
to use the vibrancy of a larger city to boost a smaller one, 'bring(ing) together cities that are in geographical proximity but where there is a gap
in their pace of development' (*Gujarat: A Glorious Decade of Development*
Gujarat Official State Portal). Discovering two urban economies to be 'complementary' seemingly also enhances motivation, or at least justification,
for twinning. However, what this means apparently varies. Geographical
distances between actual and potential twins also vary greatly. Some are
very close – like Pimpri-Chinchwad and Sarendranagar–Wadhwan, both
3 km apart; others are more distant but, like Sangli–Miraj, sprawling into
indistinguishable proximity by the point of local-governmental merger,
or (like Hubli–Dharwad) are increasingly so doing. Here, twinning might

seem unproblematic, even natural. Others are more distant with considerable non-urban intervening space: Surat–Navsari (28 km by crow; 38 km by road); Morbi–Wankaner (25 and 31 km); and one Modi pair, Vadodara–Halol (35 km by crow and 40 km by road). If only because distance probably enhances the number and complexity of interests needing appeasement, this likely also enhances twinning problems.

And this relates to the greatest contrast. Our planned twins vary dramatically in comparative size, fitting into what has been seen as the most common characteristic among twin cities generally, in India and elsewhere: dominance and subordination. Two of our pairings have roughly equivalent populations: Hubli–Dharwad and Sangli–Miraj. However, of Gujarat's real or projected twins, all but one (Bharuch–Ankleshwar) comprise twins (on 2011 figures), where one heavily outweighs the other. Sarendranagar (177,851) is over twice Wadhwan's size (75,755); Morbi (246,008) is 4.6 times bigger than Wankaner (43,881), while, for Vadodara–Halol (1,670,806 vs 59,605), Surat–Navsari (4,467,797 vs 160,941) and Ahmedabad–Gandhinagar (5,577,940 vs 292,167), the first partner is over 27 times the latter's size. Some Gujarat officials raised this when Modi announced the Ahmedabad–Gandhinagar link-up: 'we need to ensure that the big does not eat away the share of the small, because being big, its needs would also be big' (*DNA* 4 December 2011). Put another way, Gandhinagar might appear just a temptingly taxable snack for its vast and voracious 'partner' – India's fastest-growing city (https://www.forbes.com/2010/10/07/cities-china-chicago-opinions-columnists-joel-kotkin.html#7cdc602854e2). However, such objections were apparently overlaid by Modi's counter-feeling that the larger, more vibrant twin would boost its 'neighbour' to their mutual benefit.

In fact, as with our customary twins, the situation is more complicated than just dominance-subordination, even though the descriptor remains useful. Being smaller does not necessarily entail becoming subsumed if the smaller has distinctive functions and character. Thus, Ahmedabad is certainly vast and vibrant: a massive producer of cotton and cotton goods, 'the Manchester of the East'. It has thriving heavy, chemical and pharmaceutical industries, it is a major higher education, IT and research centre and it possesses its own stock exchange, yet also manages a low crime rate, and the reputation as 'the best place to live among India's megacities' (*TOI* 11 December 2011). However, though named by Modi as a new twin in November 2009, Gandhinagar was planned and created from its start (1965) to become the newly created state of Gujarat's capital, taking over that role from Ahmedabad once ready. Implicitly, it seems an early example of our type 3 category of twin. Gandhinagar was Indian-designed and intended as a fit inheritor of the name and philosophy of the man most closely associated with India's foundation, Mahatma Gandhi who was born in what became the state. It also has its own vibrant civil society and retains its own municipality.

This points to the frequently used justification that, outwardly unequal or not, planned/deliberate Indian twins are supposedly natural partners in that they 'complement' each other, most frequently economically. What this means varies. It can entail both partners having similar, and thus, mutually reinforcing economic bases. A key justification for twinning Morbi–Wankaner has been that both possess thriving ceramics industries, such that something that had already 'given global identity and glory to Morbi' (C.M. Vijay Rapani, *DeshGujerat* 15 August 2016) could only get better. Given Morbi is already supposedly producing 90% of India's ceramics products (*Vibrant Ceramics* www.youtube.com/watch?v=0B3jCSxVC0I) and the world needs tiles and toilets, hyperbole is understandable. Navsari might seem miniscule alongside Surat (the world's 'fourth fastest growing city' according to the City Mayors Foundation), but it too is growing rapidly and the partners' industrial strengths particularly in 'chemicals, pharmaceuticals and food-processing' were seen as mutually reinforcing, especially given their location on the Delhi-Mumbai Industrial Corridor (www.navsari. in/2009/12/20/navsaris-economy-and-industry-profile/). Similar justifications appeared for Bharuch–Ankleshwar (chemicals) and Vadodara–Halol. The latter twins were Special Economic Zones, and the interconnecting Expressway across their long hinterland is seen as 'a prominent development corridor' (*Real Estate Times* 13 August 2017) with the twin-city project seen as producing a real-estate boom.

But 'complementing' each other can also entail useable contrasts – economically or in other ways. Planned twins should not simply merge, rather use pooled resources to flag and exploit their supposed complementary distinctiveness. As noted, Hubli and Dharwad were seen as natural partners as early as 1925. As the *Revised City Development Plan 2041* (RCDP) noted, the British had turned Dharwad into 'an administrative and educational centre' and 'identified Hubli as a commercial and trade centre for the entire Northern Karnataka region' – roles continuing and expanding ever since, doing so apparently in mutually useful ways, for example, with Dharwad's educational/higher educational strengths helping support Hubli's industrial base (ibid. 34). Modi's much later argument for twinning Ahmedabad and Gandhinagar rested not just on Ahmedabad's ability to economically ignite the latter, but also on its capacity to provide Gandhinagar's governmentally related citizens with a social and cultural life outside of office hours, and its role as a haven of 'rurban' peace for weary Ahmedabad counterparts. All the more, once it had consented to become a 'green city' (Twin-City Model www.narendramodi.in/ twin-city-model-to-be-replicated-in-many-other-cities-of-gujarat-4464).

Before exploring consequences, we can usefully draw together what a typical twin-city plan might look like, given what we have said about the factors underpinning the general urge to plan urban places, and the aims particularly of planned twin-citydom as they have emerged by the early twenty-first century. The plan should be long-term and 'integrated' based upon a 'vision' of where the pair and their intervening, even surrounding, hinterland should be in 20–30 years of time; this will likely be updated every ten years or so. It should build upon and enhance their respective economic, educational and other strengths, exploit mutual synergies and enhance their linkages via 'state-of-the-art'/'smart' infrastructural improvements, particularly in road and rapid-transit terms. This in turn should partly aim to ensure that the hinterland develops rationally rather than 'haphazardly', taking into account both what and who is already there and the anticipated economic and demographic pressures. The plan should be 'inclusive' and consultative, embracing the full spectrum of interests/'stakeholders', with provision for self-contained 'integrated townships' for the middle classes and 'affordable housing' for the multiple ranks below. Given federal-state involvement via the Ministry of Urban Development, plans also carry expectations 'to enhance overall development in the state' (BAUDA). Increasingly, plans will also aim to make the twins 'smart' and preferably 'green'.

Impact and consequences

Exploring consequences is difficult, particularly because there seem few if any studies of how far twin-city urban development matches long-term aims outlined in development plans. It would be difficult to start here, given limited verbal space. What follows is just a start, and it mainly focuses on type 2 twins.

Short-term impacts are relatively easily traceable. They occur in the wake of publicly promised twinning. Very common are signs of anticipatory excitement in the real-estate sector located in and around the hinterland between projected twins. Thus, as twinning plans hardened for Surat–Navsari during September 2013, the *TOI* reported that their hinterland seemed 'to be turning into a goldmine for developers' with district authorities subject to 'a deluge of applications for conversion of huge chunks of agricultural land' for 'commercial and residential purposes' (*TOI* 3 September 2013). *Magicbricks* (7 May 2014), a real-estate publication, showed comparable salivation in May 2014, reporting that 'Ahmedabad's property picture is very bright ... the city is dragging its sister city on the same lines of boom, making Gandhinagar ... one of the most viable options for property investment in recent times' – doubly so as rapid-transit plans began making the hinterland evermore appetising. However, other groups show less enthusiasm about twin-city inclusion. Agriculturalists are often understandably ambivalent – as evident in Bharuch–Ankleshwar, where a conflict with building interests has paralysed twinning plans at the point of writing. Equally in Morbi–Wankaner: here, rural perceptions that urban inclusion brought few of the promised benefits like affordable housing, sewage disposal, etc., while rendering agricultural pursuits harder led the Gujarat state government to exclude several villages from the joint Urban Development Authority's territory (https://counterview.org/2016/05/06/gujarat-villages-resist-urbanisation-as-it-has-failed-to-bring-in-basic-amenities-for-which-they-struggled).

One almost invariable medium-term consequence of planned Indian twinning is the development, at a minimum, of major highways through or around the twinned area. Besides being bicycle-friendly, this is often accompanied by the emergence of plans for bus and/or tram-based rapid-transit systems (BRTS) linking the two; seen as crucial if the projected benefits of partnership regarding labour mobility, hinterland development, access to each other's facilities, etc., are to become real. Indeed, without this, citizens have no reason to think the twinned neighbour more important than anywhere else nearby. And if twinning benefits are to extend downwards, BRTS must be cost-accessible.

Equally important are governance changes. These vary, perhaps confirming a World Bank criticism of urban India for the unpredictable variability of its local-governmental provision (www.worldbank.org/en/news/feature/2011/07/04/indias-urban-challenges). Sometimes recognising interdependence has entailed the early formation of joint municipalities: as with Hubli–Dharwad (1925 reinforced in 1962), Pimpri-Chinchwad (1970 and more powerfully 1982) and Sangli–Miraj (1998–99). These may be supplemented by joint urban development authorities, as with Hubli-Dharwad from 1987. Sometimes separate municipalities have been retained, but supplemented by joint urban development authorities. Sometimes joint municipalities alone appear to have been deemed sufficient to oversee twin-city relationships.

All these authorities are involved in producing and implementing twin-city planning. The rough principle seems to be that UDAs are responsible for constructing long-term plans, negotiating details with municipalities and all levels of stakeholders, then submitting the agreed result to relevant state ministries for final approval.

Planning is central to Indian contemplations of urban areas, but three aspects may have special pertinence to twin-city planning. First is the need to exploit the perceived strengths of both partners for the general communal good. This is most easily illustrated by the Vision statement in Hubli–Dharwad's RCDP. Contemplating a twin city wherein Hubli was the major commercial and business centre and Dharwad the major administrative (for its District) cultural and educational centre, it envisaged the two becoming North Karnataka's 'commercial, cultural and educational hub' and creating 'an opportunity for growth to all segments of people' (*RCDP Executive Summary, 2*).

Second, planned twins need to exploit and develop their mutual hinterland. Thus, the aforementioned plan aimed primarily at 'growth along the Hubli–Dharwad highway' particularly in the light of BRTS development (ibid.). This is often seen as central to such aims – not just by planned twins, but by the evolving unplanned twins we have classified as type 1, though, in both cases, this partly depends on distance apart.

Third, given the above need, twin cities experience often heightened needs to negotiate plans not just with urban groups, but also with the many rural groups and individuals located in their frequently extensive hinterland of villages. The relationship is ambivalent. As noted with Surat–Navsari, some villages can successfully resist inclusion within a projected joint UDA on the grounds that urban benefits are not all they are claimed to be. If included, the dilemmas on both sides become clear. Sangli–Miraj contains much agricultural land which, as of 2007, comprised 48.59% of the total municipal area. Furthermore, 'this land is precious and productive ... hence it cannot be brought under ... urban use'. Yet, while limiting urban expansion, agriculture engaged just 10% of the workforce (*IJRFR* 3, 7 March 2016). Around Hubli–Dharwad, attitudes are more equivocal still. The *RCDP 2041* prepared for the federal Ministry of Urban Development August 2014 saw agriculture as playing an important role in the twin-city area (57/8): another observer noted, 'There are many peri-urban villages where agriculture is important' (Brook nd., 8). Yet, the period 1975–2011 saw huge losses of agricultural land (4941 down to 2320 hectares), mostly for building purposes. This observer described urban authority attitudes as comprising 'studied indifference ... (but) disapproval of open hostility' (ibid. 10).

Such ambivalence points to agriculture's ambiguous role in urban, perhaps particularly twin-city, areas. There are two kinds here: agriculturalists in their own right 'waiting to be taken over by urban development' (ibid.), and often back-yard and part-time agriculturalists who crucially exist to serve the immediate needs of urban dwellers: urban dairies and wastewater-irrigated vegetable producers (ibid.). Even planned-twin inhabitants need both.

Some indication of how even deprived rural, rural/urban and urban groups can become part of the twin-city planning process, perhaps especially in the light of hard-earned experiential learning, can be gained from a *Resettlement Plan* drawn up in December 2013, catering for people affected by a projected trunk sewer and its tributaries (65 km) on behalf of the Hubli–Dharwad Municipal Corporation (HDMC) by Karnataka state's Urban Infrastructure Development Finance Corporation (UIDFC), along guidelines set by 2007 federal legislation. The UIDFC's plan seemed to rest partly on its view that the twin cities 'have a lot of Charm and the people who live there an Intrinsic Goodness ... that touches one and all' (UIDFC, 6). It followed extensive consultations with, and detailed surveys of, the 64 people (17 families in eight households) needing temporary resettlement if the plan was to proceed. All were below the poverty line, and included people who were effectively squatters without legal entitlement to the land they farmed, even where they had continued cultivation after being compensated (inadequately they claimed) for loss of earnings when HDMC acquired the land back in 1980. Initially very hostile, their co-operation was obtained by promising resettlement in an area of Dharwad, training and agricultural equipment, and assisted by trained Resettlement Officers (*Resettlement Plan* North Karnataka www.adb.org/sites/default/files/project-document/79475/38254-043-rp-01.pdf).

Part of what is involved here, indeed creating the learning process underpinning the events above, is the sheer difficulty of urban planning (twin city or not) amidst desperate poverty levels and very high levels of rural–urban migration. One indication of this is the persistence of slums and 'gunthewari' (unauthorised settlements) even in the midst of the most successfully planned twins.

In the foregoing pages, we have said nothing about twin-city links with India's growing conurbations. Many of our cases either constitute conurbations in their own right or have increasingly become part of one. Perhaps, the best indicator is the highly planned twin city of Pimpri-Chinchwad. It is highly successful economically, and as a place to live, but, with its 'weak city identity', it is increasingly viewed as Pune's twin city, just 7 miles away (Pimpri-Chinchwad Civic Body Plans to Engage Citizens in Smart City Mission, *TOI* 10 March 2017).

Conclusion

Contrasted with other twin cities outlined in the book's introduction, Indian twins have borrowed features from each type and mixed them with their own traits. Geographically, many are akin to intranational twin cities, while their constructive co-operation and underpinning legal basis resemble international twins. Flexible attitudes to twin-able distance and the aim for better infrastructural and transport connectivity brings them closer to engineered twin cities.

Traits specific to Indian twins derive from alternative spatial forms of integrated settlements. Instead of self-centredness with little or no concern about their hinterlands usual for cross-border twin cities, Indian twins often exhibit responsible attitudes to theirs, a trait increasingly found in conurbations. The emphasis upon economic complementarity of Indian twins, with one city clearly leading, both in attracting commuters and investments and in acquiring (inter)national significance, is reminiscent centre–peripheral and core–satellite urban relationships. Recalling the term inter-urbations, coined by a Swedish geographer Nils Björsjö in 1960 to denote interdependent proximate cities with substantial age differences (Beaujeu-Garnier and Chabot 1967, 231), we find this concept overlapping with Indian twin cities too, particularly regarding types 2 and 3. The initial care from the older, more developed city for the younger one, and assistance in becoming self-sufficient and successful in performing the desired function are key to planned and instrumental Indian twins like Ahmedabad–Gandhinagar and Vijayawada–Guntur–Amaravati.

By classifying and describing Indian twin cities, this chapter seeks to start a discussion about them and indicate some promising research avenues for studying this hybrid of the twin-city extended family: a curious blend of (almost) all possible urban relationships.

References and abbreviations

Beaujeu-Garnier, J., and Chabot, G. (1967) '*Traite de geographie*', Progress Publishing House, Moscow [Russian edition].

Brook, Robert (nd.) *Urban Agriculture in Hubli-Dharwad* www.ruaf.org/sites/default/files/econf4_submittedpapers_brook.pdf (accessed 13/03/2018)

Dahl, R. A. (1963), '*Modern Political Analysis*', Prentice Hall, Upper Saddle River, NJ.

History of Urban Planning Kolkata (ref) www.studio-basel.com/assets/files/13_atlas_web.pdf (accessed 19/03/2018)

IJIFR: International Journal of Informative and Futuristic Research.

TOI: Times of India.

Urban Development Policy in Karnataka, Urban Development Department Bangalore 2015 www.indiawaterportal.org/articles/urban-development-policy-2009-department-urban-development-government-karnataka (accessed 17/03/2018)

8 Women's everyday travel experiences in the twin cities of Islamabad and Rawalpindi, Pakistan

Waheed Ahmed, Muhammad Imran, and Regina Scheyvens

Previous chapters have largely studied twin cities through the lenses of urban history and human geography, focusing on relations between twins and of twins with broader entities like conurbations, and state and national governments. This chapter combines social anthropology with development studies, and by dint of extensive interviewing, explores the experience of individuals and groups, more particularly those of women travelling within and across Islamabad and Rawalpindi in Pakistan. As the authors note, twin cities, because of their proximity and, here, their contrasting urban design and character, provide ideal venues wherein to explore this, within the context of enduringly traditional male attitudes. As expected, Islamabad, the planned and 'modern' capital city, offers multiple opportunities for women, especially in government and generally 'more acceptance for women in public space, work and transport'. However, Islamabad's layout importantly hampers realising those opportunities in travel terms, while unplanned Rawalpindi offers the liberating opportunity of crowds. However, travel within and between the twins provides multiple opportunities for expanding horizons and expectations.

Introduction

'Yeah, the same old gender question': after a sarcastic smile showing the routineness and non-seriousness of the question, the Director of one of the government's top planning offices replied 'transport doesn't have a gender thing … It is just about building roads, putting buses/vans on the roads … [so that] everybody including women can use them'. He was responding to a question from one of us about whether Pakistan's transport policy was safeguarding women's needs. Suddenly, he rang the bell and asked the minion, who instantly entered the room, 'why is the generator still not operating?' I could see sweat on his forehead as the electricity supply ceased. While considering what to ask in this gloomy environment, the director's friend, who was present, said 'you carry on [asking questions]'. I asked the same question but with a different tone hoping for something different but, before the director could answer, his friend said 'there will always be problems and unrest in society if females want to come out of their homes. The best places for women are in their homes. They should not travel without dire need'. This conversation happened during 2014 fieldwork at the Planning Commission in Islamabad. It indicates how Islamabad and Rawalpindi decision-makers think of women's transport and the social and structural issues women face in travelling there.

This chapter explores women's everyday travel experiences in the twin cities. They enable us to explore how Islamabad and Rawalpindi's highly contrasting urban form and institutional structure impact women's transport choices. We investigate the social and structural issues they face while travelling and how they negotiate mobility restrictions. The research is largely qualitative, using methods like interviews, life stories and structured observations. Researchers conducted 32 in-depth interviews to explore life stories of low-income women, business women, administrators and professional women in the twin cities. Besides interviewing women as transport-system users, male viewpoints, including public-transport drivers and conductors, and other stakeholders in government and private organisations, were also taken. Participants were selected through the purposive sampling technique. Four groups of women were targeted: professional, administrators, business women and women working in the informal sector. Women residing in different geographical areas of the twin cities were selected through the rolling snow-ball technique.[1]

Islamabad and Rawalpindi: twin cities of Pakistan

Islamabad and Rawalpindi are twin cities: mutually adjacent with people commuting daily within and between them. Islamabad's population is around 1,400,000, while Rawalpindi's is 2,600,000 (United Nations 2016). Islamabad's literacy rate (87%) is Pakistan's highest (Government of Pakistan 2007), while Rawalpindi's is 70.5%. Although lower, this is high compared with Punjab province's overall 44.09% (Waqar 2006).

Islamabad reflects post-World War II urban planning principles of low density and separate land uses. It was envisioned and started from 1958 during Ayub Khan's military regime to replace Karachi, the old capital, and aimed at nation-building and transforming an indigenous culture into a modern and civilised society (Mahsud 2013). Rawalpindi is a historical city. A small settlement before 1870, located strategically on roads linking the Khyber Pass and Kashmir with Lahore and Delhi, in the late nineteenth century it became the British military headquarters for North-Western India. After Pakistan's formation in 1947, Rawalpindi became headquarters for Pakistan's Armed Forces and Punjab's second largest city. Thus, it displays dualism in its spatial construction: 'the traditional city and the exotic cantonment, unplanned bazaar in the old city, tidy rectilinear regimentations for the civil lines and the military' (Specht 1983, 37).

To design the new capital, a Greek architect planner, C.A. Doxiadis, was engaged. Doxiadis used two planning principles. First, he sought to separate communities of different orders or 'classes', dividing the city into self-sustaining, rectangular 'sectors' accommodating 20–40,000 people surrounded by a hierarchically structured road network intersecting at right angles (Islamabad the capital of Pakistan n.d.). The second principle, 'dynapolis', involved placing the city centre in an expanding axis: sectors were 'grouped at a single distance alongside this axis, meaning that while the centre itself grew in a certain direction, new sectors could be added without increasing their distance (from) the centre' (Daechesel 2013). Doxiadis's original Master Plan included Rawalpindi as an auxiliary city developing alongside Islamabad. Rawalpindi would also provide Islamabad's initial needs, supporting the capital during formation. However, this plan was never realised and formally abandoned in the late 1970s. Punjab Housing and Physical Planning Department prepared a new plan just for Rawalpindi.

In Rawalpindi's recent Master Plan (1996–2016), mixed land-use and organic patterns dominate its central area, comprising Raja Bazar, Iqbal Road, Circular Road and Kashmir Road. This is also its main shopping area. Non-compatible land uses create congestion, traffic hazards and environmental problems. Due to over-concentrated commercial activity and linear growth, problems include inadequate parking, and poor accessibility due to road and footpath encroachments. Further north are first-generation planned housing schemes (Saidpur Road scheme, Satellite Town and Khayaban-e-Sir Syed), while southwards are second-generation schemes like Defence Housing and Bahria Town. These schemes comprise areas where land-use and building-control regulations are followed. The cantonment area is on the southern side of Murree Road, likewise the main Rawalpindi–Lahore railway track. Land use is mixed here but less problematically than in the city's older/central parts (Housing and Physical Planning Directorate 1998).

As Islamabad was a new city, people arrived from across the country, having diverse ethnic backgrounds. Being the national capital, most people came to work in government offices. Over time, the private sector also expanded particularly providing service jobs. Islamabad is also the headquarters for different private and not-for-profit organisations. These higher level economic opportunities attracted people with higher literacy rates and there has been acceptance of women for education, work and travel. However, people lacked the historical social connection to Islamabad evident in Pakistan's traditional cities. Islamabad was also planned on low-density principles, making the mixing of people rather difficult. Furthermore, Islamabad's city centre, which was planned for up-market commercial activities, largely closes down in the evening. Women do not go there then since there is no social or commercial activity and public transport services are infrequent.

Medium-density Rawalpindi, on the other hand, richly displays the Punjab's traditional, cultural and social values. Like Pakistan's other traditional cities, Rawalpindi people are accustomed to spaces for human interactions stretching from internal courtyards to community spaces (*muhallah*) to streets (Mahsud 2013). Consequently, people interact frequently, getting to know each other across generations. This social bonding provides a safety net for limited-level women's mobility notwithstanding traditional values limiting women's mobility, as evident later. This contrasts with Islamabad, where social structure does not cross generations and women sometimes feel unsafe walking Islamabad's often-deserted and isolated roads, notwithstanding good infrastructure and lighting.

Key literature and theoretical framework

Although many studies explore women's mobility in both urban and rural areas, these have typically focused on single locations, and a twin-cities perspective is missing. Such studies emerged from the late 1970s, stressing that women and men have different travel needs, primarily depending on roles in the society (Giuliano 1979; Rosenbloom 1978). Some studies have documented personal safety and harassment issues in cities (e.g. Hamilton et al. 2002). Others have focused on policy-making processes, highlighting needs to incorporate women's travel needs into transport and gender policies (Levy 1997).

Women's transport or mobility is central to academic debates around empowerment and human development. Transport was originally seen as mainly 'technical', for consideration by engineers, planners, environmentalists and economists (Simon 1996). Its social aspects were side-lined (Boschmann and Kwan 2008). However, many studies have emerged recently focusing on social issues and behaviours among different social groups regarding transport. These have recognised accessible transport systems as essential to equalising opportunity (Hine and Mitchell 2001). Nonetheless, there are no studies exploring women's mobility in neighbouring cities, especially in developing countries. We try to fill this gap by exploring the similarities and differences in women's travel experiences in Islamabad and Rawalpindi.

Women in developing countries face many exclusionary mechanisms in social, spatial and policy-making processes. These are evident in transport as it is a significant cultural area where gender is constituted or enforced. We argue that transport provides both a space wherein to observe how gender is constituted and a chance to negotiate gender stereotypes, creating new sets of gender relations. This is particularly important in a twin-city context where the possibility of crossing boundaries is much higher.

We use an empowerment approach to explore women's mobility in Islamabad and Rawalpindi. This approach represents a potential strategy for women's empowerment, putting emphasis on women's ability to challenge gender inequality within a society. This approach is important because lack of empowerment leads to marginalisation, exclusion and lack of access to resources.

Women's everyday travel experience

Social values and women's travel

Cross-country variations notwithstanding, Pakistani society is generally male-dominated (Malik and Courtney 2011). A strong 'inside/outside' dichotomy exists, whereby women belong 'inside' or at home. This can also be seen in the institution of *purdah*, entailing women's physical separation from men. This symbolises women's exclusion from the public realm, constraining mobility and travel. Women who are demure, passive and dressed in culturally appropriate ways are considered 'good' women; assertive, mobile, culturally non-conformist women in terms of dress and public interaction with men are labelled 'bad' (Ahmad 2010). These values are stronger in Rawalpindi than Islamabad, as evident in the following stories (one from each city).

Nasreen Javed, 40 years, small businesswoman (I-10, Islamabad)

I am a home-based worker and sell candles and other household decorations. My husband died three years ago and I became sole breadwinner of my 2 children. I have courageously faced the bad attitude of my relatives. They believe I should not go out to work without even realising how much hardship I was facing. I work all day ... and sell products at cheap rates. Some people, like our neighbours, are really supportive. They take care of my children when I go out to buy or sell products. Going out is another struggle, especially in Rawalpindi where I buy products from the inexpensive wholesale shops. I avoid going out as much as I can because I am a female, widow and single mother, which means I am easy prey for people to label me a 'bad' woman.

Gulshan Dilshad, 32 years, married, sales-girl (Lalkurti, Rawalpindi)

I married at 18 and my husband, Imtiyaz, had a small business of *Arhat* (buying and selling grains). After 10 years of marriage we had three children. Unfortunately, my husband's business was in debt so he started part-time work at another shop in the afternoon, but whatever he was earning was not enough to run a family. Once I was shopping with one of my 'rich' friends when I happened to meet a boutique owner who had just opened his shop and asked me to work as lady salesperson. I explained to my husband that they were offering Rs. 15000/month and also giving a pick-up and drop-off service. This service was handy because otherwise it was impossible for me to change two vans and then walk half an hour to reach the shop. My husband was a bit reluctant, not because he did not want me to work, but was worried what others would think. Ever since I started working in the boutique my neighbours have stopped visiting. They thought I was involved in a *du number* business (red-light profession) as every evening one car picked me up and then dropped me home late in the night. I can't stop working there because now I am earning Rs. 25,000/month and involved in designing dresses. I cannot remove their suspicions about the car ... and don't want to go in the public transport because that is more humiliating, people stare at you, the driver touches you, and you wait a long time at bus stops.

These cases highlight women's need for bravery in challenging social norms restricting their urban mobility. Although both face difficulty, the attitude of their neighbours towards them being 'workers' is quite different. Nasreen's neighbours in modern Islamabad even look after her children when she is away; however, Gulshan's neighbours in Rawalpindi think that she has become a prostitute and have stopped visiting. This gives a glimpse into the stark differences between these two cities regarding people's perception about women's mobility and ability to work outside their homes.

Women in public spaces

Women respondents complained about harassment in public spaces in both Islamabad and Rawalpindi. All said they had been harassed verbally, and faced verbal comments when alone or with female friends in markets, streets, parks and public transport. Such comments create insecurity among women, in turn influencing decisions to travel or not in certain areas, at certain times, in certain ways (e.g. in groups) or using certain modes of transport (Buiten 2007). It is relevant to mention here that, while Islamabad's gender attitudes might be 'modern', its planned layout of the city form makes life difficult for women than in more 'chaotic' Rawalpindi.

This can be seen in women's walking patterns in each city. Islamabad's design discourages women from walking despite its well-constructed and well-maintained walking-tracks and footpaths, running within and between sectors. Women were rarely seen walking in Islamabad from sector to sector, alone or in groups. When asked, most respondents said that they did not feel like walking the long distances from one place to another with nothing to see or enjoy. Others said they preferred walking through crowded areas like in Rawalpindi. They suggested they were scared to walk in isolated areas because of safety issues, notwithstanding Islamabad's good walking infrastructure.

Meanwhile, many Rawalpindi areas lacked walkways/footpaths altogether. The areas without footpaths were in very bad shape: footpaths with open holes or occupied by vendors, causing people to walk in the middle of roads. Yet, what was interesting was finding that women felt more secure walking in these crowded areas of traditional Rawalpindi rather than secluded spaces of modern Islamabad, even though they might risk physical touching and verbal comments. Nasreen Bibi's case illustrates this point. We should also note that men's attitudes, whether in traditional Rawalpindi or in modern Islamabad, towards women were not significantly different. Misogyny was evident in both.

Nasreen Bibi, 27, Policy Officer in an NGO, Islamabad
One day I was not feeling well at the office so I left for home early. My office was in a street in F-8, Islamabad … it took me 10 minutes to walk to a nearby bus stop. It was a very hot day in summer. As I left the office I felt someone was chasing me. I stopped immediately and looked back but there was no one. I thought maybe it is something in my mind. Anyways, I continued walking. I was alone in that street and there was nobody near … suddenly two men appeared from my back and started talking to me. I felt really scared; I cannot explain my feeling … They said that they wanted to take me to some place for lunch and spend some time with me. I tried hard not to show them my fear. I told them … I was not feeling well and they can take my phone number and we can set the time of our meeting over the phone. Thanks God they took the number and left. I was about to faint with fear. Then I hired a taxi and went home. From the next day onwards I went home by taxi.

This highlights again that walking was not a viable option for women in Islamabad: distances are great because of its low-density, so walking takes more time. This was coupled with the seclusion and perceived lack of safety when walking, there being no social or commercial activity on the footpaths. This was different in Rawalpindi, where footpaths and streets were overwhelmingly occupied by vendors, contributing towards already crowded streets and markets. Again this helps women in Rawalpindi move without fear for personal safety, apart from verbal comments and physical touching in crowded places.

Women's experience of using and navigating public transport was different in Rawalpindi and Islamabad. Even differences in drivers' and conductors' attitudes were evident. It being the capital city, Islamabad traffic police were seen regularly checking public transport at various bus stops; they were also vigilant about passengers' complaints, particularly female passengers. For example, in vans, the most frequent public-transport mode in both cities, only two seats were available beside the driver which were generally allocated to female passengers. Drivers tended to cramp three passengers in this space, but could only do this in Rawalpindi because police enforcement was less strong. When they entered Islamabad, they put seat belts on and allowed only two passengers in the front seats. This made the female seats safer for women, not having to press against the driver.

Modes of transport available

Public transport in both cities is dominated by the private sector with fragmented services and different service providers, comprising vans, mini-buses, mini-vans and taxis. Although in 2015, the government introduced a modern Bus Rapid Transit service called the Metro Bus Service (MBS) on the main Islamabad–Rawalpindi corridor, 15-seater vans remain dominant covering far more areas.

MBS provides very good experiences for women travellers. Many said that they felt safe using these buses. One reported, 'MBS had given her relief from the daily teasing remarks and glaring stares … the separate travelling area on buses made her feel safer' (The Express Tribune 2017). However, women's experiences travelling to and from metro stations were different in Islamabad and Rawalpindi. Many respondents said that they could easily access metro stations in Rawalpindi because the buses pass through the busiest route, Murree Road, where other modes of transport were abundant, particularly rickshaws. These rickshaws, informal transport, serve as feeders and are well suited to Rawalpindi's narrow alleys and streets, where other transport services can hardly go. Meanwhile, accessing metro stations in Islamabad is tough for women, there being no feeder services except taxis or Uber, expensive and unsuitable for most female passengers who generally work in the service sector or low-income professions. Furthermore, rickshaws are not allowed in Islamabad, being perceived as giving the capital city a 'bad image'. Walking to MBS stations is also unviable due to longer distances from residential locations in Islamabad (Figure 8.1).

MBS aside, women's public-transport experience is problematic. Women travellers reported the bad attitude of van drivers and conductors: Significantly, all said that they were physically touched by drivers or conductors. Many drivers and conductors said that a girl giving a little smile in response to conversation or physical touch meant that 'she wants it too'. Four of seven drivers said that they can judge from appearance whether women are 'good'

Figure 8.1 Metro bus and station in Islamabad by Waheed Ahmed (2015).

or 'bad', thereby stereotyping them by mobility and dress. There was no difference in drivers' attitudes in Islamabad and Rawalpindi. Interestingly, most van drivers and conductors lived in Rawalpindi, even though running vans in both cities as allowed by their route permits.

Furthermore, male city-planning officials were little different, as evident in this chapter's opening dialogue. One said travelling on public transport is very difficult for women, adding that van drivers 'force three females to sit in the front seats where only two passengers can sit. This puts female passengers at the mercy of drivers who touch and harass them' (EDO Spatial planning, TMA Rawal Town, Rawalpindi). Although highlighting women's travel problems, he laughed sarcastically, commenting 'it is driver's skill how he handles and manages three females in a small place'. Other government officials present (all male) laughed loudly, showing their misogyny and indifference.

Women drivers of private vehicles

Strong gender differences prevail in both cities about women driving motorcycles or bicycles. It is perceived as culturally inappropriate for women to drive motorcycles, although legally permitted. In both cities, women are seen sitting behind male drivers (mostly brothers, fathers or relatives). Some regard this as symptomatic of women's oppression and male power because, if women could ride motorbikes like men, 'this would increase their mobility' and 'choices in life'. Limiting 'women's freedom to travel was therefore invented to preserve … [male] power' (Hoodbhoy 2013).

Regarding car driving, quite a few women respondents could access a family car. Of our 22 respondents, 6 both had access and could drive. Mostly, these were professionals, like doctors, lawyers and university teachers. They had their own problems: Who drives the car (she or her husband)? Who takes the car to the office? Can a woman drive with her husband as a passenger? These questions were key issues for such women. The following story highlights some of the underlying tensions of car ownership, sharing and social norms around female drivers.

> **Uzma Tahir, 44 years, professional (Chaklala Scheme 3, Rawalpindi)**
> I work as a Project Manager at an International NGO. I married in 1997. I am lucky for not having children … it would be difficult to raise them without a father … I had to divorce my husband. It was my driving to my office that he never liked. I never understood why his mood suddenly changed [negatively] in the mornings when we would go to our offices. He was a doctor in the Combined Military Hospital (CMH) and I always dropped him first. He never drove a car before as he was not confident, even though he learnt driving. Things slowly got worse and he started abusing me like – 'my friends think you are overriding me … my family thinks driving is a man's job'. He asked me to sell the car or leave it in garage until he became an expert driver. But I said

'no'. He started abusing me even in front of my family, friends and colleagues. You couldn't tell that he was a well-educated person. Anyways, I got the divorce. Now I live with my parents ... and am happy ... life is far better.

Besides family issues, women drivers also face annoying and unwanted behaviour from male drivers of public transport and private cars. This is mostly in Rawalpindi rather than Islamabad. Most female respondents said that, in Rawalpindi, other cars including vans following them gave beeps, wrong indicators and sometimes braked suddenly in front. This kind of annoying behaviour intended to intimidate female drivers has some roots in the social belief that females make poor car drivers.

Women's agency and empowerment

Despite these barriers to women's travel in urban areas, women are exerting themselves in all spheres of life in Islamabad and Rawalpindi. Their labour force participation has increased from 16.0% in 2000 to 24.9% in the country (The World Bank 2017). This is evident in sectors traditionally considered 'male-only' professions, like transport. One of us interviewed two female professional drivers: a taxi driver and a driver in an international NGO. Both shared their life struggles in the twin cities by highlighting how they stood up for their rights, braving the odds in a male-dominated society. These two quotations highlight this:

> **Rubina Zahid, 60 years, taxi driver (Lalazar, Rawalpindi)**
> Overall, my experience as a taxi-driver in Islamabad and Rawalpindi is full of struggles and hard-work. I lived in Rawalpindi but travelled mostly in Islamabad. I never gave up what I wanted to achieve. At the end of the day, this is my life and I believe I made good decisions. I never looked back when I started this profession ... I was able to send my children to school and they are all married and leading a good life.

> **Nargis Younas, 39 years, female driver (G-7, Islamabad)**
> I think there is a lot of potential for women ... as professional drivers. If women start working as drivers, society will accept it eventually. I myself convinced a lot of females to learn driving ... poor women with limited resources. Now they are working as instructors ... earning a good amount of money for their families. Even at 'Save the Children', I have taught driving to many women employees. Some now come to the office in their own cars. I feel good in my role.

Pakistani norms and social values around gender relations and power are changing (Weiss 2001), fuelled by debates within families, increasing girls' school enrolment, and women's frequent appearance in public spheres, producing a 'constant renegotiation of power relations within the social order' (ibid.). Almost all respondents said that, despite problems, they were more confident than before and could influence their families on many issues, impossible a few years ago. They felt that leaving their homes to travel in public spaces (for education, work and leisure) increased their confidence. As one might expect, this seems particularly evident in Islamabad, while also spreading over into Rawalpindi, as we see below.

Some local NGOs have taken innovative steps to improve women's mobility. *Aurat* (Women's) Foundation, based in Islamabad, started a 'car–van leadership' project, wherein they selected 10–12 women from very poor backgrounds from Islamabad and Rawalpindi to train in driving skills enabling them to become professional drivers. Similarly, there are local-level initiatives, more often in Islamabad, where groups of women organise events like walks and bicycle or motorcycle rallies. Although informal in nature, they impact people's perception about women's roles and ability to travel in city spaces.

Conclusions

Situated in the frameworks of empowerment approaches, this chapter highlights the importance of studying gender issues in transport particularly in the context of the twin cities. As mentioned earlier, the two cities have different spatial and social dimensions making women's travel within and between them both diverse and challenging. Islamabad and Rawalpindi still exhibit their distinctive identity even though sharing boundaries. There are differences in city regulations, design, planning and social values. This chapter has highlighted how the different city forms, transport planning and people's perception about women's mobility impacts women differently and in complex ways.

We argue that transport could be used to analyse how gendered power relations of a society can be negotiated. In Pakistan's patriarchal society, women face norms inhibiting their ability to fully participate in social and economic life in cities. These were more evident in Rawalpindi, the traditional and historical city, than in Islamabad. These problems range from male misogyny, including public-transport drivers and conductors, to social values against mobile women, teasing behaviour towards independent women and restrictions stemming from female segregation. Clearly, women in the cities not only face societal attitudes but also have to tackle structural inequalities stemming from poor public transport and state indifference about women's mobility, prevalent in both twins. The perceived sense of safety was also important in women's ability to travel particularly in Islamabad's low-density and separate land uses notwithstanding good infrastructure and lighting.

However, there is hope as many women working in these two urban areas seemed highly motivated and eager. This might be due to the higher literacy levels in the twin cities compared to rest of Pakistan. Higher literacy means women are studying and working, implying the need for travel in these cities. Almost every woman interviewed for this research said that women should travel and, instead of staying home, should participate in public life. They also said that this is not an easy task, given broad sociocultural and religious values generally and male attitudes towards public areas. Furthermore, they felt that travelling and confronting problems made them feel independent and empowered. Although full empowerment is difficult to achieve, social values are changing subtly in favour of women because of economic pressures. Women who are more mobile and thus more actively engaged in the community also reported having greater influence over decision-making within the household.

Good public transport is clearly central to much of this. Some would suggest that it is a human right, but some government planners evidently support attitudes rendering its realisation very difficult. While the MBS provides a very positive initiative connecting the twin cities, far more needs to be done to encourage more open-minded male attitudes at all levels of society to women's mobility, particularly in Rawalpindi.

Note

1 This involved selecting a few relevant participants, then asking them if they knew others suiting our research requirements.

References

Ahmad, S. (2010), 'The multiple locations and competing narratives of Pakistani women', *in: Pakistani Women: Multiple Locations and Competing Narratives*, Ahmad. S. (ed.), Karachi: Oxford University Press, 1–11.

Boschmann, E. E. & Kwan, M.-P. (2008), 'Towards socially sustainable urban transportation: Progress and potentials', *International Journal of Sustainable Transportation*, 2, 138–157.

Buiten, D. (2007), 'Gender, transport and the feminist agenda: Feminist insights towards engendering transport research', New York: United Nations ESCAP (Economic and Social Commission for Asia and the Pacific).

Daechesel, M. (2013), 'Misplaced ekistics: Islamabad and the politics of urban development in Pakistan', *South Asian History and Culture*, 4, 87–106.

Giuliano, G. (1979), 'Public transportation and the travel needs of women', *Traffic Quarterly*, 33, 607–616.

Government of Pakistan (2007), Available: www.statpak.gov.pk/depts/fbs/statistics/pslm_prov2006-07/2.14a.pdf [Accessed 20/03/2015].

Hamilton, K., Hoyle, S. R. & Jenkins, L. (2002), '*The public transport gender audit*', Transportation Study Unit, University of East London.

Hine, J. & Mitchell, F. (2001), 'Better for everyone? Travel experiences and transport exclusion', *Urban Studies*, 38, 319–332.

Hoodbhoy, P. (2013), 'Women on motorbikes - what's the problem?', *The News*, February 22, Available: http://tribune.com.pk/story/511107/women-on-motorbikes-whats-the-problem/ [Accessed 22/06/2015].

Housing and Physical Planning Directorate (1998), 'Rawalpindi Master Plan (1996–2016)', Lahore: Housing and Physical Planning Directorate, Government of the Punjab.

Islamabad the Capital of Pakistan (n.d.), Available: www.doxiadis.org/files/pdf/Islamabad_project_publ.pdf [Accessed 06/02/2013].

Levy, C. (1997), 'Transport', *in:* Ostergaard, L. (ed.) *Gender and development: A practical guide* (3rd ed.), London: Routledge. 94–109

Mahsud, A. Z. K. (2013), 'Dynapolis and the cultural aftershocks: The development in the making of Islamabad and the reality today', *in:* Bajwa, K. W. (ed.) *Urban Pakistan: Frames for imagining and reading urbanism*, Karachi: Oxford University Press, 113.

Malik, S. & Courtney, K. (2011), 'Higher education and women's empowerment in Pakistan', *Gender and Education*, 23, 29–45.

Rosenbloom, S. (1978), 'The need for study of women travel issues', *Transportation*, 7, 347–350.

Simon, D. (1996), *Transport and development in the Third World*, London: Routledge.

Specht, R. A. (1983), *Islamabad/Rawalpindi: Reginal and urban planning*, Copenhagen: School of Architecture.

The Express Tribune (2017), 'Metro Bus a safe transport for women', *The Express Tribune*, March 27, Available: https://tribune.com.pk/story/1366382/easy-commute-metro-bus-safe-transport-women/ [Accessed 12/12/2017].

The World Bank (2017), 'Labour force participation rate, female (% of female population ages 15+) (modeled ILO estimate)', Available: https://data.worldbank.org/indicator/SL.TLF.CACT.FE.ZS?end=2017&locations=PK&start=1990&view=chart [Accessed 13/03/2018].

United Nations (2016), 'The world's cities in 2016_Data booklet', Available: www.un.org/en/development/desa/population/publications/pdf/urbanization/the_worlds_cities_in_2016_data_booklet.pdf [Accessed 13/04/2017].

Waqar, A. (2006), 'Literacy rate figures still not updated', *Daily Times*, March 1, Available: www.dailytimes.com.pk/default.asp?page=2006%5C03%5C01%5Cstory_1-3-2006_pg7_22 [Accessed 12/10/2017].

Weiss, A. M. (2001), 'Gendered power relations: Perpetuation and renegotiation', *in:* Weiss, A. M. & Gilani, S. Z. (eds.) *Power and civil society in Pakistan*, Karachi: Oxford University Press, 65.

9 Hong Kong and Shenzhen

Twins, rivals or potential megacity?

Marco Bontje

This first section on internal twin cities terminates with a chapter on Hong Kong and Shenzhen – fittingly since Shenzhen was partly created as a partner for Hong Kong in order to handle the latter's incorporation into China. In this sense, these can be seen as border twins transforming into internal twins. Like many cities in section 2, the relationship was intended to be positive. However, as the writer suggests, their future may be rivalrous (rendering them 'not quite twins' on the basis of border-twin expectations) and thus resemble the history of many internal twins. Unless they amalgamate – a fate rarely shared by twin cities of any sort – their relationship will continue to be special.

Introduction

Today's Shenzhen would not exist without Hong Kong right next door. Since 1979, Shenzhen has grown spectacularly from scattered small towns, villages and farmland to a megacity. Locating China's first Special Economic Zone (SEZ) directly beside Hong Kong was a strategic, deliberate choice, resting on Guangdong's history as an international gateway to China; a 'safe' place to experiment a new economic regime, sufficiently far from Beijing to not risk endangering the Communist system, while maximising potential profits from Hong Kong's proximity (Campanella, 2008). Moreover, Shenzhen, Zhuhai, Xiamen and Shantou SEZs were seen as stepping stones towards fully integrating Hong Kong, Macau and Taiwan into China (Yang, 2005). Especially at first, Hong Kong capital largely financed Shenzhen's growth; Hong Kong's example strongly influenced its planning and urban design; many early factories moved from Hong Kong or were established by its entrepreneurs (Ng, 2003).

Meanwhile, Hong Kong and Shenzhen are both megacities, competing for the same status, (inter)national functions and investments. They are deeply inter-related by intensive cross-border traffic, investments and joint-development projects. Still, they cannot yet be considered 'twin cities'. Shen (2014) argues that cross-border integration so far mostly comprises economic integration, some institutional integration, but little social integration. The 'twin-cities' idea has often been suggested in recent decades (e.g. SCMP, 2003; *China Daily*, 2016a). Some projects like the Free-Trade Zone Qianhai or the Lok-Ma-Chau Loop, and events like the Biennale of Urbanism and Architecture, employ the term to suggest that the cities are growing closer. However, concrete policy actions so far remain limited.

Relationships are complicated and tense. Although both are part of China (Hong Kong since 1997), Hong Kong retains special status as a 'Special Administrative Region' (SAR) under the 'one country, two systems' policy. It is very anxious about mainland China's aim of full integration, despite Sino-British agreement maintaining special status until at least 2047. It may face choices: does it see Shenzhen as a twin city, thereby perhaps strengthening its prominent international status, a rival or the vanguard of mainland China's takeover? If so, collaborating and/or even merging with Shenzhen may mean Hong Kong becoming 'just another Chinese city', perhaps assuming second rank in the megacity region, much less competitive internationally. Meanwhile, Shenzhen may become less special once Hong Kong loses privileged status.

What is their future? This chapter explores the many ways wherein the two cities have become inter-connected, while remaining separate, and whether they have shared or adverse interests. Can they someday become one integrated megacity, or will border obstacles remain? Or are they better off in their current close, connected but not entirely integrated state? Their future partly depends on governments in Hong Kong, Shenzhen and Guangdong, even more how far Beijing allows them to decide their own fate.

Regional context, development history and current situation

Both cities link to a larger heavily urbanised area known as the Pearl River Delta (PRD), one of China's prime megacity regions in both population and economic importance. While the Delta's history goes back many centuries, its heavy urbanisation mostly emerged only in the last half century, especially since China began reforming and opening its economy in 1978. Guangzhou was the only significant historic regional centre, long important as a trading centre and port and remains Guangdong's provincial capital (Shenzhen is also located here). But Guangzhou's growth into megacity mostly occurred in the twentieth and early twenty-first centuries. In the late eighteenth and early nineteenth centuries, it was the sole official trading point between Western and Chinese merchants, fitting into the region's longer history as a main international gateway to China (Zhang, 2015). Hong Kong and Shenzhen hardly existed – the area mainly hosting rocks, forests, farmland, small towns and villages.

Following the Opium Wars of the 1840s and 1850s, China generally and the PRD particularly experienced a century's political turbulence, economic stagnation and foreign influence: its 'century of humiliation'. Hong Kong became a British colony in three stages: Britain colonising Hong Kong Island in 1842; adding the Kowloon peninsula's southern part in 1860 and acquiring the 'New Territories' in 1898, creating a buffer-zone between the rather isolated and vulnerable port city of Hong Kong and China (Tsang, 2007). Britain and China agreed a 99-year lease, after which the area should return to China. The Shenzhen River became the new border and remains so, even now the lease has terminated. Initially, this border was quite porous, strict controls, walls and fences only being added in 1939 as World War II commenced (Watson, 2010). In 1950, it was closed even more firmly: Hong Kong (and Britain) wanting to avoid risking Chinese Communist attack and prevent a mass refugee influx; China wanting to reduce liberal, democratic and capitalist influence on Guangdong (Smart & Smart, 2008). The 'New Territories' remained largely undeveloped until the first New Towns emerged in the 1970s, bringing urbanised Hong Kong much closer to China's border. Hong Kong grew fast, transforming from small port city into a larger, increasingly 'global', city.

In the early decades of Communist rule (1949–78), Guangdong's development stagnated – due partly to the regime's varying urbanisation policies – between 'controlled urbanisation' and 'anti-urban(ism)' (Wu, 2015). Also important was China's economic development strategy favouring industrialisation of inland rather than coastal cities (Wu, 2015). So too were disastrous revolutionary policies like the Great Leap Forward and Cultural Revolution. Guangdong was further disadvantaged by Guangzhou's history as a 'treaty port', a colonial legacy the communists disliked, and the related perception of Guangdong as a potential 'bourgeois-capitalist' risk to communist rule because of its international orientation, and trade and kinship networks with the emigrant diaspora in Europe and North America (Vogel, 1990; Smart & Smart 2008). Furthermore, many political refugees and economic migrants attempted to cross the border, risking lives and liberty. Numbers are uncertain and disputed, there being no official statistics of how many fled, migrated successfully or failed to reach Hong Kong; estimates range between 500,000 and 2,000,000; some claim that 20%–30% of Hong Kong's population growth in 1961–81 consisted of Mainland Chinese and/ or economic migrants (*China Daily*, 2011; *SCMP*, 2013).

Watson (2010) describes the border of those days as a Cold War frontline, like the Berlin Wall, suggesting that most migration into Hong Kong occurred by sea or perhaps over the mountains between Shenzhen and Hong Kong. However achieved, the massive mainland Chinese influx contributed significantly to Hong Kong's emerging low-cost mass industrial production in the 1960s and 1970s. Alongside many refugees and migrants from Guangdong, another influential group were merchants and industrialists from Shanghai, who rapidly created a competitive textile industry – many escaping mainland China during the 1945–49 civil war, moving factories and investments from Shanghai to Hong Kong (Tsang, 2007; Zhang, 2015).

Guangdong's stagnation and border tensions changed radically after 1978, with China's 'opening' and economic transformation. Guangdong, especially the PRD, was where the first experiments involving 'capitalism with Chinese characteristics' began (Yang, 2005; Campanella, 2008). Its earlier disadvantages became advantages, or were interpretable as returning to its pre-communist international orientation and entrepreneurial history. Guangdong became the region where China started opening to the world. Several SEZs were created to kick-start China's economic modernisation and transformation. Shenzhen, an early SEZ, became the largest and most successful. In line with China's modernisation strategy, SEZs like Shenzhen initially specialised mostly in industrial low-cost mass production. Hong Kong was crucial to Shenzhen's early development in multiple ways. Many Shenzhen manufacturing firms moved from Hong Kong and/ or were financed by its entrepreneurs; much foreign direct investment (FDI) establishing these firms came from or via Hong Kong. Especially early on, Hong Kong represented an inspiring example of how capitalism worked and how to plan and develop a metropolis. Since the early 2000s, Shenzhen's economy and regional context have transformed once more – from 'workshop of the world' to a twenty-first-century metropolitan economy mainly based on high-tech production, innovation and advanced producer services (Zhang, 2015). Almost everything in present-day Shenzhen was built post-1980. However, recent as this built environment largely is, many parts of the city built in the 1980s or 1990s were already transforming to adapt to the city's changing economy and inhabitants' changing needs and preferences (Vlassenrood, 2016).

Meanwhile, Hong Kong also changed dramatically, influenced by China's transformation and Shenzhen's rapid emergence. After changing from a mainly port and trade city into an industrial city in the 1950s and 1960s, it mostly lost its industrial mass production to Shenzhen in the 1980s. Since then (while remaining a globally important port city), it has mainly specialised in advanced producer services, especially finance, insurance and real estate. This economic transformation paralleled radical change in its political future, with the 1984 UK–China agreement that Hong Kong would return to China in 1997. However, this was conditional upon the 'Basic Law' becoming the constitutional document establishing Hong Kong as a SAR of China. Effectively, this implied that most pre-existing colonial laws and regulations were maintained after handover. The Basic Law and the 'one-country-two-systems' principle theoretically give Hong Kong considerable executive and legislative autonomy, and its citizens rights unavailable to mainlanders, like freedom of speech, press, association and assembly. Also, the Basic Law establishes some limited democracy, with universal suffrage promised for the future. This SAR status should last until at least 2047.

However, several times already, China has apparently interpreted the Basic Law differently from Hong Kong's government and citizens, enhancing their growing anxiety about its autonomy (Keatley, 2016). Especially the denial of universal suffrage produced repeated mass citizen protests, most recently in the Occupy Central Movement and the Umbrella Revolution in 2014 (Ortmann, 2015). Significant steps towards full integration had already occurred, like the Closer Economic Partnership Arrangement (CEPA) and the 'Individual Visitor Scheme' both in 2003. Meanwhile, many Hong Kong residents have become apparently more aware of what distinguishes them from mainland Chinese. Instead of moving towards one Chinese identity, 'localism' and hostility towards mainland China has apparently increased since 1997 (Kwong, 2016; Xiyuan, 2016).

The Hong Kong–Shenzhen 'borderscape'

Breitung (2004) mentions four unique features of the border between mainland China/Shenzhen and Hong Kong:

- it is not cross-national;
- it is currently undisputed;
- it is internal, heavily guarded, but extensively crossed;
- it expires in 2047.

It also looks quite different according to one's vantage point. While Shenzhen's built-up area nearly reaches the border, most of Hong Kong's border-zone comprises fishing ponds, farmland or nature-area (Figure 9.1); the few exceptions include villages established long before the British colonial era (Watson, 2010). Equally contrasting is the infrastructure: several Shenzhen roads end at the border, connecting with nothing. The reasons

Figure 9.1 Shenzhen–Hong Kong border: built environment vs farmland and nature-conservation area by Marco Bontje (2013).

mainly connect to Hong Kong's colonial history, where British colonial rulers saw the New Territories as a buffer-zone against China. When Hong Kong Island and Kowloon appeared too small to accommodate Hong Kong's fast demographic and economic growth, New Towns were developed in the New Territories. However, so far these are all distant from the border and mainly surrounded by country parks, nature reserves and reservoirs. Another planning measure maintaining distance between Hong Kong's New Towns and Shenzhen is the 'Frontier Closed Area' at the Hong Kong side, installed in 1951 as a 'buffer-zone' against possible Chinese invasion and discouraging illegal migration and smuggling. Before 1951, border controls were much less intensive. Hong Kong's return to China in 1997 did not immediately change the Frontier Closed Area's status. Recently, however, plans have emerged to decrease the area, enabling new development plans, as we shall see later.

In total (including people from the rest of the world), an impressive 290,000,000 passengers crossed Hong Kong's borders in 2014, against c277,000,000 in 2013 (Information Services Department, 2015). Longer term growth is equally impressive: passenger numbers were around 153,000,000 in 2003 against 209,000,000 in 2005–6 (Smart & Smart, 2008). Currently (2018), there are five border-crossings. These differ in several ways. The Luohu/Lo Wu and Futian/Lok Ma Chau crossings can be reached by metro on both sides; passengers then walk across the Shenzhen River. These are the busiest: 87,150,000 people crossing between Luohu and Lo Wu in 2014, and 28,540,000 using the Futian/Lok Ma Chau crossing (Information Services Department, 2015). These include many crossing the border multiple times; official statistics provide no means of discovering numbers of different individuals involved. Private cars, taxis and buses can cross at Huanggang, while only cars use Wenjindu. The fifth border-crossing, Shekou, could only be reached by ferries from Hong Kong Central, Hong Kong Airport or Macau, until the Shenzhen Bay Bridge opened in 2007. To further facilitate cross-border traffic, two ambitious railway connections are planned.

Although the border-crossings differ, they share one unusual feature: though Hong Kong is formally part of China, progress from one to the other resembles crossing an international border. Crossing either way, you must pass through two sets of customs, separated by small pieces of 'no-man's land'. Moreover, Hong Kong and China each have their own visa regimes: these have little effect on Hong Kong or Shenzhen/Chinese citizens, who can access separate and quicker channels, but they do affect foreign travellers, doing so differently according to direction travelled and thus whether Hong Kong or Shenzhen visa regulations apply. Furthermore, even internal border-crossers are affected: Hong Kong and Macau travellers to mainland China find it easier than those proceeding in the opposite direction (Smart & Smart, 2008; Shen, 2014), although Shenzhen residents can cross more easily than others from the mainland.

Wanted and unwanted migrants and visitors

Hong Kong–Shenzhen migration is quite complex: many migrants travel back and forth, some keep homes in both places and many cross the border for work, study or family visits. As evident from the survey for Hong Kong SAR Planning Department and Shenzhen Statistics Bureau (2008), at that time about 62,000 people with Hong Kong identity cards or residence permits resided in Shenzhen. However, about half were apparently returning migrants, born in Shenzhen or elsewhere in mainland China. Furthermore, almost half of Hong Kong migrants to Shenzhen belonged to 'mixed households' with Hong Kong and mainland household members. About 10,000 were students, with 41% studying in Hong Kong and 59% crossing to study in Shenzhen. About 27,000 migrants were working, with 66% working in Hong Kong. Thus, alongside many tourists and day visitors, commuters and students also contribute significantly to the busy daily cross-border traffic. The most frequently mentioned reason to move from Hong Kong to Shenzhen was reunion with parents, partner and/or children (70%), and also lower living costs (25%), work or study (25%) or expecting a better living environment in Shenzhen (20%). Smart and Smart (2008) highlight Hong Kong's ambiguous rules, laws and policies regarding mainland China migrants and visitors. Especially in the first years after 1997, Hong Kong feared large numbers of mainland migrants and tried regulating entry: investors, tourists (especially wealthier ones) and talented workers were welcomed to contribute to Hong Kong's economy, but it struggled with the right of all children of Hong Kong residents to enter and reside in Hong Kong, as stipulated in the Basic Law. This right did not exist pre-1997, and involved around 1,670,000 people then living in mainland China. Despite its highest court confirming this right, Hong Kong's government negotiated a quota system with Chinese central government.

Despite these attempts to limit migration, since 1997 many children of mainland Chinese parents have been born in Hong Kong hospitals, thereby enabling access to education in Hong Kong. Such parents often made strategic choices: China's one-child policy did not apply in Hong Kong; Hong Kong's health care is reputedly better; its education and welfare higher and Hong Kong passports enable children to travel more easily outside China. Furthermore, parents often see it as an investment in the general family future. However, although such children get 'permanent resident' status, their parents do not. Hence, a peculiar group of daily cross-border migrants has emerged: 'border-crosser children' (Reinstra, 2015). These cannot attend mainland Chinese public schools, and private schools are very expensive for 'non-locals', but their home-base is mainland China, especially Shenzhen. Therefore, many cross the border from home to school and back, escorted by 'nannies'. Numbers have grown from 12,865 in 2011–12 to 28,106 in 2015–16. Children and families alike face complex situations, trying to arrange their lives and struggling with identity (Xiyuan, 2016).

Numbers will eventually reduce because of another measure: a 'zero-delivery quota' for expectant mainland mothers in Hong Kong hospitals in 2013 (Harbour Times, 2016). This followed a wave of 'birth-tourism' 2001–13, causing overcrowded hospitals and Hong Kong citizens' protests. The effect is already clear: mainland births in Hong Kong first increased from 620 in 2001 to over 35,000 in 2011 and 2012, and then dramatically dropped to 173 in January–September 2013 (Xiyuan, 2016).

Another contentious issue problematising cross-border traffic and Hong Kong–mainland relations is 'parallel trade'. Historically rooted in smuggling and small-scale cross-border trade between relatives, this has grown dramatically and become 'big business', especially since 1997 and the 2003 CEPA and Individual-Visitor Scheme. Many mainland visitors enter Hong Kong as 'tourists', but are really parallel traders, buying high-quality consumer goods in Hong Kong and very profitably reselling them in mainland China. These are unavailable, more expensive or of poorer quality in mainland China, so demand is high. Mainland food-safety scandals, like the use of poisonous melamine in milk powder, have further contributed to parallel trade. This is so extensive that Hong Kong retailers run out and locals can hardly buy the products or cope with rapid price rises. Moreover, especially in Hong Kong's North District opposite Shenzhen, parallel traders crowd trains and roads and claim public spaces for their operations, producing daily chaos and disturbance, protest and sometimes violent confrontation. Determined Hong Kong customs and police efforts to curb this semi-illegal trade have so far had little success (Cheung et al., 2015). Indeed, drastic measures affecting all mainland visitors, like the 'one visit per week' limit for Shenzhen visitors, may even have adversely affected mainland visitor groups Hong Kong wants to attract.

Economic connections, complementarities and/or rivalries?

Although, after transforming from 'factory of the world' to 'world city', Shenzhen's economy is becoming more like Hong Kong's, the two retain contrasting positions in the 'world-city network'. In analysing the office-networks of leading companies in advanced business services, Taylor and Derudder (2016) present Hong Kong and Shenzhen as exemplifying, respectively, 'globalism' and 'localism'. 'Globalist' world cities have strongest relations with other leading world cities beyond their region. 'Localist' world cities instead have strongest relations with other world cities within their region. The authors rank Hong Kong second among 'globalist' world cities and Shenzhen tenth among 'localist' ones. While Hong Kong's strongest inter-city links are with cities outside Pacific Asia, Shenzhen's are inside Pacific Asia, especially other Chinese cities. In this respect, we could see the two cities' advanced-service sectors as complementary. Possibly, this complementarity will eventually change as Shenzhen's advanced-service sector strengthens and matures and maybe becomes more internationally orientated. However,

Taylor and Derudder also highlight Hong Kong's extraordinary situation as an 'exterior power' of China: under China's political control, but with considerable economic autonomy. This 'exterior power' is 'built upon transactions that are necessary but not possible in China itself' (181). While Hong Kong retains its SAR status and mainland China is 'less free' economically, Hong Kong will retain significant advantages over Shenzhen, which cannot easily compete in the advanced-services sector.

Trujillo and Parilla (2016) develop a global-city typology, wherein the two cities are categorised as different types. Based on 35 competitiveness variables (assessing 'tradeable clusters', 'innovation', 'talent' and 'infrastructure connectivity'), this typology locates Hong Kong as an 'Asian Anchor' and Shenzhen an 'Emerging Gateway'. 'Asian Anchors' function as command-and-control centres, whose prominent position in the world economy rests on connectivity, a talented workforce and high capacity to attract FDI. 'Emerging Gateways' like Shenzhen, though fast-growing, still lag behind leading global cities on most key competitiveness factors. This suggests hierarchical relations between Shenzhen and Hong Kong, with Hong Kong clearly leading. Yet, if the development of 'Emerging Gateways' like Shenzhen continues, the gap with 'Asian Anchors' like Hong Kong could lessen.

Alternatively, are the two cities' advanced-producer service sectors gradually intertwining, eventually integrating them into one larger metropolitan whole? There seem recent signs of such gradual development. One is the Shenzhen–Hong Kong Stock Connect Programme, starting in December 2016. Part of the Chinese financial system's ongoing liberalisation, the programme allows overseas investors to trade 881 stocks on the Shenzhen Stock Exchange, and mainland Chinese investors to trade 417 stocks at Hong Kong's equivalent (SCMP, 2016). However, Chan and Zhao (2012) note the obstacles to further collaboration and integration of the two cities' advanced producer service sectors. Next to the border, Hong Kong and Shenzhen businesses necessarily operate in two quite different political–economic contexts. Thus, the authors argue, stakeholders' interests on either side of the border match poorly. Hong Kong businesses and government mainly seek access to new markets in Shenzhen and the PRD; Shenzhen businesses and government mainly want advice from Hong Kong businesses to improve their advanced-producer services.

Strategies to encourage or discourage collaboration and integration

According to Yang (2005), Zacharias and Tang (2010) and Shen (2014), in the years around Hong Kong's return to China, Shenzhen, Guangdong and China's national governments were more eager to collaborate with Hong Kong and encourage the two cities' further integration than *vice versa*. The reasons for Hong Kong's reticence include retaining 'a unique legal and cultural environment ... attractive to globalised business, while offering easy access into the China market' (Zacharias & Tang 2010, 215). Shenzhen may also prefer to maintain the internal border and profit directly and indirectly from Hong Kong's semi-autonomous status. Still, both cities also acknowledge possible benefits from intensifying collaboration. Their policy-makers and stakeholders meet frequently to discuss collaborative projects in the yearly Cooperation Meetings and twice-yearly Cooperation Forums (Shen, 2014). Several projects have recently started to encourage or facilitate collaboration, though some may not proceed beyond good intentions. Here are three prominent examples.

The Lok Ma Chau Loop (LMCL) is often seen as promising. This innovation and technology park is one of the ten 'major infrastructure projects' mentioned in the strategic-development vision, *Hong Kong 2030*. Its Planning and Engineering Study ambitiously seeks,

> to develop ... into a HK/SZ Special Co-operation Zone and a hub for cross-boundary human resources development within a Knowledge and Technology Exchange Zone under the principle of sustainable development ... benefit(ing) the long-term development of HK, the Greater PRD and South China region.
>
> (HKSAR Planning Department and Civil Engineering and Development Department 2015, 7)

Unfortunately, the site remains undeveloped. Though there are quite detailed spatial plans, possible urban designs and assessment studies, actual construction remains little beyond 'virtual reality'. One obstacle is the question of who owns the land. LMCL was originally on Shenzhen's side of the Shenzhen River, until its realignment in 1997 (Shen, 2014). Each city claims ownership. Understandably, Hong Kong refused to agree to the mainland Chinese government's request that it pay all development costs while recognising Shenzhen's ownership.

Qianhai and Shekou Free Trade Zone (FTZ) is one of Shenzhen's most recent new developments. It is situated west of Shenzhen, adjacent to Shekou, one of the places where Shenzhen's rapid growth began in the early 1980s. It is strategically located between the two city airports and well connected to Shenzhen's current Central Business District in Futian. In 2010, China's State Council approved the 'Overall Development Plan' issued by Guangdong province. The fact that Guangdong rather than Shenzhen created the plan and needed the Chinese government approval highlights its regional, even national importance. Such direct higher government involvement in Shenzhen's affairs happens more often than in most other Chinese cities because Shenzhen was China's first SEZ; some even say Shenzhen is China's *de facto* fifth 'directly-controlled municipality' after Beijing, Tianjin, Shanghai and Chongqing (Polo 2016).

Alongside neighbouring Shekou, Qianhai is Shenzhen's part of the Guangdong FTZ, with other parts situated in neighbouring Guangzhou and Zhuhai. FTZs are the next generation of SEZs, and another government attempt to open parts of its economy to FDI. Like the earlier SEZs, China has chosen to experiment in a few locations, and then maybe spread throughout the country. This is increasingly happening: seven new FTZs emerging in 2016 alone. FTZs are areas where goods may be landed, handled, manufactured and re-exported without customs intervention. They can also experiment with financial models and have more possibilities to attract investment than elsewhere in China. Qianhai is sold as a success story in Chinese media, supported by probably overblown statistics, for example, claims that, within a year of this FTZ's creation in 2015, over 61,000 companies registered there, producing an annual growth rate of 265% (*China Daily*, 2016b).

Although it is notoriously difficult to discover the 'real' intentions underpinning prestigious Chinese urban development projects like Qianhai, Polo (2016) argues convincingly that the project has changed course drastically in recent years. Though initially planned in 2010 to enhance Shenzhen–Hong Kong collaboration and integration, it now seems to be morphing into just another Central Business District for Shenzhen. And the perceived future relationship between Qianhai and the international financial centre of Hong Kong looks competitive rather than collaborative or integrative, a competition Qianhai cannot win. Despite initial good intentions, the Qianhai project has not broken through the institutional barriers, and apparently the three governments involved (Hong Kong, Shenzhen and the Chinese government) are insufficiently supportive for the project to succeed (Chan & Zhao, 2012).

The Bi-City Biennale is equally ambiguous. In 2005, Shenzhen organised its first *Urbanism and Architecture Biennale*. The organisers said that they wanted to develop this event with Hong Kong in future. Hong Kong joined in 2007, formally rendering it a *Bi-City Biennale*. However, the term seems just a label for two separate events with little in common beyond 'urbanism and architecture'. Each has its own location, opening period (same year, different months), particular theme, curators, promotional materials, etc. The homepage of the most recent Hong Kong edition (2017; http://uabbhk.org/uabbhk/) hardly mentions Shenzhen. The Shenzhen homepage (http://en.szhkbiennale.org/) does note that the event is 'co-organized by the two neighboring and closely-interacting cities of Shenzhen and Hong Kong'. However, beyond that, Hong Kong hardly features. In some past biennales, the two cities' inter-relatedness featured more: the 2013 Shenzhen Biennale focused on 'Urban Border', while Hong Kong's 2011 Biennale was about 'Tri-ciprocal Cities'. However, in 2013, given its sensitivity, few contributions dared address the Shenzhen–Hong Kong border or related issues around how long Hong Kong's autonomous status could survive (*Guardian*, 2013).

Longer term development strategies

Despite these recent attempts at collaboration, integration and barrier erosion, mutual anxieties and tensions remain. Interestingly, both cities now apparently face comparable challenges, having run out of suitable development land, even though their populations and economies are expected to grow considerably. Hong Kong has long struggled with this dilemma; Shenzhen has hit it far more recently. Will the two cities find joint solutions, or have both reached their growth limits? We now 'zoom out', to explore the two cities' longer term strategic planning and the possible future vision of one integrated megacity.

Since Shenzhen became a SEZ, spatial development has largely been determined by three master plans, each having its own spatial development model. The 1986 Masterplan followed a 'clustered linear' model, with three development clusters connected by road and rail. The 1996 plan switched to a 'network model' to better connect Shenzhen's central city (the original SEZ area) with adjacent areas (those formally added to the SEZ in 2010; they had actually become part of Shenzhen long before). The most recent 2010 Masterplan, governing development until 2020, and the longer term strategy *Shenzhen 2030* rest on a 'polycentric model'. This evolution of planning and strategy also reflects Shenzhen's rapidly changing development context and the challenges faced while progressing from countryside with scattered urbanisation to megacity. The 2010–20 plan and the Shenzhen 2030 strategy emerged when Shenzhen faced new realities, like land-shortage and environmental problems. These challenges, alongside Shenzhen's transformation from a city of mass-industrial production to a twenty-first-century high-tech and advanced-services city, have seemingly produced a shift from continuous extension to redeveloping existing urbanised areas (Zacharias & Tang 2010; Vlassenrood 2016). The 2010–20 Masterplan includes clear ambitions about relationships with Hong Kong: explicitly identifying 'twin city' as a long-term planning goal, Shenzhen is presented as 'national service-base to support Hong Kong's development' aiming at 'a world-class city-region' through collaborating with Hong Kong and relying on southern China (Ng, 2011).

The *Hong Kong 2030 Planning Vision and Strategy (HK2030)*, developed in 2001, planned additional new towns, closer to the border, plus development-corridors, including two strengthening Shenzhen: a 'central development-corridor' from southeast to north, and a 'northern development-corridor', connecting several existing and future new towns, partly along the border. The 2030 plan describes the northern corridor as 'Non-intensive technology and business zones and other uses ... capitalis(ing) on the strategic advantage of the boundary location'. *HK 2030* also included plans for strengthening/developing 'regional transport corridors', four to Shenzhen and two to Macau–Zhuhai–Guangzhou (Planning Department, 2007). *HK 2030* has been the leading spatial development strategy until 2015, when its partial revision began. *HK 2030+* should be completed in 2018.

The North East New Territories plans are heavily disputed: residents of nearby settlements object to displacing indigenous villagers and fear the new development becoming Shenzhen's 'backyard', possibly a step towards Hong Kong's 'mainlandisation' (Kwong, 2016). Critics also question the need for such large-scale development; will Hong Kong's population really grow as fast as the government expects, or is this development rather serving the interests of Hong Kong's real-estate tycoons and/or Shenzhen's elite? Still, the new strategy will largely build on the earlier 2030 vision. For Hong Kong–Shenzhen relations, this implies that attempts to bridge the gap between the two cities will continue, though this will probably remain a slow step-by-step process, while the 'one country-two systems' principle and the border remain. If it someday disappears, however, a very different and unpredictable story will emerge. Meanwhile, what will be the impact of connecting Hong Kong, Macau and Zhuhai via the bridge currently under construction? Will it release pressure on the Hong Kong–Shenzhen border? Will it integrate Hong Kong into mainland China via another route, bypassing Shenzhen and lowering its importance for Hong Kong?

Conclusions and future perspectives

The answer to the chapter title's question right now is: Hong Kong and Shenzhen have far to go before we could really call them twins; they are not heading for one integrated megacity anytime soon; and they both compete and collaborate (so rivals as well as partners). Relations are complex; future development is unpredictable. They are closely inter-related in many ways. Shenzhen might hardly exist without Hong Kong. Much of Hong Kong's recent growth would probably not have happened without China's opening since the late 1970s, wherein Shenzhen SEZ remains crucial. Hong Kong planners, developers and investors inspired and enabled much of Shenzhen's early development. Most of Hong Kong's industrial mass production migrated to Shenzhen in the 1980s and 1990s. Most FDI financing Shenzhen's development came via Hong Kong. Once Shenzhen grew and matured, the inter-city relationship became reciprocal in several dimensions: investments, consumption, visitors, migrants, etc. Shenzhen's economy may be catching up, though Hong Kong still retains clear advantage as a leading East-Asian 'world city'. Collaboration has apparently increased in recent years, though remaining limited while the border survives. And that is unlikely to disappear soon. Whether Hong Kong remains a SAR within China, and how autonomously, will eventually be determined in Beijing. Recent events make tightening control likely; 'one country-two systems' may not last until 2047. Beijing's reactions, for example, to the 2014 Umbrella Revolution and Hong Kong's 2016 Legislative Committee elections rightly concern Hong Kong. Still, complete 'mainlandisation' seems unlikely anytime soon. Both cities (and mainland China as a whole) still profit too much from Hong Kong's special status to relinquish entirely.

References

Breitung, W. (2004) 'A tale of two borders'. *Revista de Cultura* 9: 6–17.
Campanella, T. (2008) *The concrete dragon: China's urban revolution and what it means for the world.* New York: Princeton Architectural Press.

Chan, T.G. & S.X.B. Zhao (2012) 'Advanced producer-services industries in Hong Kong and Shenzhen: struggles towards integration'. *Asia Pacific Viewpoint* 53 (1): 70–85.

Cheung, H., T.-H. Chiew, C. Tsoi, T. Wan & F. Wong (2015) 'Policing parallel-trading activities and the associated public disorder in North District'. Master thesis, Social Sciences/Criminology, Hong Kong University.

China Daily (2011) 'The great exodus'. www.chinadaily.com.cn/hkedition/2011-04/20/content_12358785.htm (accessed 17/11/2016).

China Daily (2016a) 'Rebuilding trust between the mainland and Hong Kong'. www.chinadailyasia.com/opinion/2016-05/16/content_15433724.html (accessed 22/11/2016).

China Daily (2016b) 'Hong Kong enterprises revving up in Qianhai'. www.chinadailyasia.com/hknews/2016-05/12/content_15432153.html (accessed 28/11/2016).

Guardian (2013) 'Shenzhen Biennale: glowing stairs and metal-free bras are the Chinese dream'. www.theguardian.com/artanddesign/architecture-design-blog/2013/dec/19/shenzhen-biennale-industrial-heritage-china-development-urbanisation (accessed 2/12/2016).

Harbour Times (2016) 'Cross-border students up 118% in five years'. http://harbourtimes.com/2016/06/23/cross-border-students-up-118-in-five-years/ (accessed 01/12/2016).

Hong Kong SAR Planning Department and Civil Engineering and Development Department (2015) *Planning and Engineering Study of Lok Ma Chau Loop, executive summary.* www.pland.gov.hk/pland_en/p_study/comp_s/lmcloop/pdf/FR%20ES/LMC%20Loop_Final%20Report_ES%20-%20Eng_20150209.pdf (accessed 22/11/2016).

Hong Kong SAR Planning Department & Shenzhen Statistics Bureau (2008) *Survey of Hong Kong people living in Shenzhen.* Hong Kong / Shenzhen: Hong Kong SAR Government / Shenzhen Municipal Government.

Homepage 'Bi-City Biennale of Urbanism /Architecture'. http://uabbhk.org/uabbhk/ (accessed 06/12/2017).

'Hong Kong 2030+, Conceptual Spatial Framework'. www.hk2030plus.hk/conceptual.asp?form=80 (accessed 01/12/2016).

Information Services Department, Government of Hong Kong SAR (2015) *Hong Kong: the facts. Immigration.* www.gov.hk/en/about/abouthk/factsheets/docs/immigration.pdf (accessed 17/11/2016).

Keatley, R. (2016) 'Can Hong Kong keep its autonomy?', *The National Interest*, 29 January 2016. http://nationalinterest.org/feature/can-hong-kong-keep-its-autonomy-14946 (accessed 05/12/2016).

Kwong, Y.-H. (2016) 'State-society conflict radicalization in Hong Kong: the rise of "anti-China" sentiment and radical localism', *Asian Affairs* 47 (3): 428–442.

Leal Trujillo, J. & J. Parilla (2016) *Redefining global cities.* Washington: Brookings Institution. www.brookings.edu/research/redefining-global-cities/ (accessed 23/11/2016).

Ng, M.K. (2003) 'City profile: Shenzhen', *Cities* 20 (6): 429–441.

Ng, M.K. (2011) 'Strategic planning of China's first Special Economic Zone: Shenzhen City Master Plan (2010–2020)', *Planning Theory and Practice* 12 (4): 638–642.

Ortmann, S. (2015), 'The Umbrella Revolution and Hong Kong's protracted democratization process', *Asian Affairs* 46 (91): 32–50.

Planning Department, Government of Hong Kong SAR (2007) Hong Kong 2030, planning vision and strategy. Final report. www.pland.gov.hk/pland_en/p_study/comp_s/hk2030/eng/finalreport/ (accessed 17/11/2016).

Polo Sainz, I. (2016) 'Reform versus co-optation, again: Investigating the purpose of a new development area in Qianhai, Shenzhen'. MSc thesis, MSc Human Geography, Amsterdam University.

Reinstra, M. (2015) 'What sets them apart? Parents of border- crosser students, Shenzhen'. MSc thesis, MSc Human Geography, Amsterdam University.

South China Morning Post (hereafter *SCMP*) (2003) 'Twin-City concept for HK, Shenzhen'. www.scmp.com/article/424356/twin-city-concept-hk-shenzhen (accessed 22/11/2016).

SCMP (2013) 'Forgotten stories of the great escape to Hong Kong'. www.scmp.com/news/china/article/1126786/forgotten-stories-huge-escape-hong-kong (accessed 17/11/2016).

SCMP (2016) 'China gives the nod for Shenzhen-Hong Kong Stock-Connect to commence on December 5'. www.scmp.com/business/companies/article/2045194/china-gives-nod-shenzhen-hong-kong-stock-connect-commence (accessed 25/11/2016).

Shen, J. (2014) 'Not quite a twin city: Cross-boundary integration in Hong Kong and Shenzhen', *Habitat International* 42: 138–146.

Smart, A. & J. Smart (2008) 'Time-space punctuation: Hong Kong's border regime and limits on mobility', *Pacific Affairs* 81 (2): 175–193.

Taylor, P.J. & B. Derudder (2016) *World-City Network: a global urban analysis (second edition)*. Abingdon, Oxon/New York: Routledge.

Tsang, S. (2007) *A modern history of Hong Kong, 1842–1997*. London/New York: I.B. Tauris.

Vlassenrood, L. (ed.) (2016) *Shenzhen: from factory of the world to world city*. Rotterdam: nai010/ International New Town Institute.

Vogel, E.F. (1990) One step ahead in China: Guangdong under reform, Harvard University Press.

Watson, J. (2010) 'Forty years on the border: Hong Kong/China', *ASIA Network Exchange* 18 (1): 10–23.

Wu, F. (2015) *Planning for growth: Urban and regional planning in China*. New York/Abingdon, Oxon: Routledge (RTPI Library Series).

Xiyuan, L. (2016) 'Seeking identity: the trans-border lives of mainland Chinese families with children born in Hong Kong', *Global Networks* 16 (4): 437–452.

Yang, C. (2005) 'An emerging cross-boundary metropolis in China: Hong Kong and Shenzhen under "Two Systems"', *International Development Planning Review* 27 (2): 195–225.

Zacharias, J. & Tang, Y. (2010) 'Restructuring and Repositioning Shenzhen, China's new Megacity', *Progress in Planning* 73, 4, 209–49.

Zhang, X. (2015) 'From "workshop of the world" to emerging global city-region: Restructuring the Pearl River Delta in the advanced-services economy'. PhD thesis, Amsterdam University.

Part II

International twin cities

Until now, we have focused on borders of the municipal sort. We now shift attention to international ones and thus to cross-border twins. Here, we find far more positive expectations of twinned relationships, often embodied in formal written agreements. Such positivity often stems from their acquiring, even being allocated, cross-border functions, and even becoming allocated, the role of 'laboratories'; occasionally their respective nation-states give them a mandate and resources for being more active and for carrying out para-diplomacy. Over the next few chapters, we focus on twins straddling borders within the Americas, Asia and Europe. In some, we begin seeing the emergence of trans-border communities, raising interesting comparisons with the equally mixed loyalties of modern internal twin citizens.

10 Tabatinga, Leticia and Santa Rosa

Emergence, transformation and merging of paired and triple cities in the Amazon

Carlos Zárate Botía and Jorge Aponte Motta

The discussion of cross-border twin cities, in English at least, has rarely focused upon Latin-American examples. This chapter remedies this gap somewhat by exploring urbanization processes and trans-frontier interactions on the borders of Brazil, Colombia and Peru. Based on the investigation of local social, spatial and economic dynamics, it argues that residents of the Amazon cross-border urban continuum tend to identify themselves as part of cross-border communities – a greater or lesser feature of other cross-border twins in some of the chapters that follow; doubly so perhaps because this produces characteristic tensions with other higher levels of government.

Introduction

Currently, the Amazon border between Colombia, Brazil and Peru comprises a complex network of cities that include paired and triple settlements, belonging to a trans-border region with around 300,000 inhabitants, which vibrates and changes constantly in response to demographic, economic, social and cultural factors (Zárate et al. 2017, 51). The urban cross-border continuum of Tabatinga and Leticia, along with the little Peruvian village of Santa Rosa and other small settlements, are part of this. They gather slightly over 105,000 inhabitants according to the latest census data. Using a historical standpoint, this chapter aims to present the key moments in the emergence and transformation of this urban cross-border continuum.

Our pairs are unexceptional. In Latin America, particularly straddling national borders, there is a huge network of settlements. Many have emerged as paired, and even triple, border cities. Their existence and dynamics, indeed the border-world generally, are largely unknown to local and central governments and researchers placed in distant capital cities. However, in recent years, academic interest at least has begun changing, with increasing desire to understand what is happening in border cities in South America and in the Amazon.[1]

At the outset, we should explain that we do not find terms like 'twin' or 'mirror' cities helpful in understanding border-area settlements. The term seems to denote equality and similarity, whereas we see such settlements, at least in Latin American borders, as essentially dissimilar and asymmetric. Instead, following Buursink's definition (2001, 7), we prefer the notion of 'paired cities' denoting distinctive places located on either side of a boundary or dividing river between two states and dependent on that boundary for their existence, like those spanning the Amazon borders.

The national and transnational character of such settlements is a main trait of these areas. They bear indelible marks from complex historic, social, economic, environmental, political and cultural factors that affirm the singularity of the border environment. In addition, as Martínez (1994) noted, they have very particular characteristics like the fact that they are places of conflict, ethnic and international interactions accompanied by the sense of separation, otherness and exclusion expressed by their inhabitants.

Within this context, we focus on the cities of Leticia, Tabatinga and Santa Rosa. These urban settlements are examples of the making of paired and triple cities, very like other border cities in the Amazon and South America, for example, those of Asís, Bolpebra and Iñapari along the border of Brazil, Bolivia and Peru in the Amazon, and the cities of Puerto Iguazú, Foz de Iguaçu and Ciudad del Este near the famous Iguazú Falls between Argentina, Brazil and Paraguay.

The making of paired cities in the Amazon

Most current Amazon border cities originated from the dissolution of the Hispanic–Lusitanic colonial order, and the simultaneous emergence of new nation-states. This produced the division between the Brazilian and the Andean Amazon countries (Venezuela, Colombia, Ecuador, Peru and Bolivia) over a prolonged conflictual period from the late eighteenth to the mid-twentieth centuries when the borders of these nations finally stabilized.

The basic outlines of the current paired cities of Leticia and Tabatinga emerged from the treaties of Madrid (1750) and San Idelfonso (1777), and from work undertaken by the expeditions organized by the border mixed commissions[2] in charge of turning these treaties into reality on the ground (Zárate 2012). These imperial agreements sought to demarcate the Amazon space and regulate conflicts between Spain and Portugal over access, control and natural resources and labor in this vast jungle region. However, these delimitation attempts in the colonial period failed, leading to the dismantling of the expeditions. Nevertheless, a century later, such attempts resulted in the founding of paired border towns along the river boundaries of the new nation-states.

Nuestra Señora de Loreto de Ticunas was founded on the Amazon River in 1760 as a Spanish missionary village, while Tabatinga, 60 km downstream, was a Portuguese military fort established in 1766. This confirms that Portuguese colonial expansion was invariably supported by the military, whereas its Spanish counterpart was heavily supported by the church, in this case, the Jesuits (Zárate 2012, 32–33). *Loreto de Ticunas* marked the border on the

Spanish side, and depended on the missionaries ruling over the so-called *Maynas* missions from Quito.

By contrast, Tabatinga served as the border on the Portuguese side. There was a buffer zone between these two communities with no geopolitical control by either empire, which did not prevent it from being inhabited or transited, mainly by indigenous *Ticunas* people. This zone marked a non-delimited boundary between the two empires.

The Jesuits' expulsion from Portuguese domains in 1759 and from their Spanish counterparts in 1767 had important consequences for these border settlements, particularly the Spanish ones, due to their dependence on Catholic missionaries to guarantee their territorial dominance. Consequently, after 1767, the *Loreto de Ticunas* mission was reduced to almost nothing, leaving the Spanish side of the border unprotected from the opposite military forces placed in Tabatinga, its Portuguese counterpart.

At the end of the colonial period, these border towns almost disappeared. Their restoration occurred almost 100 years later in 1866, within the framework of the new agreements of delimitation and configuration of nation-states. Then Brazil and Peru marked the border at the mouth of the *San Antonio* creek. According to Peruvian official sources, it was located '2,410 meters from the Brazilian fort of Tabatinga' over the north band of Amazon river (Zárate 2012, 39).

These first delimitation attempts in the republican period had begun in 1851 when Brazil and Peru settled their Amazon border and river-navigation rights through a treaty of limitation and navigation, thereby relaunching the delimitation processes suspended at the end of colonial period. As a result, Brazil and Peru established a border binational commission and consequently, in 1866, the expedition of limits marked out the Apaporis–Tabatinga line taking advantage of the San Antonio creek.

A year later in 1867, this enabled Leticia's foundation as a border city built by the Peruvian government. Initially, Peru tried to build a fort called *Ramon Castilla*, whose name recalled the president who tried hardest to incorporate the Amazon into Peru in the mid-nineteenth century. This was devised as a counterweight to the Brazilian fort of Tabatinga. However, the building of this fort failed and the camp they established to make it became the border city, since known as Leticia (Zárate 2012, 40).

Nevertheless, we should note that these agreements were unrecognized by other Amazonian countries like Colombia and Ecuador, who also demanded rights over the riverine lands where Tabatinga and Leticia were placed, and, no less important, demanded direct access to the Amazon river waters.

The appearance of thousands of rubber gatherers employed by national and transnational extractive companies related to commercial and navigation ones, mainly from England, in these still-disputed border zones in the final decades of the nineteenth century,[3] seriously disrupted delimitation endeavors. They forced the weak state apparatus into becoming a mere spectator or even servant of the rubber entrepreneurs, who greatly benefitted from the lack of established boundaries, customs or taxes. This also explains why paired cities like Leticia and Tabatinga derived no benefit from their potential as border-checking and tax-collection points, and remained poverty-stricken witnesses to the trade in 'black gold' and its huge ships.

The Amazonian rubber crisis during the twentieth century's second decade gave states the opportunity to restart delimitation processes. This process was characterized by bilateral negotiations between Brazil and its neighbors in the Amazon region that, inasmuch as they were finalized, created new conflicts because they affected third countries' interests or simply because those countries were excluded. Colombia was no exception: under the 1922 Lozano-Salomón treaty, it acknowledged the Peru territories already granted to Ecuador under a 1916 treaty. This was the confirmation and continuation of bilateral negotiations initiated by Brazil and Peru under the agreement of 1851.

Under the 1922 agreement, Peru recognized Colombia's title over the so-called Amazonian Trapezium, the geometrically shaped territory that allows this last country access to the Amazon River and includes the city of Leticia (see Figure 10.1 upper left box). The Congress of Peru and Colombia finally approved the agreement in 1925 and 1927, respectively, and then proceeded to create a mixed commission by 1928 in charge of delimitations in situ. The task ended in 1929 allowing the Trapezium's delivery to Colombia in August 1930.

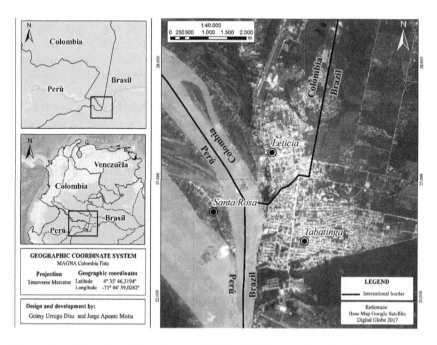

Figure 10.1 Map of triple border cities (1930–1950) by Angela López Urrego, Carlos Zárate and Jorge Aponte Motta.

However, this did not produce the expected resolution of border contro-versies between the two countries. Instead, it intensified them, producing the 1932–33 armed conflict. Many Peruvians from the Loreto region, espe-cially those involved in big rubber business and leaders of the regional state power established in Iquitos, Loreto's capital, the Peruvian Amazon region, did not agree with the delivery of Leticia and Trapezium to Colombia and decided upon its recovery.

Conflict began on 1 September 1932 when a group of civilians and troops from Loreto took Leticia and overthrew the Colombian civilian authorities. In reaction, the Colombian president decided to contest this Peruvian af-front. Then, in a hasty and disorganized way, he mobilized the Colombian army, and a full-scale three-month long war started five months later. The struggle between two armies lasted from February 15 to May 25, 1933, when a ceasefire agreement was signed in Geneva by the Society of Nations, to which both countries belonged, after many controversies and disappoint-ments. That Society created a mixed commission, established in Leticia for a year to administer the Trapezium in the name of Colombia. Once Colom-bia resumed control of the Trapezium, in 1934 the Amazonian border space was finally demarcated and reorganized, continuing with few changes until the present.

Since then, Colombia has officially become the third country in the Amazon basin with direct access to the Amazon River, alongside Brazil and Peru. Then, the Peruvian government decided to transfer its civilian and military inhabitants in Leticia to the south side of the Amazon River, founding – with nostalgia for a former glorious past – the settlement of Ramón Castilla. Later, this became the village of Santa Rosa.

Leticia, Tabatinga and El Marco: border resettlements in the post-war era

When Colombia first received the Amazonian Trapezium in August 1930, Leticia was just a hamlet of barely 24 'miserly-built' houses with about 150 inhabitants. This was fewer than the numbers living in *Hacienda La Victoria*, about 20 km up-river and lodging 477 people (Abdón Villareal 1930, 332–335). These were mostly indigenous agricultural workers dedi-cated to wood-processing as well as growing, cutting and converting sugar cane into alcoholic beverages. From a demographic and geographical viewpoint, this meant that *Hacienda La Victoria* was more important than Leticia, the capital city of the *Comisaria del Amazonas*, the new territorial entity the Colombian government created provisionally to administer the territories granted by Peru.

On the other hand, Tabatinga's situation on Brazil's side of the border was not economically or demographically very different from Leticia's since its inhabitants mostly belonged to a border military garrison. Nevertheless, a few dozen inhabitants congregated in El Marco, a small settlement on

Figure 10.2 Map of Leticia–El Marco–Tabatinga (1930–1950) by Geimy Urrego Díaz and Jorge Aponte Motta.

the Brazilian side of the San Antonio Creek, equidistant between Tabatinga and Leticia (see Figure 10.2). El Marco pinpointed the new border between Columbia and Brazil fixed by an agreement in 1928. These inhabitants consisted of riverbank, indigenous and *caboclos* people, surviving by selling fish, other foods and supplies to Tabatinga's military fort and providing manpower to build Leticia from the moment it became a Colombian city (Picón 2010, 45).

According to demographic data available on the studied area, by 1934, a year after Leticia's takeover and the end of the war, the relative proportions of the three nationalities, at least on the Colombian side, were starting to change and continued so doing. Leticia then had 402 inhabitants compared to *La Victoria*'s 496. By 1937, 1,343 people were living in Leticia, of whom 738 – the majority for the first time – were Colombian, alongside 328 Brazilians and 277 Peruvians. During that year, only 180 lived in *La Victoria* (Convers Pinzón 1937, 58). This shows how Colombia had managed to insert a majority of its nationals and, at the same time and more importantly, how Leticia was already becoming a border tri-national community, not just a binational one.

The relatively numerous presence of Brazilians and Peruvians in Colombian Leticia during these early years primarily resulted from its becoming a city of workers in a region without employment sources. Their arrival resulted from the construction of local, regional or national government infrastructure, plus the influx of Colombian civil and military institutions and the arrival of civil servants and their families to meet their needs.

Consequently, Leticia had started growing by the late 1930s, reinforced by investment from the Colombian government, which conducted the first proper urban delimitation process and other important construction works. These included the military headquarters, the first school, a customs building, a military hospital, plus market stalls for Peruvian, Jewish, Lebanese, some Colombian and other European merchants, who founded settlements and trading houses in the late 1930s and early 1940s (Picón 2010, 59).

By contrast, even in the late 1950s, Tabatinga remained a small, shabby military post, far from the border and attached only to the afore-mentioned El Marco. Its precarious situation justified Brazilian military complaints comparing it unfavorably with the construction of military facilities and the increase in Colombian army officers in Leticia. Captain-lieutenant Aécio Pereira Souza, officer-in-charge of patrolling and providing logistic support to border patrols, remarked:

> The harsh situation of Tabatinga's platoon was even more distressing because they were stationed right next to the Colombian city of Leticia, a modern and well-designed city with public utility services, electricity, banks, businesses, etc., military base of the Colombian Military Command of the Amazon with more than 1,000 troops, with a naval base with three modern gunboats recently built in England, and one air base operating large aircraft.
>
> (de Souza 2000; Cited By Steiman 2002, 88)

By then, the defensive character and dominance of military establishments in both border communities was evident. Civilians were less important, notwithstanding the early development of equipment, infrastructure and urban public utilities. Several kilometers of creeks, swamps, paddocks (in the rainy season becoming mud), plus the settlement of El Marco, then separated Tabatinga and Leticia.

By the 1960s, that settlement, due to its location half way between Leticia and Tabatinga, and because of the laboring and productive activities of its inhabitants, serving both cities, facilitated the beginning of the spatial fusion of the border cities of Leticia and Tabatinga. It also produced changes in the inhabitants' communication and perceptions of the border, now seen as one co-joining trans-border place.

The different places of commerce and leisure located in El Marco, together with the activities developed through them, increased cross-border mobility along pedestrian trails, as well as the occasional actions carried out by both states to 'control the border' with border patrols. That allowed this settlement to become a crucial communication point for Colombian and Brazilian people from both border cities (Aponte Motta, 2012). So, while Leticia and Tabatinga symbolizes the military, religious and civil urban spaces of the Colombian and Brazilian nation-states, El Marco expresses the space created by the trans-border relations of the people who cross it in their everyday lives, rendering it a border regime beyond conventional structures of sovereignty.

At a military/political level, efforts made by Brazilian and Colombian local authorities facilitated the regulation of the fluxes of people and goods between the border cities of Tabatinga and Leticia and the settlement of El Marco. Accounts by Brazilian witnesses suggest that, in the early 1950s, the Brazilian army started road construction linking the military garrison and El Marco to 'control the Colombian border' (Personal interviews, February 2008, with Luiz Altaíde, a political leader of El Marco in the 1970s). This was strongly related to government policies from 1955, creating military colonies to reinforce and protect border settlements (Steiman, 2002, 23).

Almost simultaneously, in 1956, Colombia began constructing the road to Brazil, in addition to the former and lonely trail named *Caminito de Didimo* (Picón 2010, 64) (see map 1). This new road connected Leticia to Brazil, 200 metres above El Marco. It resulted from the policy of the Colombian military regime of Gustavo Rojas Pinilla to build border infrastructure. Constructing these roads stimulated communication between the cities and promoted the territorial occupation of the borderland. So, we see the birth of two main socio-spatial results: first, the emergence of an urban continuum embracing the border cities of Leticia, Tabatinga and the settlement of El Marco, and second, the progressive abandonment of the Amazon river as a main communication between Leticia, El Marco and Tabatinga people.

Leticia and Tabatinga: an unplanned urban cross-border continuum

Between the 1970s and 1990s, the dynamics of trans-border articulation and the development of the three settlements of Tabatinga, El Marco and Leticia, along with Santa Rosa (former Ramon Castilla), underwent decisive and profound change. This resulted from the intrusion of external factors: macroeconomic forces, and regional and even global policies.

One economic force leading this change was undoubtedly the trading of jungle fauna and fur and their massive export to the USA. This was rapidly followed by the irruption into this border region of businesses associated with the illegal drug trade, mainly the processing and transport of cocaine across the Amazon River and on to Europe and the USA. These activities permeated and greatly stimulated the border economy. Moreover, in those years, Brazil, Colombia and Peru, among other South American countries, undertook political and economic changes to modernize and decentralize their states. This coincided with the fall of governments and military dictatorships, and for Colombia with the reform of its centenary constitution, updating by unfinished attempts of decentralizing its administration and territorial order.

Brazil and Colombia thus became simultaneously but unco-ordinatedly interested in undertaking physical adaptations and investments in the border. Thus, the former road connecting Tabatinga's military fort to El Marco started becoming, on the Brazilian side, a modern double-lane highway with a tree-decorated median strip and sidewalks which in 1988 was named *Avenida da Amizade* (Friendship Avenue). In 1976, the Colombian government has already started a similar, if more modestly laned, endeavor: the construction of the *Autopista Internacional* (International Highway) to connect Leticia to the Brazilian border (El Leticiano 1976, 8). Both avenues finally connected up and became the main corridor for this border conurbation, which runs parallel to the Amazon river, symbolizing the desire that earlier rhetoric about sovereignty and border control could ultimately give way to border integration.

For the first time, the Amazon River no longer constituted the main border corridor: the cities of Tabatinga and Leticia focused attention upon, and displaced, border life towards *Avenida Internacional* on the Colombian side or *da Amizade* on the Brazilian. Thus, Brazil started locating its main public and business entities like supermarkets along, or close to, *Avenida da Amizade*, thus making it a political, administrative and business center. Colombia behaved similarly in Leticia: drawing in commercial businesses, while slowly changing or relocating leisure facilities like bars, nightclubs and brothels, established too close to the Brazilian border, along *Avenida Internacional*.

Finally, these two cities met at the border through avenues and neighborhoods that had been located beside the international boundary during the 1970s and 1980s. From this moment, they began to grow in disorderly ways in different directions. Leticia in the 1990s grew outwards to the north, following the international borderline. Tabatinga did so to the east, away from the border, a movement that intensified during the 2000s.

This growth was driven by rapid increases in the number of inhabitants in both cities. According to the projections of the Brazilian census of 2010, Tabatinga had 63,635 inhabitants in 2017 (*Instituto Brasileiro de Geografia e Estadistica* 2010). Leticia, according to the projections in the Colombian census of 2005 had 41,957 for 2017 (Departamento Nacional de Estadistica 2005). Meanwhile, the Santa Rosa Island Minor Populated Center had 776 inhabitants in 2013 (Fondo de Inversión en Telecomunicaciones 2013). The other engine of urbanization was the emerging action of urban agents involved in the appropriation and the market of land, accompanied by erratic urban policies in both cities. These elements have continued to put pressure on the growth of cities, generating new dynamics around the border (Figure 10.3).

In the mid-2010s, Tabatinga began building houses in rural areas to the east and northeast of the city, coinciding these last with the international limit. This meant that the international border again became the most dynamic areas of urban expansion in the north ends of both cities (Aponte Motta 2017). This is generating complex cross-border dynamics yet to be studied.

This situation has occurred within the framework of intense limitations on urban expansion in both cities. In Leticia, the restriction is marked by the international limit with Brazil to the south and east of the city. To the west, it is prevented by the Amazon River. Thus, the only possible direction of growth is towards the north, where military areas and indigenous lands severely limit the city's expansion. This situation has pushed Leticia to expand its urban perimeter, allowing it to build new neighborhoods and housing at the north end of the city along the Brazilian border.

Meanwhile, Tabatinga also has very limited expansion possibilities. To the west, there is the Amazon River and, to the south, military and indigenous areas. To the north lies the part already urbanized against the border with Colombia, while to the east and northeast there are lands managed by the *Instituto Nacional de Colonização e Reforma Agraria*, INCRA (National Institute of Colonization and Agrarian Reform), a federal institution that manages rural areas and assigns lands for agricultural uses. Therefore,

Figure 10.3 The Columbian border post between Leticia and Tabatinga by Jorge Aponte Motta (2017).

Tabatinga's formal territory is almost full. That is why it is growing up outside its municipal boundary, in the northeast area, inside the lands administered by INCRA.

These northern extra-border areas of both cities over the borderline attract various urban agents. Some are speculative land-buyers who make investments in the urban land market of both cities. They buy parcels of land and await the future legalization of their use for urban purposes, and consequential increments in land prices. Others are various low-income inhabitants looking for better lives – new regional migrants from the three countries who originally lived in wild settlements distant from Leticia and Tabatinga.

On the other hand, many of the border inhabitants between Brazil and Colombia live along the San Antonio creek, over the borderline, occupying flood-prone lands with wood stilt houses in irregular settlements. Close to the mouth of the creek on the Amazon River, there are other settlements that comprise floating houses tied to the land by ropes. Such 'houses serve as parking places for boats from the three countries' (Vergel 2006, 40) and as places to sell fuels, to buy or sell fish, or as small food markets or as points for boarding and landing boats.

At the north of San Antonio Creek in another point along or on the joint border between the cities, inhabitants live in more formal settlements located in lands that are not flood-prone. Here are brick houses, organized neighborhoods with paved streets, urban services and facilities. However, at the most northern area of the border, self-built wood houses emerge again, mixed with plots of non-urbanized land (Aponte Motta 2017). That suggests a particular dynamic of the land market, whereby different agents take advantage of the possible price increment of the urban lands, mixed with the needs of low-income inhabitants to find a place to live inside these rapidly growing cities. Such inhabitants occupy the cheapest or non-legally usable lands, like the ones adjacent to the borderline, or flood-prone lands as Vergel (2006) and Aponte Motta (2017) have shown.

Among the economic trends that, in the immediate future, will continue dynamizing the urban transformation particularly in Leticia, though also (to a lesser extent) in Tabatinga and Santa Rosa, is tourism. In just over a decade, tourists have increased tenfold from 5,215 visitors per year in 2000 (Banco de la República de Colombia 2001, 31) to 46,195 in 2013 (Banco de la República de Colombia and Departamento Nacional de Estadística DANE 2014, 42). This increased not only flights and travel agencies, but also hotel infrastructure, restaurants, transports and other related businesses.

Overall, tourism is becoming the key economic axis of the trinational border. Recent investments expanding and modernizing Leticia and Tabatinga international airports will support an intense expansion of this sector. Nevertheless, it is unclear how the cities are going to incorporate these major new elements into their traditional dynamics, alongside their increasing traditional floating population. It is also unclear whether local administrations will find ways to co-ordinate their plans and public agendas, related to demographic, economic or environmental urban tourist effects across the border. Moreover, we should remember that transnational companies devoted to tourism and travel, particularly international airlines and large hotels, capture most benefits of this activity by controlling the flows of national and foreign visitors. Therefore, to positively impact local economies in the tri-national Amazon region, touristic development requires special plans and rules ensuring the sharing and co-ordination of its management on both or three border sides through supervision by authorities and civil society. Unfortunately, none of this is really evident.

Conclusion

We have shown why and how these border cities emerged and became located in the triple Amazonian border between Colombia, Brazil and Peru. They resulted from interactions between national political and economic processes alongside local ones developed in the more peripheral spaces of the three nations, where past and present state weaknesses are evident and uncontested. Undoubtedly, these cities bear the marks of the particularities of the nations involved in border interactions and of the various generations of people who have arrived in recent decades mixed with the original populations or indigenous peoples. We do not forget that the emergence of this special urban border continuum should be explained furthermore according to cultural, riverine and Amazonian rainforest conditions and interactions.

We have also shown how, between Leticia and Tabatinga, a transborder urban continuum formed. This is today in the midst of intense transformation. These cities are growing across formerly rural lands in the context of various jurisdictional and territorial tensions and pressures, which are changing the urban expression of the cities and generating new forms of relationship through the border. Furthermore, it is evident that Leticia and Tabatinga have two simultaneous but wholly unco-ordinated neighborhood construction processes converging along the borderline, subject to different and incompatible regulatory regimes not just about borders but also urban planning standards.

The growth and configuration of this urban continuum in future decades will likely maintain the chaotic dynamics organizing urban space and the urban land market. This will affect housing availability for the growing population. Tourism and its impact will probably increase due to the expansion of the two cities' international airports. But it seems unlikely that the co-operation agreements or local policies will respond to the local situation. Neither national politicians of the two countries nor local governors seem interested in finding ways to solve the needs of border inhabitants, although they are very sensitive to the desires and trans-border activities of real-estate capital.

Without doubt, these cities have potential to lead the border integration process proclaimed by politicians and economic leaders in the three nations. However, the current reality contains no indication that the growth and transformation of these border cities will start being planned anytime soon. Although Tabatinga and Leticia both have territorial municipal plans, these are not designed to interact with each other, nor to organize the territory in ways taking account of the border situations and interactions. So, even if those plans are implemented one day, they will probably prove unsuitable to the trans-border context. If this continues, inter-government co-operation and inter-municipal partnership between these two cities and their little Peruvian neighbor will remain asymmetric and disordered, or at best, just spontaneous and occasional. The fate of the urban cross-border continuum, along with the lives and fortunes of its inhabitants, may well remain uncertain and accident-prone.

Notes

1 Excluding the immense current bibliography concerning the border of the USA and its neighbors as well as the cases inside Europe, we can mention that there is disagreement about the use of notions like 'twin', 'pair' or 'mirror' cities in Latin America. Authors active in this discussion include Carrión and Espín (2011), Dilla (2008), Valero Martínez (2002), Vergel (2006) and Zárate (2012).
2 The delimitation commissions refer to the diplomatic and technical agents who define and mark the limits or borderlines, in situ, through expeditions.
3 Rubber exploitation, for European markets to support the increasingly growing car industry, originated by 1830 near the mouth of the Amazon River and, by the end of the nineteenth century, had covered the entire Amazon River system (Hemming 1995).

References

Abdón Villareal 1930. *Informe al Ministerio de Gobierno.* Comisaría Especial del Amazonias, Leticia. Archivo General de la Nación. Fondo Ministerio de Gobierno, 1931. f. 332–335. Bogotá.

Aponte Motta, J. 2012. Comercio y ocio en la transformación del espacio urbano fronterizo de Leticia y Tabatinga, in: Zárate, C. (Ed.), *Espacios Urbanos y Sociedades Transfronterizas En La Amazonia.* Universidad Nacional de Colombia Sede Amazonia, Leticia, Amazonas, CO, 205–235.

Aponte Motta, J. 2017. *Leticia y Tabatinga. Construcción de un espacio urbano fronterizo. Hacia una geohistoria urbana de la Amazonia.* (Tesis Doctorado en Geografía). Universidad Autónoma de Madrid, Madrid.

Banco de la República de Colombia 2001. *Informe de Coyuntura Económica Regional del Amazonas*, IV trimestre de 2000. Banco de la República, Bogota.

Banco de la República de Colombia, Departamento Nacional de Estadística-DANE 2014. *Informe de Coyuntura Económica Regional Departamento del Amazonas.* Departamento Nacional de Estadística/ Banco de la República de Colombia, Bogotá, CO.

Buursink, J. 2001. The binational reality of border-crossing cities. *GeoJournal* 54, 7–19.

Carrión, F., & Espín, J. 2011. *Relaciones fronterizas: encuentros y conflictos.* Flacso-Sede, Quito Ecuador.

Convers Pinzón, R. 1937. El Trapecio Amazónico colombiano en 1937. *Boletín de la Sociedad Geográfica de Colombia* 4(1), 54-64.

de Souza, A.P. 2000. Fronteiras da Amazônia: uma guerra silenciosa. Rio de Janeiro: Razão Cultural.

Departamento Nacional de Estadistica, 2005. *Estimación y proyección de población nacional, departamental y municipal total por área 1985–2020.* [WWW Document]. URL www.dane.gov.co/censo/files/cuadros%20censo%202005.xls (accessed 10/08/14).

Dilla, H. (Ed.) 2008. *Ciudades en la Frontera: Aproximaciones críticas a los complejos urbanos transfronterizos.* Manati, Santo Domingo.

El Leticiano 1976. *Rescate Urbano de La Autopista.* El Leticiano 37, 8. Archivo Biblioteca Banco de la República. Leticia.

Fondo de Inversión en Telecomunicaciones 2013. *Localidades incluidas en proyectos de telecomunicaciones región Loreto 2013.* [WWW Document]. URL www.fitel. gob.pe/archivos/FI51ae409a0060f.xls (accessed 30/05/16).

Hemming, J. 1995. *Amazon frontier. The defeat of the Brasilian Indians.* Papermac, London.

Instituto Brasileiro de Geografia e Estadistica 2010. *Informaciones completas. Tabatinga.* Censo 2010 [WWW Document]. @cidades. URL https://cidades.ibge. gov.br/brasil/am/tabatinga/panorama (accessed 14/01/18).

Martínez, O. 1994. *Border people: Life and society in the US- Mexico borderlands.* University of Arizona Press, Tucson-London.

Picón, J. 2010. *Transformación urbana de Leticia. Énfasis en el periodo 1950–1960.* Gente Nueva, Leticia.

Steiman, R. 2002. *A geografia das cidades de fronteira: Um estudo de caso de Tabatinga (Brasil) e Letícia (Colômbia)* (Pós-Graduaçao em Geografia). Universidade federal do Río de Janeiro, Rio de Janeiro.

Valero Martínez, M. 2002. *Las fronteras como espacios de integración.* Universidad de los Andes, Caracas.

Vergel, E. 2006, *Twin Cities in Amazonian Transnational Borders, an Appropriate Cross Border Approach for Squatter Settlements on flood pronelands located on border's fringe: The Case Study of Leticia and Tabatinga* (Master in Urban Managment), University of Rotterdam, Rotterdam.

Zárate, C. 2012. Ciudades pares en la frontera amazónica colonial y republicana, in: Zárate, C. (Ed.), *Espacios Urbanos y Sociedades Transfronterizas En La Amazonia.* Universidad Nacional de Colombia Sede Amazonia, Leticia, 21–44.

Zárate, C., Aponte, J., & Victorino, N. 2017. *Perfil de una región transfronteriza en la Amazonia.* La posible integración de las políticas de frontera de Brasil, Colombia y Perú. Universidad Nacional de Colombia, Leticia.

11 The resiliency of Los Dos Laredos

John C. Kilburn and Sara A. Buentello

In this chapter, we move northwards to the US–Mexican border, but stay with the issue of cross-border identity and active trans-boundary networks of people. By exploring the flows of people and commodities across the border, as well as the use of services and languages on both sides of the border, the authors conclude that a specific pattern of interactions and behavior, and indeed, a vibrant cross-border economy has been created over time in 'Los Dos Laredos'. Again, we come across tensions between these border communities and the changing agendas of higher levels of government.

Traditionally, Laredo, Texas, USA, and Nuevo Laredo, Tamaulipas, Mexico, have existed harmoniously for over a century. Divided by the Rio Grande River, the four international bridges are only a short 320-m walk to cross. They serve as entryways into and out of the USA and Mexico. With a combined population of over 600,000, 'Los Dos Laredos' have survived multiple setbacks jointly and separately, but nevertheless affecting each in some way – most notably the closing of the Laredo Air Force Base in 1973, and the rise of drug cartel-related violence in the last 12 years. Despite this, the twin cities continue trading and people continue taking advantage of easy crossings to enjoy the best of each side, as will become evident in the pages that follow.

Of the 57 ports between Mexico and the USA, the Laredo–Nuevo Laredo crossing handles the most commerce. The relative ease of crossing the Rio Grande in Laredo is facilitated by the infrastructure of one hybrid pedestrian-automotive bridge, three additional motor-vehicle bridges and one dedicated rail bridge. This ensures that commerce continues flowing across the border notwithstanding regional economic and political challenges. Interstate 35, from Mexico, covers more than 2,500 km through major US cities. At the Mexican border, Interstate 35 becomes Mexico's Federal Highway 85, traveling through major population centers like Nuevo Laredo, Monterrey, Ciudad Victoria, Pachuca and the southern edge of Mexico City. Over time, the cities' relationship has changed. Informal interaction and shopping in the respective downtowns have significantly decreased, while the port has actively grown with international trade.

Contemporary issues around restricting immigration with Mexico, renegotiating trade deals and building a wall crossing the entire US–Mexican border have been under constant attention. At Donald Trump's official campaign announcement, he stated,

> When Mexico sends its people, they're not sending their best. They're not sending you. They're not sending you (sic). They're sending people that have lots of problems, and they're bringing those problems with us. They're bringing drugs. They're bringing crime. They're rapists.
>
> (Washington Post, 2015)

This inflammatory language is unusual given the customary discourse of international diplomacy, but the intense accusations are clearly divisive. In the first week of his presidency, Trump tweeted that the North American Free Trade Agreement (NAFTA) was unfair and one-sided in favor of Mexico, leading to repudiation, renegotiation and potential threats of placing a sunset provision on the entire agreement (Swanson, 2017).

> The U.S. has a 60 billion dollar trade deficit with Mexico. It has been a one-sided deal from the beginning of NAFTA with massive numbers of jobs and companies lost. If Mexico is unwilling to pay for the badly-needed wall, then it would be better to cancel the upcoming meeting.
>
> (Abdullah, Gamboa and Alexander, 2017)

Additionally, as of this writing, Trump continues insisting that his campaign promise of building a wall across the US–Mexican border will come true, and Mexico will pay. This contention produces serious concerns from both nations, threatening to hinder commercial trade and interpersonal interaction in a region with a long history of operating as twin cities with common interests. Additionally, NAFTA is being renegotiated.

Texas has a long border with Mexico, and Laredo's four crossings represent only a small number of the total number from the USA to Mexico (Table 11.1).

Table 11.1 Mexican–USA border crossings

State	Vehicle / pedestrian	Rail	Border length (miles)
California	9	2	140.4
Arizona	9	1	372.5
New Mexico	3	0	179.5
Texas	28	5	1,241.0

Source: US Department of Transportation, Bureau of Transportation www.bts.gov/content/border-crossingentry-data (accessed 20 February 2018).

All the areas around the crossing ports face similar issues and have some interdependence with the city across the Rio Grande. They have their local, state and federal governmental relationships to build on. All the land ports must cooperate with each other due to the many people crossing and sharing Rio-Grande water. However, the various border-crossing areas are unique. This chapter argues that Laredo and Nuevo Laredo have the strongest claim to be called twin cities because this is where the two economies and peoples are most enmeshed along the US–Mexican border.

US and Mexican governmental authorities must cooperate and coexist. Just one example is the sharing of the Rio Grande as the primary source of drinking water for both US and Mexican communities. The bi-national *International Boundary and Water Commission* regulates boundary demarcation, national ownership of waters, sanitation, water quality and flood control in the border region. Sharing resources, monitoring and information is essential to ensure that the local supply is safe.

There are several locations where people may cross from Mexico to Texas, and goods and services flow. All have trading importance. The western crossing of the Rio Grande is led by the City of El Paso. While having the largest population among border cities (US Census estimate: 835,593 in 2015), its diverse economy and the presence of Fort Bliss (a large military base) makes the region less dependent on international trade. Additionally, there are few geographic advantages to building a land port in El Paso because it is not located near the major Mexican population and manufacturing centers. Looking south along the Rio Grande, the crossings at Presidio, Del Rio and Eagle Pass are very small due to lack of population on both sides. South of the Laredo crossing, there is 'the Rio Grande Valley'. While the metropolitan region boasts over one million residents, the valley actually consists of various small cities with their own governmental systems.

Meanwhile, Laredo is connected to Interstate 35 permitting rapid transportation of goods. This is frequently referred to as the NAFTA Corridor (see map 1). While the Laredo crossing has been active for hundreds of years, significant expenditures accompanied the passage of NAFTA. This greatly reduced tariffs and regulation on goods imported and exported between the USA, Mexico and Canada. According to a recent Texas Department of Transportation Border Corridors and Trade Report (2015), Laredo's World Trade Bridge alone accounts for more truck border-crossings than the next three bridges combined. Laredo's rail bridge also sees more traffic than rail bridges elsewhere. In 2015, $204.43 billion worth of trade crossed into Mexico through Laredo (approximately $92.74 billion from the USA to Mexico and $112.65 from Mexico to the USA). Nearly all border trade (96.84%) is with Mexico. While there are 57 US–Mexican Entry Ports, Laredo accounts for 36.9% of the total trade and 21.96% of the total of Mexican International Trade (Schaffler, 2017). The automotive industry is highly dependent on US–Mexican trade (McBride and Sergie, 2016), and one of the most significant forms of trade comprises automotive parts exported to Mexico and motor vehicles entering the USA from Mexico. In spite of a cross-border effect risking delays and disruptions, supply-chain management continues to be a common way to produce goods between the USA and Mexico (Cedillo-Campos et al. 2014). Thus, trade continues growing notwithstanding concerns about drug cartel violence (Kilburn, San Miguel and Kwak, 2013). Of course, trade alone does not make Los Dos Laredos twin cities. The Laredo–Nuevo Laredo region emerged over hundreds of years. It represents one of the narrower parts of the Rio Grande, and makes crossing easier. So, alongside commerce in both directions, Aguilar (2015) reports that, since 1978, about 45% of business for Laredo retailers comes from Mexican shoppers (McAllen has about 35%, Brownsville 25% and El Paso 15%). In context, shoppers from Tijuana and Baja California spend approximately $4.5 billion annually (Guerrero, 2017) among the nearly $450 billion dollars of annual retail sales (San Diego Source, 2017). Recent US government threats to withdraw from or renegotiate NAFTA have produced significant concerns regarding the rebuilding of trade agreements, contracts between international corporations, intergovernmental partnerships around security and crime, indeed simply crossing for shopping and tourism. Long-standing border relationships among local environments like border communities are now being restructured by the two national governments.

Road and rail transportation from the USA through Mexico to Central America is most efficient through Los Dos Laredos. Currently, they are home to 510 freight-forwarding companies, 210 trucking companies, and 105 US Customs Brokers, plus one rail bridge and four vehicle bridges for commercial and/or private vehicle traffic (Laredo Development Foundation). Over 1.5 million loaded trucks crossed the bridges northbound and nearly 1.4 million crossed the bridge southbound; so did over one million bus passengers and more than five million personal vehicles with over ten million passengers. Additionally, nearly 3.5 million foot passengers crossed the border. With such substantial cross-border traffic, this region has benefitted greatly from the rapid overall trade increase between the USA and Mexico. In terms of unadjusted US dollars, US–Mexican trade via the Port of Laredo rose from $19.4 billion in 1994 to $192.7 billion in 2014.

At one time, the only barrier to crossing into Laredo from Nuevo Laredo was the river and its bridge. Little scrutiny occurred; almost nobody was denied entry or exit on either side. Relationships between both cities and residents were familial. This was challenged after 11 September 2001. U S border agencies became hyper-vigilant after the World Trade Center attacks, and many more commercial and passenger vehicles were subject to extensive searches in the name of counterterrorism (Kilburn et al., 2011). The rhetoric of building a substantial wall spanning the entire US–Mexican border was greatly elevated throughout Donald Trump's 2016 US Presidential campaign. Casual pedestrian bridge-crossers also experienced longer processing and searches. These additional precautions could have discouraged people from crossing to the USA, but traffic has continued steadily.

Over the past decade, the drug wars have dramatically impacted both economies. Garza (2008) describes how US warnings, fear of crime and narco-terrorism have lowered American tourism. In consequence, in 2006, Nuevo Laredo shut down 700 businesses expunging 3,000 jobs. This has resulted in more property crimes like burglary, further harming business. This in turn produced lower tax revenue for local government and the loss of 200 governmental jobs. All these issues create the impression that Nuevo Laredo is unsafe.

Rios and Shirk (2011) claim that 41,648 homicides from December 2006 to June 2010 are attributable to Mexican drug cartels. While many homicides occurred elsewhere throughout Mexico, Los Zetas of Laredo, known as one of the cartels fighting for dominance in the region, were targeted in the violence and credited with many of the murders. Fear of crime in the region altered many behaviors; however, there are still many people who will cross into Nuevo Laredo despite the warnings (Kilburn et al., 2013). In addition to work, many Laredoans travel to Nuevo Laredo to visit family, to shop, to eat or to receive health care.

One of the challenges this region faced was the boom-and-bust cycle around the building and subsequent closure of the Laredo Air Force Base. Laredo has had a long military affinity beginning with the establishment of Fort McIntosh in 1849. According to Thompson and Juarez (n.d.), Laredoans first became familiar with 'aero planes' in 1914. The Wright Brothers' biplane called 'Scout' was first tested at Fort McIntosh. The plane, which succeeded in its initial maiden flight to Eagle Pass, nosedived into the Rio Grande on its return flight two days later. During World War I, small landing fields were needed in South Texas as Kelly Air Force Base in San Antonio reached capacity. Due to the warm climate, Laredo was chosen as a site for a small landing field. In 1940, this was converted into an airport. In 1942, with World War II in full swing, the government expanded the airport from 320 to 2,085 acres, and established an air-corps gunnery school. By 1945, the small landing field had grown into what was then known as Laredo Army Air Base. It quickly became a hub for military training that included, for the first time in history, training women to fly military aircraft. These women were engineering test pilots of B-26, P-40, and P-63 aircraft (Thompson and Juarez, n.d.).

After a long shutdown, the government reactivated the base in 1952, renaming it Laredo Air Force Base, providing basic training to American and foreign pilots. By 1973, 9,000 pilots had been trained and the base employed 450 civilian workers. While the base expanded, bars and restaurants, clothing stores, car and motorcycle dealerships, hardware stores and other businesses sprouted. Residents discuss the 'glory days' of crossing from the base to Nuevo Laredo for a great night out or to shop for jewelry and other bargains across the river. Much of the labor in building Laredo businesses came from Mexicans.

When the Laredo Air Force Base closed in 1973, nearly all civilian workers lost their jobs; the economic impact on the city was devastating. However, as air base property reverted to the city, its officials began looking for ways to stimulate the local economy by offering airport properties for economic ventures. By so doing, the city created an influx in the import–export business. Acreage of military property was distributed to the US Postal Service, the Department of the Interior, the National Guard, the Texas Department of Transportation, the Laredo Municipal Housing Authority, the City of Laredo and Webb County. Flynn Investments purchased the base housing units and the Catholic Diocese purchased the non-denominational church property (Thompson and Juarez, n.d.).

In February of 1975, the air base was conveyed to the city and became the Laredo International Airport, now controlling the landing strips and offering daily domestic and international flights. While the nature of the relationship changed from military personnel purchasing goods and services in Nuevo Laredo to one of importing and exporting goods, Los Dos Laredos began cooperating on joint commerce more than ever before.

Throughout the 1980s, much of this region's fate rested on the local oil boom-and-bust cycle. However, the local economy changed substantially in 1994 when President Bill Clinton signed into law the NAFTA between the USA, Mexico and Canada. It eliminated tariffs on imported and exported goods crossing these borders, and with no labor standards applying, large US companies moved their manufacturing plants to Mexico, where products were produced at substantially lower costs. While there have been numerous criticisms of NAFTA harming the overall US and Mexican economies (Esquivel and Rodriguez-Lopez, 2003; Kletzer, 2008; Scott, 2011), it brought immediate benefit to Laredo and the lower South Texas region. The communities went from offering only retail jobs to offering more lucrative transportation, logistic warehousing and distribution jobs. Laredo's population went from 72,000 to 250,000 as job seekers flooded the region seeking work. According to a Texas Department of Economic Development report, Texas is the principal conduit of US trade with Mexico. Over 70% of all US exports to Mexico are handled by Texas ports, bridges and airports. The Laredo border-crossing processes well over one-third of all US exports to Mexico. While oil prices and the state's per capita income fell in the 1980s, the 1990s brought highly successful international trade with NAFTA.

As the foregoing suggests, the relationship is not uni-directional, involving just Mexican dependence on the USA. There is strong interdependence. Nowhere is this more evident than in health care. Many Americans, particularly women, cross to Nuevo Laredo in this connection. Over 40% of residents on the US side of Los Dos Laredos cross the border for this purpose (Kilburn & Leon, 2012; Landeck and Garza, 2003). None reported having health insurance and, due to American medical-care costs, obtaining medical services in the USA is impossible (Uribe-Leon, 2016). And a doctor visit in Nuevo Laredo currently costs only around 35 pesos ($2.50 US). Patients can see a physician, and receive medications usually in a pharmacy right next door to the doctor's office. Medical care is quick and easy without an appointment or long wait. In a recent paper on health care at the border, Uribe-Leon (2016) noted that health-care costs in terms of doctor fees, diagnostic tests and treatment are significantly cheaper in Nuevo Laredo. Many residents speak of better quality primary care there due to shorter waiting times, and more interaction with physicians. They explain the attention they receive is more personalized and they can ask questions without feeling disrespected or rushed. Above all, they prefer primary-care treatment from a native Spanish speaker. As current governmental policies project increased numbers of uninsured US residents, we expect to see health-care-related crossings multiply.

Curious, but worth noting, are perceptions by some that Mexican medications, particularly antibiotics and pain relievers, are more potent than comparable medication prescribed in the USA. One respondent stated that she thinks the medication they get in Mexico is stronger than the comparable US brand. She said Mexican doctors usually administer a shot that works immediately, while American counterparts prescribe mild antibiotics and medications that are ineffective and require additional visits (Buentello, 2015).

Laredo is known as the Gateway to the Americas. Import and export of commercial goods and, to some degree, retail sales, are possibly the region's main source of economic prosperity. Retail brings Mexican shoppers to the USA to purchase better American-made products.

American residents traveling to Mexico daily do so for employment. Many Laredoans work in large American corporate plants or *maquiladoras*, like Sony and Delphi Automotive, in Nuevo Laredo. Many US citizens live in Nuevo Laredo because housing and living costs are cheaper than in Laredo. So are rents and utilities, and the money employees earn working in the USA goes further and lasts longer than in Laredo.

Concern about crime has led established Mexican businesses to relocate from Nuevo Laredo to Laredo, in quest of a safer environment (Garza and Landeck, 2009). Local shops and restaurants have closed because of the city's violence and insecurity. Well-established specialty shops like *Marti's*, originally the place to shop in Nuevo Laredo, closed their doors. Garza (2008) documented 23 established businesses completely closing down in Nuevo Laredo and relocating to Laredo. News reports suggest that thousands of affluent Mexicans have quietly moved their families and businesses to the USA and being welcomed wholeheartedly because they bring millions of dollars into the American economy via expensive home- and vehicle-purchases, and establishing businesses.

Nuevo Laredo's economy continues suffering as Mexico loses the drug war. In recent years, Laredo has seen more Mexican-inspired restaurants and bakeries opening up. These popular establishments have been welcomed by Laredoans who miss daily river crossings to dine and shop. While Nuevo Laredo's economy suffers, Laredo businesses have flourished.

Each Convention and Visitor's Bureau now acts independently in advertising each city's attractions. The Sister Cities may be slowly distancing themselves from each other in subtle, yet obvious ways. For example, the Texas Department of Transportation modified a mile marker showing travel distance to Laredo *and* Nuevo Laredo to only Laredo. Additionally, Laredo began a major campaign advertising that 'Laredo is Safe' when it emerged that people throughout the USA believed that Nuevo Laredo's violence was spilling over into Laredo.

According to the US Customs and Border Protection (CBP), border protection has been present in Laredo since 1924. Currently, CBP's Laredo Sector covers 110,000 square miles and 116 counties, and has one sector-complex and nine stations. Over 1700 agents are employed in the Laredo Sector, bringing $80 million dollars into the community annually.

CBP primarily seeks to deter illegal entry by undocumented aliens into the country while apprehending and deporting those establishing themselves without proper documents. Its mission today also includes preventing terrorist acts and weapons entering the country, and the illegal transportation of narcotics and other contraband. The usual means to achieve their objectives continue, for example, line-watching (placing agents in strategic locations along the border to stop illegal entry) and sign-cutting (a form of tracking, looking for signs of environmental disturbance left behind). However, CBP also now uses high-tech military-surveillance equipment they say makes them more effective.

While the CBP's presence creates concerns about criminality among people crossing the border, we cannot deny the economic impact of having significant numbers of federal agents. On the positive side, Laredo greatly benefits from the thousands of federal jobs existing to address these concerns. However, the security professions have expanded due to painting this region as dangerous, and this creates anxiety for citizens and legal residents who feel more scrutinized than usual and their legality evermore questioned. It has become obvious to many that making immigration a crime is causing havoc across the nation as Americans are divided about appropriate action. A recent Gallup poll of the American public showed 36% supporting a border wall with Mexico with 56% opposing, despite President Trump's repeated claims that the USA will build a wall across the border (Newport, 2017). Many positive contributions made by Mexican immigrants have been overshadowed by concerns about illegality. This discourse has spilled over with negative consequences for legal immigrants as well as American-born citizens including children, families and the general community (Lykes, Brabeck and Hunter, 2013).

Between October 2015 and September 2016, over 102,105 children from Mexico and South America were apprehended by the US Border Patrol. In Laredo, 2,953 unaccompanied children were apprehended. According to reports, they were fleeing unsafe conditions in Central America, especially El Salvador, Nicaragua and Honduras. Based on false information, children without parents were being sent to the USA expecting that they would be allowed to stay. This sudden influx produced a humanitarian crisis, overwhelming CBP and catching local communities unprepared. Nevertheless, church groups immediately mobilized and alongside community non-profit organizations created humanitarian relief teams to welcome and assist refugees with food and clothes, and help in contacting family members living in the USA. The Holding Institute, an affiliate of the United Methodist Church, took the lead and became ground-zero the distribution of thousands of donated items. As Border Patrol was apprehending, processing and releasing these individuals into the community, social services agencies were welcoming and providing assistance. While this activity is considerable, we must recognize that Laredo regularly handles a considerably larger proportion of US–Mexican trade. This is a large amount of commerce relative to the number of Border Patrol apprehensions occurring in the region.

During this humanitarian crisis, many Americans assumed that these migrants were from Mexico, demanding they be sent back. However, since they were not, other rules applied. Children arriving without parents are considered vulnerable and have certain protective rights under US immigration policy. They are processed under specific relief policies like asylum or offered Special Immigrant Juvenile Status. Unlike adults or families, arrested children must be processed and protected and cannot be returned to their country of origin without due process (American Immigration Council, 2015).

As Central Americans continued penetrating US borders, Mexico was blamed for not helping stop the influx through its southern borders. Kahn (2014) wrote that, when Mexico eventually began establishing road-blocks and check-points to seal its borders and stopped the dangerous practice by Central Americans of hopping onto a freight train named La Bestia (the beast) to enter the USA, the flow of immigrants began slowing. According to Mexico's interior minister, over 30,000 Central Americans were stopped in this way. According to the Department of Homeland Security Office of Immigration Statistics (2017), border-patrol arrests dropped by 61% from 1,189,000 in 2005 to 409,000 in 2016. The drop according to DHS could be for various reasons including stricter border enforcement, the US economic downturn and harsher employer sanctions.

The history of Los Dos Laredos began in 1836, when the Republic of Texas seceded from Mexico and a group of native Laredoans, not wishing to be Texans, crossed the river and established their own Laredo naming it Nuevo (New) Laredo. This divide not only carried over certain cultural traditions but also strengthened ties between the two cities even though each sat in different nations. The distinct yet similar cities, Los Dos Laredos, continued merging and combining though traditions steeped in culture and ethnicity. Since then, there have been several symbolic rituals taking place demonstrating the bonds of people from the twin cities.

Among the symbolic markers reinforcing the US–Mexican relationship is the long-standing tradition of an *abrazo* or hug, which began in 1898 during what was known as the 'Laredo Celebration'. US and Mexican local and military officials gather yearly to walk arm-in-arm in a show of unity into downtown Laredo. In 1969, the Washington's Birthday Celebration Association (WBCA) and the International Good Neighbor Council introduced the 'Abrazo Children' to further promote awareness and understanding between the people of the Americas. An abrazo occurs annually during the month-long Washington Birthday Celebration, when four children known as the Abrazo Children, two representing the USA and two representing Mexico, along with US dignitaries and their Mexican counterparts, gather at their respective side of the bridge and walk to the center to exchange a symbolic hug. This demonstrates good will, friendship and mutual respect between Los Dos Laredos and the two nations. The locally televised ceremony begins with the first hug between each pair of children, followed by individual embraces between the two sets of dignitaries. This marks the start of the grand parade immediately following. For the past 119 years, the WBCA has celebrated the birthday of the first US President with a month-long agenda filled with celebratory activities embracing both sides of the border (Figure 11.1).

Another continuing ritual is the Silver Rose tradition, also signifying unity between the peoples of the Americas, dating back to 1960. The International Catholic Church organization, Knights of Columbus and designated squires carry one of the four silver roses to various locations across 38 states of USA and 5 Canadian provinces. Each year, the silver roses cross at the pedestrian bridge across the Rio Grande before returning to the Basilica of Guadalupe in Monterrey. The ceremony draws hundreds of formally dressed organizational members marching to the bridge and sharing prayers as they return the

Figure 11.1 Annual Abrazo Ceremony between Laredo and Nuevo Laredo by Washington's Birthday Celebration Association (2015).

roses. Similar meanings can be found in the Noche Mexicana: a Presentation of Señor and Señora Internacional established in 1977 as a bi-national and bi-cultural event honoring Latinos for various achievements on each side of the border. Meanwhile, the International Sister Cities Festival for the past 13 years has promoted Mexican artistry by offering the citizens of Los Dos Laredos and surrounding cities a taste of Mexico by promoting the old style Mexican 'Mercado' to the US consumer, proclaiming it a world of shopping under one roof. Over 200 exhibitors from throughout Mexico's interior converge for a week-long event in an air-conditioned venue to show and sell their handcrafted artisan products.

There are several long-term bonds making Laredo–Nuevo Laredo twin cities that predate their role as international border twins. First, there is a long tradition of people crossing back and forth for hundreds of years. People have moved at will from one side to the other. Secondly, as the City of Laredo grew, there was even more commerce in Nuevo Laredo, and as Nuevo Laredo grew, there was more commerce on the Laredo side. The cultural customs remain and, although international boundaries produce evermore complex barriers, more goods are transferred across this border each year. The tradition of calling the region 'Los Dos Laredos' emphasizes that residents of this region do not solely identify with one nation. As Michael Dear states in *Why Walls Won't Work* (2013), borders are areas of connection. People find their ways to pass through, around or over any wall.

References

Abdullah, H., Gamboa, S. & Alexander, P. 2017. 'White House Floats 20 Percent Tax on Mexican Imports to Pay for Border Wall', *NBC News.* www.nbcnews.com/news/us-news/white-house-floats-20-percent-tax-mexican-imports-pay-border-n712461 (accessed 22/12/17)

Aguilar, J. (2015) 'Plummeting Peso Hurting Border Economy', *Texas Tribune.* September 2, 2015.

American Immigration Council. (2015) https://www.americanimmigrationcouncil.org/research/guide-children-arriving-border-laws-policies-and-responses (accessed 26/01/17)

Buentello, S. (2015) 'Twin Cities', Paper at the 12 Annual Pathways Research Symposium. Corpus Christi, Texas.

Cedillo-Campos, M.G., Sanchez-Ramirez, C., Vadali, S., Villa, J.C. & Menezes, M. (2014) 'Supply Chain Dynamics and the "cross-border effect": The U.S.-Mexican Border's Case', *Computers and Industrial Engineering* 72, 261–273.

Dear, M.E. (2013) *Why Walls Won't Work.* New York: Oxford University Press.

Department of Homeland Security Office of Immigration Statistics. 2017 Efforts by DHS to Estimate Southwest Border Security between Two Ports of Entry. www.dhs.gov/sites/default/files/publications/17_0914_estimates-of-border-security.pdf (accessed 04/12/17)

Esquivel, G. & Rodriguez-Lopez, J. (2003) 'Technology, Trade and the Wage Inequality in Mexico before and after NAFTA', *Journal of Development Economics* 72(2), 543–565.

Garza, C. (2008) 'The New Refugees: Mexican Businesses Moving to Laredo', *Border Research Reports*, 9: Texas Center for Border Economic and Enterprise Development, Laredo, Texas.

Garza, C. & Landeck, M. (2009) 'Mexican businesses moving to American side of the U.S. Mexican border', *Southwest Review of International Business Research* 20(1), 167–173.

Guerrero, J. (2017). 'Tijuana Reaping Profits That Once Flowed to San Diego', *KPBS*. www.kpbs.org/news/2017/may/17/tijuana-retail-reaps-profits-once-flowed-san-diego/ (accessed 15/12/17)

Kahn, C. (2014) 'Mexican Crackdown Slows Central American Immigration into the U.S.', NPR Article September 12, 2014. https://www.npr.org/sections/parallels/2014/09/12/347747148/mexican-crackdown-slows-central-american-immigration-to-u-s (accessed 13/12/17)

Kilburn, J.C., Costanza, S.E., Metchik, E. & Borgeson, K. (2011) 'Policing Terror Threats and False Positives: Employing a Signal Detection Model to Examine Changes in National and Local Policing Strategy between 2001–2007', *Security Journal* 24, 19–36. Advance online publication May 18, 2009 doi:10.1057/sj.2009.7

Kilburn, J.C. & Leon, G. (2012) 'A Study of Barriers to Health Care and Adaptations of a Border Community Population', Paper presented at the Annual Meeting of the Southern Sociological Society in New Orleans, LA.

Kilburn, J.C., San Miguel, C. & Kwak, D. (2013) 'Is Fear of Crime Splitting the Sister Cities?' *Cities* 34, 30–36.

Kletzer, L.G. (1998) 'Job Displacement', *The Journal of Economic Perspectives* 12, 115–136.

Landeck, M. & Garza, C. (2003) 'Utilization of Physician Health Care Services in Mexico by U.S. Hispanic Border Residents', *Health Marketing Quarterly* 20(1), 3–16.

Laredo Development Foundation. www.ldfonline.org (accessed 21/12/17)

Laredo International Sister Cities Festival. http://www.laredociudadeshermanas.com/festival.html (accessed 16/12/17)

Lykes, M. B., Brabeck, K.M. & Hunter, C.J. (2013) 'Exploring Parent-Child Communication in the Context of Threat: Immigrant Families Facing Detention and Deportation in post-9/11 USA', *Community, Work and Family* 16(2), 123–146.

McBride, J. & Sergie, M. (2016) 'Nafta's Economic Impact', *Council on Foreign Relations*. www.cfr.org/trade/naftas-economic-impact/p15790 (accessed 21/12/17)

Newport, F. (2017) 'Building a Wall Out of Sync with American Public Opinion', *Polling Matters.* www.gallup.com/opinion/polling-matters/209384/building-wall-sync-american-public-opinion.aspx (accessed 17/12/17)

Rios V. & Shirk D. (2011) 'Drug Violence in Mexico Data and Analysis through 2010', Special Report for TBI, USD, San Diego.

San Diego Source. (2017) Retail Sales. www.sddt.com/Economy/indicators.cfm?Report_ID=63&_t=Retail+Sales#.Wia_ObBryUk (accessed 22/12/17)

Schaffler, W.F. (2017). *Impacts in Los Dos Laredos in A World Without NAFTA.* Laredo, TX: Texas Center for Border Economic and Enterprise Development.

Scott, R. (2011) 'Heading South: U.S.-Mexico Trade and Job Displacement after NAFTA', Economic Policy Institute briefing paper # 308.

Swanson, A. (2017). 'How the Trump Administration Is Doing Renegotiating NAFTA', September 28. *New York Times.*

Texas Department of Economic Development. www.texasedc.org (accessed 04/12/17)

Texas Department of Transportation Border Corridors and Trade Report. (December 2015) https://ftp.dot.state.tx.us/pub/txdot-info/iro/border-trade-report.pdf (accessed 09/12/17)

Thompson, J. & Juarez, J. (n.d). A Brief History of Laredo Air Force Base. https://www.webbcountytx.gov/HistoricalCommission/HistoricalMarkers/AirforceBase.pdf (accessed 11/12/17)

Uribe-Leon, M. (2016) 'First-time Mothers, Latin-American Women Experiencing Motherhood in a New Sociocultural Context.' Lamar Bruni Vergara Academic Conference. Texas A&M International University, Laredo, Texas.

U.S. Department of Transportation, Research and Innovative Technology Administration, Bureau of Transportation Statistics, based on data from the Department of Homeland Security, U.S. Customs and Border Protection, Office of Field Operations.

U.S. Custom and Border Protection (CBP). www.cbp.gov/border-security/along-us-borders/border-patrol-sectors/laredo-sector-texas (accessed 12/12/17)

Washington Birthday Celebration Association. www.wbcalaredo.org (accessed 06/12/17)

Washington Post. (June 16, 2015) 'Donald Trump Announces a Presidential Bid', www.washingtonpost.com/news/post-politics/wp/2015/06/16/full-text-donald-trump-announces-a-presidential-bid/?utm_term=.5c50b7bfce2d (accessed 11/01/17)

12 Border-city pairs in Europe and North America

Spatial dimensions of integration and separation

Francisco Lara-Valencia and Sylwia Dołzbłasz

With this chapter, we start the themes of integration or disintegration and strengthen the comparative dimension of our book by contrasting twin-city experiences on the US–Mexican and the Polish–German borders. By exploring the outward-specialization of border-cities the chapter contributes to studying cross-border mobility and shopping, as well as the developing sense of familiarity among borderlanders. One theme emerging again is the importance of borders – as points where both the weakening and the strengthening of cross-national differences can be occurring simultaneously, with implications also for the way places appear on the ground.

Introduction

Historically, international borders' most fundamental function lies in enforcing national sovereignty and control. Borders emerged to regulate movements of people and things, little considering their consequences for local economies and communities. In particular, location along national borders imposes unique conditions on cities and towns shaping development and spatial structure (Buursink 2001; Arreola 1996). In fact, borders express their power via physical, legal, economic and even mental barriers directly influencing cities' functions, extent, form, centrality and stability (Agnew 2008). Thus, traditional location theorists have described border-cities as deviant urban places where borders create functionally fragmented spaces, characterized by scant economic activity and small market thresholds (Hansen 1977).

However, newer spatial theories see border locations as competitive advantages; part of cities' territorial capital (Sohn and Lara-Valencia 2013). This perspectival shift was triggered by the late twentieth-century popularity of globalization theories predicting the world's debordering, a view galvanized by intensifying cross-border movement of goods, services, technologies and capital. Thereby, border regions and their urbanized cores necessarily become locii for intense integration and interdependence. Regional experts and policy-makers began visualizing development paths where border-cities could become central nodes of a networked global economy (Keohane and Nye 1998).

Border-city pairs can be seen as gateways connecting markets in different countries. Here, goods, traders, tourists and shoppers traverse international boundaries, creating cross-border economies fluctuating with changing price differentials, exchange rates and transaction costs. As is well known, economic and political forces regulating such cross-border flows affect the urban space of border-city pairs by assembling material and symbolic landscapes dominated by buildings, roads, bridges, walls and signage regulating cross-border mobility and used by tourists, travellers, shoppers and residents to navigate and comprehend the city. Such assemblages have long interested geographers and urban planners seeking to understand how border-city pairs conflate national and international elements to create distinct urban borderscapes (Arreola and Curtis 1993; Nugent 2012). They concentrate most intensely around international crossing facilities, often anchoring the city's central business district.

Visually inspecting downtown areas in many border-city pairs often reveals the importance of trade-related facilities, tourist-oriented business, shopping and hospitality services directly related to border adjacency. Border-induced trade economies are evident on both sides, though significant differences commonly occur between cities resulting from competition, complementarities and specialization.

This chapter examines the spatial form and structure of Nogales/Nogales (USA–Mexico) and Gubin/Güben (Poland–Germany), two paired border-cities impacted directly by integration policies due to co-adjacency to international boundaries. It combines visual survey data and geospatial analysis to explore the outward/inward orientation of the service sector (retail and hospitality industries) in both pairs' central business districts. We argue that, even when openness towards the 'other side' reflects complex political, historical, cultural and economic factors, it also indicates local responses to intensifying processes of rebordering/debordering. We compare both material and symbolic expressions of spatial and functional asymmetry, complementarity and competition in each border-city pair to highlight the spatial dimensions of border-induced integration and separation forces. Although identifying differential patterns attributable to historical, geographical and other place-specific factors, our analysis tends to support notions of rather unified urban border experiences across continents.

Cross-border shopping

Among many interactions, cross-border shopping is one of the most noticeable, with economic and spatial consequences for border twins. It is a modality of market interactions distinct from those occurring further inside national boundaries, impacting mainly in downtown districts. To start with, consumption markets along open international borders allow consumers and retailers to make commercial decisions around price structures and differentials particular to each side, creating economic opportunities not easily available to people far from the border. Price differentials and thus consumer choices will depend on various factors, including exchange-rate stability and relative sales taxation. Others include transportation costs and check-point waiting times. Border twins seem natural places for cross-border shopping because border adjacency reduces transport costs to points where shoppers become indifferent between shopping in either market, given equal prices and quality. However, limitations of border-crossing infrastructure, protracted check-point inspections or difficulty navigating shopping areas can increase the border's frictional effect, thereby increasing cross-border travelling costs and discouraging out-shopping. Citing medical towns[1] along the US–Mexican border, Arreola (2010) suggests that smaller border-cities have better chances of evolving into competitive service hubs by shielding shoppers from having to navigate larger distances and longer inspection lines.

Non-economic factors' role in cross-border shopping has been increasingly analysed (Spierings and Van Der Velde 2008, 2013). Shoppers and merchants differ in ability to interact with people from the other side, depending on knowledge of language and customs. Obviously, this tends to be higher among border-city residents. To reduce the negative impact of intercultural barriers and promote cross-border shopping, border-cities merchants sometimes adapt to customers' needs by advertising in their language, accepting foreign currency, creating downtown tourist circuits and, generally, adopting marketing strategies reducing the cultural transaction cost of border crossing. These strategies try to lessen those elements making shoppers feel uncomfortable or confused, without neutralizing national differences rendering cross-border shopping worthwhile.

As suggested above, what cross-border shoppers consider familiar/unfamiliar depends upon exposure to the border. Obviously, border proximity enhances likelihood of engagement in cross-border activities, as well as familiarity with resources and opportunities on the other side. People also have developed familiarity from participating in multiple cross-border activities, not just shopping. Cross-border shoppers fall into two broad categories with unique implications for border-cities' commercial and spatial structure.

First, we have utilitarian cross-border shoppers guided mainly by rational desires to exploit price differentials. Their shopping is frequent, product-oriented and discriminating between suppliers based on location, quality, convenience and other factors. Utilitarian shoppers cross to buy goods and services they normally buy in their own city, behaving as typical, rational consumers. They are attracted to shops frequented also by local consumers. On the other hand, we have leisure cross-border shoppers, usually tourists and travellers crossing borders to experience the other side's life and culture. They are attracted by a place's uniqueness, seeking escape from daily routines. Thus, the price may not guide their shopping decisions.

They mainly confine themselves to tourist corridors within border-cities' central business districts and greatly help create cross-border shopping landscapes often found in border-cities.

Many businesses near border crossings cater to clienteles comprising varying mixes of local and cross-border customers: utilitarian and leisure-driven. Thereby cities along open borders have retail and service sectors that seem overgrown compared to their population base. The 'excess' is because the market threshold of each city extends beyond the border, creating commercial systems that complement and provide unique goods and services to clients on both sides.

With clienteles comprising extra-local shoppers with different income levels, cultural dispositions or preferences, some businesses sell goods and services often unaffordable to typical local consumers, or incompatible with local tastes and sensibilities. To benefit from clustering economies, such as comparative window-shopping and high shopping traffic, some businesses congregate together, creating distinctive, outwardly oriented commercial districts. Such downtown clustering visually distinguishes border-cities due to the density of similar facades, signage, sounds and transient people, thereby helping produce unique urban landscapes.

One element clearly reflecting the combined outward/inward orientation of border-city pairs is distinctive commercial signage. As in any city, border retailers attempt to call tourists and regular shoppers' attention by displaying distinctive shop-front signs in terms of size, colouring and lettering styles. One singular aspect of border-cities, reinforcing their trans-border character, is the volitional usage of foreign language in commercial communication. This increases communicative effectiveness, and assists legibility and cultural wayfinding.

We now examine the external orientation of commercial and service establishments in Gubin/Güben and Ambos Nogales city centres. This rests on data collected by visually inspecting retail and service establishments within each city pair using four indicators of outward market orientation: (1) language on signage, (2) language in marketing materials; (3) acceptance of foreign-currency payment and (4) employed personnel's bilingual language skills. These were chosen as signalling local business dispositions towards clients from across the border, and perceptions of cross-border shopping as an economic resource for local development. Such data was complemented with locational analysis of business establishments' spatial distribution and clustering.

Gubin/Güben: a divided city

Gubin and Güben form a border-city pair on the banks of the Neisse River, marking the Polish–German border. In 2013, Gubin's population was 20,049, and Güben's 21,800. Alongside Frankfurt/Słubice and Görlitz/Zgorzelec, it is one of the three city pairs along this border divided after European boundary realignment following World War II. Consequently, most of Gubin/Güben's history as paired border towns has consisted in bridging differences and building trust, especially in recent decades. The two cities disconnect until the early 1990s contrasts with many varied EU-supported local and regional initiatives in more recent years (Kulczyńska 2010).

Cross-border shopping

Gubin's centre lies at its geographic core, while Güben's retailing is mostly located near the border in its eastern part. Gubin's retail activity is dispersed across its centre, concentrating particularly around the main square. Most Güben shops lie along the corridor formed by Frankfurter Strasse and Berliner Strasse, connecting the city centre with the border crossing. Although Gubin/Güben's geographic retail distribution is typical of a small urban market, it also reflects bordering and debordering processes.

The Neisse River had no border towns before 1945, when Poland's eastern boundary with the USSR was relocated west and the Polish–German boundary was moved to the Oder–Neisse line (Dołzbłasz and Raczyk 2012). Partitioning mediaeval Güben created Gubin and Güben on the east and west banks, respectively. It left two-thirds of the old city in Poland, including the centre and two large suburbs (Matthiesen and Bürkner 2001). Consequently Güben had to develop from scratch some urban functions lost to Gubin, including a central business district.

As Kulczyńska (2010) noted, the new border functioned as a barrier restricting inter-communal interactions, making commercial exchange and cooperation difficult. This changed after German reunification in 1989 and the 1991 German–Polish treaty. Opening the Polish–German border created new retail opportunities in the area, promoting commercial activity focused on servicing customers from the other side. In the early 1990s, Gubin experienced significant multiplication of small retail establishments focused on German shoppers. Around 2000, retail expansion based on traditional shops was supplanted by a rise of big-box stores with large supermarkets appearing in the city centre. Güben's supermarkets emerged in the late 1980s, and opening the border helped support an already successful large-format retail model based on national and multinational retailers like Lidl. Unlike smaller retail shops, big-box stores require large buildings and abundant parking space, thus favouring suburban locations.

Cross-border mobility

A recent survey among Gubin residents indicated that 80% have visited the city across the river (Dolińska et al. 2013): to shop, visit friends and engage in other social activities. This increases with education level, wealth, having friends and relatives across the river, and German-language proficiency; it decreases with age. Among many reasons for visiting Güben, shopping was undoubtedly most important (around 93%). Others were strolling (about 76%), visiting acquaintances (40%), and entertainment and cultural activities (about 33%). Other less common activities were utilizing personal-care services, using transport facilities or participating in religious celebrations and family holidays.

The same survey indicates that Gubin residents perceive their Güben neighbours mainly as tourists and customers (about 80%), although also as friends and neighbours (about 60%). Overall, both the type of dominant cross-border interactions and perception of the neighbours mainly as potential customers indicate that the connection is primarily pragmatic: access to goods, services and economic opportunities, while altruistic and communitarian feelings are less widespread (Dolińska et al. 2013).

German–Polish price differentials best explain Gubin/Güben interaction. In 2012, Polish prices of essential foodstuffs were among the EU's lowest: overall about 60% of the EU average, while non-alcoholic beverages were around 79% lower (Kurkowiak 2013). By contrast, German prices were much higher: food 106% of EU average; non-alcoholic drinks 104%. Only alcohol bucked the trend, German prices being 82% of EU average against 93% in Poland. Conversely, German tobacco prices were the EU's highest (102%) against Poland's 58%. The overall effect on cross-border shopping flows is obvious: many Gubin shoppers cross to buy alcohol, and Güben shoppers seek food and tobacco.

Changing economic conditions on either side can affect relative prices, shifting cross-border shopping's intensity and direction. In the 1990s, traffic was one-sided: German shoppers crossed to Poland to buy food. Some estimates of Gubin retail activity calculate that 50%–60% of sales were to German shoppers. Recently, reducing price differentials encouraged Poles to shop in Güben, mainly for cosmetics and household cleaning products, while Germans remained attracted by lower prices of food and some personal-care services: mainly hairdressing and dentists (Kulczyńska 2010).

Cross-border openness

Overall, 320 retail and service establishments were inspected in Gubin/Güben. Comparing each city's business mix reveals a pattern consistent with border retailing (Table 12.1). In Gubin, clothing and shoe stores predominate among city-centre retailers, followed by grocery stores heavily concentrated near the international bridge. In particular, grocery-store location seems to indicate sales not only to locals but also to Germans attracted by Poland's food prices. Personal-care services like hairdressing shops and beauty parlours are also significant among Gubin businesses near the bridge, many being cross-border shoppers. On the German side, Güben had more restaurants, coffee shops, travel and tourism services, and gift shops reflecting higher demand for these in Germany's more affluent economy. Polish customers increasingly utilize these businesses, but the

Table 12.1 Gubin/Güben retail trade and hospitality services by activity subsector

Type of activity	Gubin (PL)	Güben (DE)
Clothing and clothing accessories	22%	13%
Food and beverage stores	17%	5%
Hairdressers and beauty parlours	12%	4%
Banks and real state	4%	7%
Food services and drinking places	7%	16%
Health- and personal-care services	3%	6%
Travel and tourism services	4%	7%
Other retail and hospitality activities	31%	42%

Source: Author's research.

clientele remains mainly German. Overall, Polish customers able to afford the prices prefer German electronics appliances, and clothing perceived as better quality than available in Gubin.

Overall, these data show that Gubin's retail economy is more reliant on cross-border shopping than Güben's. Marketing practices confirm this observation. These include language of store-front signage, printed fliers and other materials, bilingual employees, and accepting foreign currency. In Gubin/ Güben, such practices were considerably more prevalent among Polish businesses than German equivalents. Approximately 37% of Gubin businesses displayed German signs, 32% used German information materials, almost 80% allowed euro payments, and nearly all had German-proficient staff. Meanwhile, Güben businesses rarely seemed to market to Polish clients: about 90% using none of the above practices, though a few (9%) had Polish-speaking employees, mostly because they employed Polish workers.

Ambos Nogales: complementary cities

The boundary separating Nogales, Sonora, and Nogales, Arizona, dates back to 1853, after a disastrous war and negotiation forcing Mexico to cede almost half of its territory to the USA. The area where Ambos Nogales (the two Nogales) was settled was sparsely populated but already important in commerce between settlements in what is now Northwest Mexico and Southern Arizona (Flores 1987). Railroad expansion sparked initial US interest in this territory and underpinned this border-city pair's foundation in 1884. The two cities were born and planned together as places to support and regulate commerce coming from better communication between large urban centres in Mexico and the USA. However, they grew as separate but interdependent urban and economic units. Currently, Nogales, Sonora, has approximately 300,000 residents and its economy is dominated by export-oriented manufacturing (maquiladoras) and commerce. Nogales, Arizona, has a stagnant population of 20,500 with jobs heavily dependent on border enforcement, retail and transportation.

Ambos Nogales are united by numerous factors. According to the 2010 US Census, about 95% of Nogales, Arizona, residents had Mexican ancestry and almost 42% were foreign-born, mostly from Northwest Mexico. Many families have members on both sides; individuals cross the border for work and schooling, and commerce is lively. Uneven urban growth, migration and heightened border security, a powerfully and heavily consequential trend even before Donald Trump's emergence, has strained relationships and cooperation.

Cross-border shopping

On the Mexican side, the backbone of downtown Nogales is a tourist corridor, starting at the check-point and extending a few blocks along Obregon Avenue. It is dotted with curio shops, pharmacies, dentists, barber shops, restaurants and bars. English is the trading language, since potential customers are US tourists crossing the border as pedestrians and strolling the corridor after a good Mexican deal. Nearby, there is a red-light district and currency-exchange office clusters catering demand from visitors and locals alike. Relatively new in downtown is a group of health-care offices, including dentists, spas, plastic surgery offices and other specialized health-care facilities. Beyond these externally oriented retail clusters, the Mexican downtown contains government, religious, commercial and service activities serving local residents. US tourists are unusual, as is English-language usage.

About 9,600,000 border crossers entered Nogales, Arizona, from Nogales, Sonora, in 2013. Approximately 83% (7,800,000) were Mexican nationals crossing for leisure reasons, including shopping, and visiting friends and family. The remaining 17% are US residents returning after visiting Mexico. Nogales, Arizona, has no tourist district, but does have shopping areas including clothing stores, shoe and appliance stores and supermarkets visited by crowds of Mexicans crossing regularly for US goods that are cheaper and considered better quality than those in Mexico (Figure 12.1).

Studies of Mexican visitors' impact on Arizona's economy suggest that almost all are Sonora residents. They stay a few hours and cross several times monthly. (Pavlakovich-Kochi and Charney 2008). Mexican shoppers spend around $40 per person each time. This represents almost half of Nogales, Arizona's taxable sales, supporting the employment of 45% of the local labour force (City of Nogales 2010).

Fast-food and chain restaurants are located near shopping areas in Nogales, Arizona, alongside banks and gas stations. Other common border-related activities are currency-exchange offices and transportation services. With Mexican visitors supplying the main demand, Spanish is frequently used in signage and shoppers can expect Spanish-speaking attendants. Ostensibly, biculturalism and bilingualism are strong in Nogales, Arizona. The US Census shows that 89.4% of residents spoke a

Figure 12.1 Vehicular traffic entering the USA through Nogales by Francisco Lara
        Valencia (2016).

language other than English at home and approximately 57% of businesses
were Hispanic or Latino-owned.

Cross-border mobility

Cross-border movement through Nogales, Arizona's three land-entry-ports
has been intense but shifting. Three distinct periods are observable between
1995 and 2013 (Figure 12.1).

The first comprises 1995–2001 and encompasses US economic recupera-
tion and Mexican economic expansion after the 1994 North American Free
Trade Agreement. During this period, border crossings peaked at nearly
16,300,000 in 2000, 16% higher than the 14,100,000 in 1995. During 2000,
for each pedestrian crossing, 2.5 crossed in private vehicles or buses. This is
critical to understanding the economic and spatial impacts of cross-border
visiting in Nogales. Visitors in private vehicles tended to travel to shopping
malls outside Nogales' urban core, or even locations like Tucson or Phoenix
up to three hours away. They also tended to stay longer and spend more.

The second period spans 2001–2007, the post-9/11 period. This is characterized by substantially decreased cross-border activity due to prolonged
waiting times at most border entry-ports, mainly resulting from enhanced
US government inspection routines and border-control procedures. Data
indicate a drop of almost 1,500,000 crossings in 2008 relative to previous

years, then gradual recovery to pre-9/11 levels from 2003. Most significant is the shift in travel modes: the passenger-to-pedestrian ratio went from approximately 2 to 1 to almost 1 to 1 in 2001–2006. This probably indicates vehicles being more rigorously inspected than pedestrians and therefore waiting longer at entry-ports. Arguably, this forced many to forsake their cars for pedestrianism, thereby avoiding lengthy delays. Pedestrian crossings grew by almost 2,000,000.

From 2008, Nogales witnessed severely reduced cross-border activity. Crossings dropped almost to 4,700,000 during 2007–9. The lowest cross-border activity occurred in 2011, when total crossings dipped to 8,900,000, half the peak 16,500,000 crossing in 2007. The drop resulted from falls in both passengers and pedestrians. The post-2008 drop in transborder activity probably results from various factors, including Mexicans feeling that Arizona was not a friendly place after the passage of anti-immigrant legislation and xenophobia there. Cartel violence in Mexican border-cities also grew dramatically, producing negative media reports about safety in Mexico and frequent US State Department travel warnings.

Cross-border openness

In Ambos Nogales, two very different consumption markets interact continuously. Approximately 529 retail and service establishments operated downtown in both cities in 2014. Nogales, Sonora, has the most with 412 (78%), while Nogales, Arizona, had the remaining 117 (22%). The business mix in each city reveals interesting contrasts and similarities (Table 12.2).

In Nogales, Arizona, the largest percentage of businesses is clothing and clothing-accessories stores, mainly along Morley Avenue Corridor. Following these are shoe and general merchandise stores, alongside clothing stores, representing over half the business in downtown Nogales, Arizona. The Avenue's commercial landscape is completed by gift, jewellery and perfume stores. It connects directly with a pedestrian border, and is close to

Table 12.2 Ambos Nogales retail trade and hospitality services by activity subsector

Type of activity	Nogales (USA)	Nogales (MEX)
Clothing and clothing accessories stores	44%	0%
Food and beverage stores	7%	3%
Hairdressers and beauty parlours	0%	8%
Banks and real state	3%	4%
Food services and drinking places	3%	13%
Health- and personal-care services	4%	27%
Travel and tourism services	3%	20%
Other retail and hospitality activities	36%	25%

Source: Authors' research.

the largest entry-port in Ambos Nogales. The area therefore benefits from intense pedestrian traffic, mainly Mexican shoppers from Nogales, Sonora, crossing regularly, sometimes more than weekly. 95% of businesses deploy Spanish speakers.

Parallel to Morley Avenue, runs Grand Avenue. Aside from some clothing and shoe stores, its commercial mix is different: it includes used-car dealers, auto shops, gas stations, fast-food restaurants and grocery stores. Businesses use Spanish in facades and advertising, though slightly less prevalently than in Morley Avenue due to the branding requirements of the avenue's franchised retail stores. However, their clientele is also mainly Mexican: Spanish is spoken and Mexican Pesos mostly accepted. A similar situation prevails in the smaller commercial area on the west side of downtown Nogales, Arizona, even though the commercial mix is slightly different again. On the Mexican side, Nogales, Sonora's central business district shows an equal border impact. Within the 20 blocks forming the 'primer cuadro', in 2014, we found 65 dentists, 33 curio shops, 29 drug stores and 16 doctors: a third of the area's businesses and around 27% of this kind of businesses in the whole city. Curio shops, long mainstays of Mexican border towns, are being replaced by dental offices, doctor offices and pharmacies due to the many border-crossers attracted by Mexico's much lower health-care prices. Visually surveying these establishments, mostly located along Obregon Avenue and near the border, indicated all using English in their facades and marketing material, and with bilingual personnel. Unsurprisingly, all accepted US dollars, a practice commercially long embedded in Mexican border-cities. We can assume that most customers are US residents, many travelling from large Arizona cities like Tucson and Phoenix.

The *'primer cuadro'* also has numerous currency-exchange offices, restaurants, coffee shops, beauty parlours and barber shops. Unlike dentists and curio shops, their clientele is both Mexican and US Cursory observation indicates many are Mexican-origin residents of Nogales, Arizona, and neighbouring areas. Around one-third deployed English signage, all accepted dollars and half had English-speaking personnel.

Integration and separation

As shown in this chapter, borders impact border-city pairs in contradictory ways: material and symbolic features in Gubin/Güben and Ambos Nogales downtown areas show borders both connecting and dividing, not only integrating, but also separating. Although this may seem cyclical, in reality, separation and integration occur simultaneously and often overlap, reflecting borders' selective permeability, and hence their political instrumentality. This challenges dualistic representations of the border as either wall or bridge. Rather, it assumes that borders are contingent entities that can simultaneously be either barriers or connectors, depending on who or what is trying to cross and when (Böröcz 1997).

Border contingency is underpinned by the original bordering that created Gubin/Güben and Ambos Nogales as border-city pairs. Güben's partitioning created two incomplete cities, each forced to gradually recreate elements of urban structure lost via the new Polish–German boundary. For a long time, they turned away from each other. Strong ethno-cultural differences, alongside resentment resulting from the displacement of Germans and relocation of Poles greatly hindered cross-border interactions (Buursink 2001). Although the fading of European internal borders following EU enlargement augmented cross-border opportunities in Gubin/Güben, contemporary patterns of cross-border consumption still reflect disruptions resulting from the cities' birth in 1945.

Ambos Nogales are practically joined at the hip, their origin being catalysed by the boundary's location. The original settlement emerged at the first border check-point; for most of the twentieth century, their growth was anchored here. To some degree, both Nogales are mirror images, though the projected images are filtered by economic, cultural and institutional differences between Mexico and the USA. Their downtowns are still conduits for most interaction, and destinations for most cross-border tourism and shopping. Our data showed that outward orientation of these spaces is conspicuous and abundant, a reality evident since the Nogales emerged.

As Arreola and Curtis noted (1993), the mix and spatial organization of economic activities in border-cities' downtown areas are not fixed and they respond faster and better than other components of the urban system to changing national and regional forces. Downtown landscapes represent thermometers of changing cross-border dynamics. Indeed, various factors, separately or combined, determine certain activities' economic viability and greatly help shape the type, number and density of particular businesses and overall shopping landscapes. Undoubtedly, differences in disposable income on each side largely explain consumption types and levels, and each side's role in cross-border marketplaces. However, more incidental factors also contribute. For example, variations in the Polish Zloty's value might discourage some Gubin residents from engaging in cross-border consumption while stimulating visitors from across the border. Similarly, long check-point waiting times or perceived insecurity might affect willingness to cross the border in Ambos Nogales.

Ambos Nogales illustrate the point: the shopping scene has shifted constantly, recycling downtown buildings and land to accommodate trending activities like discotheques in the 1980s, pharmacies and health-care offices in the 1990s and shuttle services in the 2000s, depending on the demand by the two traversing streams of shoppers. Currently, Nogales, Sonora, specializes in medical services and drug stores, while Nogales, Arizona, still specializes in clothing/fashion.

Because specialization results from each city's objective conditions, businesses naturally take advantage to participate in the regional marketplace. In border-city pairs, this behaviour is advantageous for both sides, arguably producing complementary commercial structures supplying the needs of both sides' consumers. In Ambos Nogales, the Mexican city provides enough dentists, pharmacies and doctors to cover both sides' needs. Meanwhile, commercial activity in the US city produces sufficient clothing and grocery stores for both. In isolation, each city's commercial structure seems distorted, but the excess disappears when demands are considered in combination. This explains why both Gubin and Güben and Ambos Nogales have developed commercial landscapes contrasting so significantly. Each of them are twin cities by virtue of dissimilarity and thus complementarity.

Specialization in cross-border demand is important because it produces needs to adopt commercial practices reducing the border's barrier effect. Highly dependent on Güben customers, Gubin beauty parlours commonly advertise in German. Likewise, tourist corridors, like Obregon Avenue in Nogales, Sonora, where merchandise, architecture, colours, language and even sounds help decrease unfamiliarity, are niches where visitors can find security levels minimally matching their tolerance of otherness. As businesses specialized for cross-border shoppers agglomerate, they help create distinctively outward spaces capable of including foreign consumers.

In fact, cross-border openness is clearly asymmetrical in Gubin/Güben, with Polish businesses being more open than German equivalents in Güben. Our survey suggests the symmetry or otherwise of cross-border openness does not depend on city centres being physically proximate, or physical and administrative barriers being present. Openness seems most influenced by price differentials, alongside cultural differences. Indeed, increasing cross-border activity in Gubin/Güben after the inter-city collaboration agreements in 1991 and Poland's EU inclusion did not change the basic fact that Gubin has comparative advantages in supplying personal-care services, while Güben is more competitive in retail and high-order services. Regarding cultural closeness, bilingualism may influence spatial and socio-economic integration in border-city pairs more than policies promoting integration via border-infrastructure investment like bridges and roads.

Although shopping may not enhance closeness or spatial integration between border-city pairs, it is certainly positive for border living, increasing local prosperity and helping residents on both sides satisfy needs through interconnected market areas. It also exposes people on either side to the neighbouring city's culture and space. Arguably, these two aspects could produce interdependency and forms of cooperation and integration.

However, outward specialization among border-city pairs has risks. Over-specialization can render certain activities highly vulnerable to bordering processes curtailing cross-border consumer-access. Witness the many curio shops in Nogales, Sonora, ceasing business due to long checkpoint wait times, perceptions of the border as insecure and dangerous, and the shifting demographics of cross-border shoppers from Arizona.

Conclusion

Overall, several conclusions emerge regarding how bordering affects border-city pairs and their retail and service activities. First, their urban landscape is shaped by intersecting forces of integration and separation emanating from national borders. Certainly, due to borderline adjacency, their downtown areas show the highest density of cross-border retailing and services, yet are continuously adapting to shifting bordering levels. This shows up in asymmetrical commercial structures influenced by cross-border functionality and complementarity. Borders create patches of outwardly and inwardly directed zones connecting and separating both sides. Second, the type and form of retailing and services are clearly economically driven, mainly by neighbouring countries' price and income differentials. However, their precise impact varies greatly according to local conditions around cross-border transaction costs: prevailing business practices, culture and language on both sides can entice or discourage people from cross-border transactions. Third, shopping intensity and direction depends on the border's permeability in turn enabling or restricting flows of people and goods. However, permeability is not merely physical: political and cultural factors also play important roles. Finally, as locations of varied and frequent cross-border exchange, border-city pairs can become highly interconnected and interdependent. Commerce has long been recognized as a transformative force capable of changing local cultures and social dispositions, and is worthy of further investigation.

Note

1 Medical border towns are settlements on the Mexican side where the main business activity is health related services for US (and Canadian) tourists attracted for more affordable medical care. This is one role of Nuevo Laredo in Chapter 11.

References

Agnew, J. 2008. Borders on the mind: Reframing border thinking. *Ethics and Global Politics*, 1(4), 175–191.

Arreola, D. D. 1996. Border-city idee fixe. *Geographical Review*, 86(3), 356–369.

Arreola, D. D. 2010. The Mexico-US borderlands through two decades. *Journal of Cultural Geography*, 27(3), 331–351.

Arreola, D. D. and Curtis, J. R. 1993. *The Mexican border cities: Landscape anatomy and place personality.* Tucson: University of Arizona Press.

Böröcz, J. August 1997 Doors on the bridge; The border as contingent closure. *American Sociological Association Annual Meetings*, Toronto, Canada.

Buursink, J. 2001. The binational reality of border-crossing cities. *GeoJournal*, 54(1), 7–19.

City of Nogales 2010. *Nogales, Arizona: Shopper capture potential*. Nogales, AZ. https://imageserv11.team-logic.com/mediaLibrary/78/Shopper_Capture_Potential_Report_1.pdf (accessed 02/02/2018)

Dolińska, K., Makaro, J. and Niedźwiecka-Iwańczak, N. 2013. Instrumentalność versus autoteliczność praktyk transgranicznych mieszkańców Zgorzelca i Gubina. *Opuscula Sociologica*, 2, 31–46.

Dołzbłasz, S. and Raczyk, A. 2012. Tranborder opennes of companies in a divided city: Zgorzelec/Görlitz case study. *Tijdschrift voor economische en sociale geografie*, 103(3), 347–361.

Flores, G. S. R. 1987. *Nogales, un siglo de historia.* Hermosillo: INAH-SEP Centro Regional Noroeste.

Hansen, N. 1977. Border regions: A critique of spatial theory and a European case study. *The Annals of Regional Science*, 11(1), 1–14.

Keohane, R. O. and Nye, J. S. 1998. Power and interdependence in the information age. *Foreign Affairs*, 77(5), 81–94.

Kulczyńska, K. 2010. The Gubin-Guben transborder urban complex as an arena of consumer behaviour. Bulletin of Geography. Socio-Economic Series 14, 79–89.

Kurkowiak, B. 2013. Comparative price levels for food, beverages and tobacco. *Statistics in focus* 15. http://ec.europa.eu/eurostat/statistics-explained/index. php?_beverages_and_tobacco&oldid=207727 (accessed 04/02/2018)

Matthiesen, U. and Bürkner, H.-J. 2001. Antagonistic structures in border areas: Local milieux and local politics in the Polish-German twin city Gubin/Guben. *GeoJournal*, 54(1), 43–50.

Nugent, P. 2012. Border towns and cities in comparative perspective. In: Wilson, T. and Donnan, H. eds. A companion to border studies. Malden, MA: Blackwell Publishing Ltd., 557–572.

Pavlakovich-Kochi, V. and Charney, A. 2008. *Mexican visitors to Arizona: Visitor characteristics and economic impacts, 2007–08.* Tucson: Eller College of Management, The University of Arizona.

Sohn, C. and Lara-Valencia, F. 2013. Borders and cities: Perspectives from North America and Europe. *Journal of Borderlands Studies,* 28(2), 181–190.

Spierings, B. and Van Der Velde, M. 2008. Shopping, borders and unfamiliarity: Consumer mobility in Europe. *Tijdschrift voor economische en sociale geografie*, 99(4), 497–505.

Spierings, B. and van der Velde, M. 2013. Cross-border differences and unfamiliarity: Shopping mobility in the Dutch-German Rhine-Waal Euroregion. *European Planning Studies*, 21(1), 5–23.

13 Spatial portrait of twin towns on the Danube[1]

Máté Tamáska

In this chapter, the authors focus on the ways borders impact the often very different appearance of the previously united, and now divided, towns on either side: both on how they look and on how they are experienced by their citizens. Nevertheless, as this chapter argues, this has implications for integration: for this reason, the analysis, as with Chapter 12, is set within the frame of integration and disintegration. The fact that disintegration may come to seem more evident than integration may partly connect with the fact that the Danube is a very wide river.

This spatial portrait explores two post-socialist townscapes of two sets of mid-size twin towns, each linked by a bridge over the Danube between Hungary and Slovakia. The twins are: first, Esztergom (Hungary) and Šturovo (Slovakia); and second, Komárom (Hungary) and Komárno (Slovakia). These towns changed names several times over their history. To ease understanding, I use the contemporaneous town names in the historical parts of the study. Both towns developed in what was until 1920 the Hungarian Kingdom, which was part of the Habsburg Empire from the sixteenth century and of the Austrian-Hungarian Monarchy 1967–20. The post-1920 border divided pre-existing riverside twin communities, straddling what became Hungary and Czechoslovakia – later Slovakia after Czechoslovak split in 1993 (Sikos and Tiner 2008). Such urban divisions were common in twentieth-century Central Europe, and have consequently become major subjects of regional academic discourse (see Chapters 12 and 14 in the current volume). As with other such twins – for example, Český Těšín–Cieszyn (Czech Republic–Poland), Frankfurt (Oder)–Słubice or Görzlitz–Zgorzelec (both on the German–Polish border) – we can assume three phases of development: (1) integration; (2) disintegration and (3) partial reintegration.

Both sets of twins experienced very intensive *integration* of their urban fabric following bridge construction in the late nineteenth century. The notion of the 'urban fabric' (or 'townscape') denotes the ground structures (streets, plots) and buildings in the towns, viewed as a whole. The bridges linked the historical towns to their suburbs on the opposite river bank. The settlements pre-existed the bridges. However, only the bridges created real town–suburb relationships. After 1920, both sides became separate entities: the historical town on one bank and the newly constituted town (ex-suburb) on the other. For 70 years, each developed separately (until 1990). This *disintegrative* phase distinctively shaped the structure of the twin towns. The *reintegration* period after 1990 has been a long-term process (involving EU membership 2004, and then Schengen membership 2007). The infrastructure, cultural life, shopping and tourist facilities have begun integrating, but education, health-care and real estate, for example, have remained largely separate. The reintegration since 1990 did not return to pre-1920 conditions, but has often involved co-operation between separate and independent units.

To understand the 1920 border change, we must set it in a wider historical and geographical context. Fundamental here is the fact that the Danube today is a mosaic of jurisdictions: eight countries have Danube riversides, and four capitals are located on the river (Vienna, Bratislava, Budapest and Belgrade). This mosaic results from twentieth-century border changes. The Danube region was until 1920 mostly dominated by the Habsburg Empire. The Empire had been the winner of sixteenth/seventeenth-century struggles with the Ottoman Empire within the territory of the medieval Hungarian Kingdom. Esztergom and Komárno were located directly in the frontier zone of this near-200 year struggle. Thus, it is unsurprising that neither had served as crucial ferry points before the bridges were built (Beluszky and Győri 2005). Bridge-building produced relative prosperity not only for each town but also on the river's opposite bank, where there was a village (today Štúrovo and Komárom, the original names being Párkány and Új-Szőny). The typical form of this period was a dominant historical town opposite a rural setting.

The integration of the old towns and their rural settings peaked at the turn of the twentieth century under the Austro-Hungarian Monarchy (1867–1920), when both sets of twins were co-joined by highway bridges (1895 in Esztergom, 1898 in Komárom). In 1896, Komárno and Komárom were even unified under one government. Compared with other Danubian middle-size towns (like Baja, Paks, Tolna), we can assume that the rapid integration of urban fabrics on both riverbanks was more an exception than a general trend. Similar highway bridges were erected only in Budapest, the capital.

Before the bridges, each part of the twin towns developed relatively independently. Esztergom and Komárno became middle-sized towns with solid bourgeois dwelling houses, while their suburbs still looked like villages with very simple architecture. The bridges spurred their suburbs to become more industrial, thanks to the main railway lines passing not through the historical towns, but the opposite rural settings: Komárom and Štrurovo. The elites in the historic towns welcomed this situation. They envisaged development in which the historical center preserved its 'bourgeois character'. The workers in the upcoming industry migrated mostly into the former villages, which became evermore suburbs of the old towns on the opposite riverbank. Social segregation however was not a specialty of the twin towns but a consequence of capitalism.

The fall of the Austro-Hungarian Monarchy in 1920 produced a new situation. It started a slow disintegration of the twin towns. The new state border between them and their suburbs (Esztergom and Štrurovo, Komárno and Komárom) initiated new separate existences. The adaptation of town structures to the new borders was no simple process. People on both sides of the Danube saw the borders as temporary, doubly so because most people in Komárno and Štrurovo spoke Hungarian and had Hungarian identity (Szarka 2013). The creation of a border zone characterized the interwar decades. Before 1920, the former bridgeheads had acquired central positions in their respective townscapes, with bustling trade and some characteristic municipal buildings. Now what had been the central parts of the twin towns (the bridgeheads) became border areas, patrolled by guards. The shops, cafés, promenades emptied.

One former suburb, today's Hungarian Komárom, developed most intensively. Its population nearly doubled from 5,836 (1900) to 11,018 (1930). Thousands moved from Komárno to Komárom, where the Hungarian state tried to replace the administrative position of lost Komárno, which had formerly been the seat of regional government. The state financed building a new main square, alongside new administrative and school buildings. All these projects aimed to form a new middle-sized town in this rural area. However, Komárom's economic power was modest. Therefore, the private dwellings mostly comprised only one flat.

On the other side, 'old' Komárno lost its central position as a regional administrative center. The new Czechoslovakia did not trust this 'Hungarian town', so educated people (who previously worked in state administration) moved to Komárom, or even Budapest. Nevertheless, Czechoslovakia needed Komárno as an important harbor on the Danube. Thus, population grew in the period (1900 17,810; 1930 21,158). A new harbor was built. This investment ensured economic prosperity on the one side. On the other side, new entities (storehouses, an embankment and elevators) were installed on the riverbank. So, a barrier emerged composed of industrial structures along the river, and around the bridge. This zone was strictly separated from the town's inhabitants, other than as workers.

Esztergom and Štrurovo showed significant difference from Komárno–Komárom. The more important town was Esztergom, which remained in Hungary. There was consequently no significant migration from Štrurovo to Esztergom. Pre-1920 Esztergom was the capital of a 1000 square-kilometer county (or *vármegye,* a regional unit of the old Hungarian Kingdom) also called Esztergom. Because of the new border, the town lost half its catchment area. Inward migration slowed and so population decreased (1900, 18,206;

1930, 17,543). The result was marked stagnation in the urban fabric: very few new buildings.

Štrurovo remained essentially a village with very simple architecture, notwithstanding its population increasing from 3,793 in 1900 to 5,867 in 1930 (Table 13.1). Its only truly urban area before 1920, distinguished by multistory municipal housing, was located around the bridgehead. This remained the case after 1920. However, after 1920, even this area lost importance. On the village's non-river side was the railway station, which became an international point of traffic, but this caused no interwar expansion. The only sign of Štrurovo's emergent border functions was new barracks for the frontier police.

Summarizing these interwar processes overall, we can suggest that these Danube border towns needed to find new roles. The old historical towns suffered through losing their positions as administrative centers for wider regions. The former suburbs showed a different way of developing. Komárom became a new town and tried to replace the functions of lost Komárno. Štrurovo became an international railway cross-border point, though without developing new infrastructure. Regional planners of the time had no chance to solve the problems caused by dividing the towns. The political climate caused by enmity between Hungary and Czechoslovakia prevented cross-border co-operation over urban development. The developing projects responded only to the urgent needs for offices of state administration, flats for migrants and harbor infrastructure. Most local people waited for administrative reunion – which happened in 1938, as Hungary retrieved large territories from Czechoslovakia (losing them again in 1945). However, this brief period of administrative union could not integrate the twin towns.

The year 1945 saw rapid disintegration of the urban fabric of both parts of both twins. It became clear that the divisions between them were going to be long term. Deportations (Jews to concentration camps in 1944) and forced migrations (the Hungarian minority from Czechoslovakia to Hungary in 1945–46) broke up the social texture of local communities (Molnár and Szarka 2010; Shmuel and Wigoder 2003). For example, Komárno's population fell from 21,158 in 1930 to 17,395 in 1949. The two neighboring socialist countries organized their regional planning only within national frameworks. The border after 1945 divided people's everyday contacts very radically. Later, after 1970, some controlled bilateral international exchange in social life could emerge (including workers migrating to the fabrics on the other side of the twin towns), but disintegration and the independent reshaping of the urban fabrics on both riverbanks characterized the period (Bideleux and Jefferies 2007). The towns in each 'twin' integrated increasingly into their own national urban network. Regional plans integrated each part of the twin towns in structures that assumed the border was closed and the connections (roads, railways, education systems, etc.) existed only inland. The bridge between Esztergom and Štrurovo was blown up at the end of the war, remaining in ruins until 2001. The Komárom–Komárno bridge was reconstructed, but not for local needs: it functioned as an international Czech–Hungarian trade route, even to Poland and East Germany.

The two communist state's governments reorganized their urban networks (Gutkind 1972). Czech development shifted northwards away from the Danube, along the line to Nové Zámky and Nitra. But the regime also forced the creation of local towns able to absorb rural migrants. Komárno, with its urban character and industrial background, seemed ideal: its new (enlarged) dockyards were among the most important developments in postwar Czechoslovakia. Beside these emerged completely new quarters of prefabricated houses – mostly blocks of flats in socialist realist style. The population of Komárno doubled in this period from 17,395 in 1949 to 37,366 in 1990, thanks mostly to rural migration. New avenues and squares created a very new town plan, in which the other side of the river, and consequently, the area around the bridgehead played no role.

The same happened in Štrurovo, where the population also doubled: 1949, 5,662; 1990, 12.146. Moreover, because Štrurovo had no traditionally significant center or industrial zones, the change was felt even more deeply. The district created by the new prefabricated flats turned its back on Esztergom and faced the enlarged international railway station and the newly established industrial concern (a paper mill). Komárno and Štrurovo's development showed very similar patterns: external industrial areas, large areas of prefabricated blocks of flats and some derelict zones of historical townscape.

Comparing the patterns with the Hungarian towns, Esztergom and Komárom, one can note that they showed less intense population growth (Esztergom: 1930, 20,363; 1990, 29,841; Komárom: 1930, 12.681; 1990, 19.532). This allowed them to preserve more features of their historical townscapes.

Esztergom, the main seat of Hungarian Catholicism, was punished by the Communist administration. The period's main axes of urban development by-passed the town. The new transport routes avoided Esztergom, the region's favored location being Tatabánya where mining was central (Laki 2016). Administrative staff were relocated from Esztergom to this 'miner city'. Fundamental change in the region was delayed until the 1960s, when the central government decided to create a 'new Esztergom of workers'. On Esztergom's outskirts, there developed new residential districts for people working in the fabrics industry and mines in and around Esztergom, first in Dorog. The former Catholic town's social structure changed fundamentally. Most inhabitants in 1990 were newcomers, culturally unconnected to its Catholic past (Bánlaky 1992). The special connections across the Danube also lost their importance. The bridge was not reconstructed – hence Esztergom people had little contact with Štrurovo.

Communications between Komárom and Komárno were much better, thanks to their bridge. However, inter-twin relationships were limited to shopping and culture. Work-based migration remained small. In Komárom,

Table 13.1 Population change in twin towns on the Danube 1850–2011

	1850	1869	1900	1930	1949	1990	2011
Esztergom (HU)	11,454	14, 512	18,206	17,543	20,363	29,841	28,926
Štrurovo (SK)	1,174	missing	3,793	5,867	5,662	12,146	10,828
Komárom (HU)	2579	3677	5836	11.018	12.681	19.532	19.284
Komárno (SK)	13,000	12, 812	17,810	2,.158	17,395	37,366	34,607

Source: based on official censuses – the Data for 1850 was borrowed from Fényes, E. (1951) Magyarország … 1851. (Geographical Dictionary of Hungary), statistics for other years came from Hungarian and Slovak Statistical Offices.

a new main square and 'boulevard' with shopping facilities defined the new urban center. Meanwhile, prefabricated housing followed the old street lines; even the churches (built around 1900) were incorporated into the new urban design as symbols of the history of a very young town (as Table 13.1 showed).

The year 1990 saw a further change in the twin towns' history. The fall of the Communist regimes and EU enlargement progressively opened the borders. After Czechoslovakia split in 1993, Komárno and Štrurovo grew closer to Slovakia's new state capital, Bratislava (Slavik 2000). The gradual opening of borders has enhanced communication between the two river-sides. However, these Central-European geopolitical trends did not enhance a north–south connection (between Slovakia and Hungary), where the twin towns could have exploited their bridge positions. In the new geopolitical situation, the region of Central Europe oriented definitely to the West (to-wards the EU countries: work-based migration for better salaries). It formed an international region, wherein Vienna was central. Vienna was competing in the urban networks with the nations' centers (Budapest and Bratislava). Hungary's and Slovakia's western territories belong today partly to the Austrian capital's catchment area. The twin towns lay relatively far from Vienna. However, a secondary pulling effect of commuting can be observed. Because people commute to Vienna from Bratislava and Győr, the work-places in those towns need people. Komárom and Komárno residents now easily find jobs in Győr or Bratilsava (journey time approximately 30–60 minutes, depending on type of transport) (Enyedi 1998).

After 1990, a very important economic sector has been cargo: the multinational companies need storage capacities for the goods. In Slova-kia, investors preferred the regions around, and to the north of, Bratislava. Thus, towns like Komárno and Štrurovo lost out. The industrial concerns of the communist period could not compete, so most closed their doors. The traditional waterways and railways were no longer utilized. Komárno's harbor and Štrurovo's international goods station declined (Gajdoš and Pašiak 2006).

The transition particularly harshly affected the Slovak side on the twin towns, because they retained communist industrial capacities unable to integrate into the new economic system. On the other side, international capital favored their Hungarian counterparts because those counterparts had good connections to the emerging international economic axis between Vienna and Budapest (Barta 2005). Global firms, like Nokia or Suzuki, chose Komárom and Esztergom as their base. Recently, Nokia has left the region (2014), but new firms arrived. Consequently, over the past quarter century, a north–south difference came into existence among the old twins. The southern parts integrated in the international economy, and the northern ones remained in the shadows of the region's new economic structure.

This trend has been expressed in population and migration trends and recorded by National Statistical Offices of Hungary and Slovakia. Komárom (HU) could attract new inhabitants (thanks to inward migration) between 1990 and 2011. Meanwhile, 2,100 people left Komárno between 1996 and 2014. Moreover, the birth rate also declined from 370 in 1996 to only 240 in 2014. Overall, at the point of writing, Komárno now has around 3,500 (c10%) fewer inhabitants than in the 1990s. Similar trends are evident in Štrurovo.

Meanwhile, the Hungarian towns attracted new inhabitants. The trend of nearly 20 years was that inhabitants of Slovak twin towns migrated or commuted daily into the Hungarian twins. This commuting and migration (change of residence) was also culturally grounded due to the Hungarian-speaking minority in Komárno and Štrurovo. However, migration's economic rationale can always change. Since Slovakia joined the eurozone in 2009, Hungarian salaries have become less attractive for people in Komárno and Štrurovo (Katalin 2015).

The euro's effects on Slovakia's growing economy are hard to grasp in the context of twin-town statistics. Thus, we can only assume that the period 1990–2011 was characterized by inward migration on the Hungarian side and outward migration on the Slovak side. However, because of low birth rates, Esztergom and Komárom's population remained stable.

Our local interviews suggest that open borders generated a new wave of shopping and entertainment activity between the riverbanks. This has followed price differentials between Hungary and Slovakia. Very good examples are the restaurant and spa services in Esztergom and Štrurovo. After the bridge re-opened in 2001, Štrurovo's main street became a gastronomic attraction for Esztergom's inhabitants because of low prices. However, these restaurants mostly closed after 2009, as Slovak euro-based prices increased for Hungarian consumers. Moreover, Slovak citizens now prefer going to Esztergom, mostly to enjoy low-priced spa services.

The very variable economic situation between the Slovak and Hungarian twins has made it hard to undertake common urban planning (Hardi 2017). The only urban area showing some reintegrated fabric is around the bridge-heads. The bridgeheads comprise not only the entrances to the bridge, but the wide areas around those entrances. It is impossible to precisely estimate the areas affected by the two bridgeheads, but they are approximately 2–300 meters from the bridge entrances (Figure 13.1).

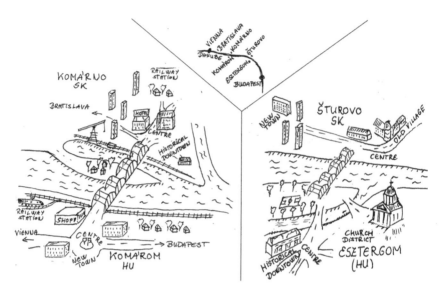

Figure 13.1 Urban structure of twin towns on the Danube by Máté Tamáska.

In many ways, the Danube is too wide for middle-sized twin towns. Esztergom's central area is a 1-km walk from Šturovo's main street. During this walk between the two central areas, we leave Esztergom's built-up areas and cross the island. The island is today the recreation zone of the modern town. This zone developed from the 1960s, during the period of disintegration. Consequently, it offered recreation possibilities mostly attractive to people from Esztergom. Šturovo built its own recreation zone in the same decades, far from here.

Very similar things happened in Komárno (SK), again on an island in the river. Here, the town has used the territory for inner suburbanization, especially since 1990. The suburb links only to Komárno, has no significant connection to Komárom and cannot provide a common urban space. What is the situation directly at the bridgeheads? As noted, the Komárno–Komárom's bridge was quickly re-established after 1945, developing as an international check-point. The former check-point area has no function today, just a brown-field relic of the former socialist regime and border police (unused offices and resting places). It has become a 'non-use' zone. For any pedestrian/visitor, the distance between the twin towns seems even larger than it really is.

In Esztergom–Šturovo, the bridge was not restored after the Second World War; for decades, only ferries served those wanting to cross. Consequently, the border control did not create any infrastructure. When the bridge was reconstructed in 2001, the twin-town structure again became important. However, a noise and air pollution problem arose, due to buses, cars and trucks crossing Esztergom's former recreation area (the island).

The bridgeheads' changing role can be seen also in the character of the buildings. Komárno–Komárom has only the historical town, a characteristic historic urban area around the bridgehead. Its architecture dates from around 1900. During 1900–20, the bridgehead became the main place for establishing hotels, cafes and promenades serving the bourgeois culture of both towns. The divide in 1920 progressively destroyed this square. Visitors today see only a large road with very heavy traffic. There are some important historical buildings here, but the whole aspect of the area is not typical for a historic town, where the urban design preserves what is historically valuable. It is also outside the town's listed monument zone[2]. On the other side, in Komárom, the bridgehead never had much distinguished architecture. Forty years of border control shaped the urban fabric. Urban development documents of the 1970s defined the bridgehead as a 'green zone without urban usage': it belonged to the border police, so urban planners had no input. After 1990, a warehouse store occupied this very important area. It has become a vibrant point of urban life: people from both sides meet here. They talk volubly but just while shopping or in the parking area. There are no cafes or other ways of spending time. Most people in the twins retain a rural life-style. The warehouse fulfills the function of the old markets, where people met spontaneously and chatted.

Esztergom and Štrurovo have enjoyed a more fortunate situation. The bridgeheads are more or less in the same position they were in the age of integration in the early twentieth century. In Esztergom, there are immediately two imposing gates into the town: one leads into the archbishop's quarter (Figure 13.2), and the other to the historic market town. Štrurovo was always poorer than Esztergom. Thus, its bridgehead is also modest. The main street of historical settlement has for hundreds of years run down to the Danube crossing point. But the late nineteenth-century permanent bridge was located some way from this point. As a result, the town built a new street between the bridgehead and the traditional main street. This street became an important highway when the developments of the 1960s and 1970s created a new quarter from prefabricated block of flats. The street became the axis between the new and old towns. Nowadays, the bridgehead is a typical traffic zone.

Viewed overall, these two sets of twin towns show how borders can change such towns over time. The relationship between both pairs before 1920 was very asymmetrical: the historical town occupied the leading position; the other riverside functioned as a completion of the town structure. The railway station and some industrial undertakings comprised two suburban townscapes. Later, the two acts of division, together with the socialist border regimes, helped turn former suburbs into small towns. As a result, the townscapes of both riverside pairs are very different. The historic parts of Esztergom and Komárno have clear monumental zones, which serve as identification points also for people on the other riverside. Meanwhile, the two former suburbs look like new modern towns established in the second half of the twentieth century – as indeed they are in many ways.

Figure 13.2 Esztergom–Štúrovo – a view from the Royal Castle on the Hungarian side, photo by Jaroslaw Jańczak (2013).

The reintegration of both parts after 1990 seems to have been easier in the case of Esztergom–Štrurovo. It is surprising that even Komárom–Komárno functioned as one administrative unit before the border change in 1920. On the other side, Komárom–Komárno gained an international traffic function in the socialist period, therefore the twin towns accommodated new infrastructure: check-points, military buildings and control zones. After border control had ended, there remained the now functionless buildings of the border police. This non-use zone hinders the reunification of the townscape. It would need a new development project to make the space around the bridgeheads attractive. The separation of Esztergom and Štrurovo was in the socialist period stronger but did not produce new infrastructure (it was not a major international check-point). Thus, the integration process could refasten onto the heritage of the time when the first bridge was erected around 1900.

As the introduction noted, division of towns was quite common in Central Europe. Geo-political instability started with the First World War and lasted through the century, occurring in three waves. First, the division of Komárno/Komárom and Esztergom/Štúrovo, and other previously undivided places. The second wave happened after 1945. The division of towns along the German–Polish border was extreme: not only was the border changed but also the population. The third wave occurred in the years following 1990, with the collapse of Communism and of Yugoslavia (for example, producing Brod and Slavonski Brod). With so many cases, we can construct a model of the creation of riverside twin towns:

1   They begin with integrated town structures. The historical town center and the far less urbanized suburb live in functional symbiosis. The suburbs offer mostly economic back-services for development: railway stations, industry (factories) and cheap living conditions for workers. In this phase, the bridge and bridgehead is the most important urban space. Most important new buildings are established here (like multi-storied mansions, hotels, business headquarters, etc.)

2   Disintegration is sudden and dramatic, often with human traumas. Both twins are confronted with critical situations. The historical part has to find a new catchment area, and the former suburb needs to create a new urban center. Therefore, architectural changes are more dramatic in the former suburbs, effectively becoming new towns. Due to the need for a center and other markers of urbanity, the sites become attraction points for contemporary architectural projects and display windows for contemporary architecture. In this period, the former common places of the twins, most notably the bridgeheads, became strange spaces in urban structure: either police control zones or even unused periphery.

3   The reintegration phase faces the problem caused by the twins' urban structures becoming much less functional for common urban life during the decades of disintegration: the local communities follow different social patterns, first in terms of language, but also everyday culture – how people deal with family life, jobs and consumption. This is most influenced by national policy frameworks set by German, Hungarian, Slovak or other governments and less dependent on the locality. Thus, reintegration impacts only some areas of life: festivals, tourism and culture. It also depends on non-human factors like how wide the river is at the point where the reintegrating twins are located. Thus, the twin towns do not become united, cross-border entities, but parallel towns duplicating urban functions.

Notes

1   This chapter results from research led by the author entitled 'Structure of Space and Society in Divided Towns', Nr. OTKA PD 108532, 2015–17 (National Research, Development and Innovation Office).

2   Komárom historic centre has special status as a listed monument zone. In Slovakia, there exist two categories for historical towns: reservation zones with very important historical buildings (around 30 such places) and monument zones with buildings without any special historic value (around 120).

References

Bánlaky, P. (1992) *'Esztergom. A szent és gyámoltalan város'* [Esztergom: the sacral and helpless town], Budapest: MTA.

Barta, Gy. et al. (ed.) (2005) '*Hungarian Spaces and Places: Patterns of Transition*', Pécs: Centre for Regional Studies.

Beluszky, P. and Győri, R. (2005) '*The Hungarian Urban Network in the Beginning of the 20th Century*', Pécs: Centre For Regional Studies of Hungarian Academy of Sciences, Discussion Paper 46.

Bideleux, R. and Jefferies, I. (2007) '*A History of Eastern Europe: Crisis and Change*', New York: Routledge.

Enyedi, Gy. (ed.) (1998) '*Social Change and Urban Restructuring in Central Europe*', Budapest: Akadémiai Kiadó.

Gajdoš, P. and Pašiak, J. (2006) '*Regional Development of Slovakia from the Perspective of Spatial Sociology*', Bratislava: Slovak Academy of Sciences, Institute for Sociology.

Gutkind, E. A. (ed.) (1972) '*Urban Development in East-Central Europe: Poland, Czechoslovakia, and Hungary*', New York: The Free Press.

Hardi, T. (2017) 'Asymmetries in the Formation of the Transnational Borderland in the Slovak-Hungarian Border Region', in Boesen, E. and Schnuer G. (ed.), '*European Borderlands Living with Barriers and Bridges*', Abingdon: Routledge, 176–192.

Jańczak, J. (2018) 'Central European Cross-Border Towns: An Overview', in Garrard, J. and Mikhailova, E. '*Twin Cities: Urban Communities, Borders and Relationships over Time*', Abingdon: Routledge.

Katalin, K. et al. (2015) '*Cross-Border Migration between Slovakia and Hungary*', Budapest: Kopint Foundation for Economic Research Project and Kempelen Institute.

Laki, I. (2016) 'A Short History of Hungarian Industrial Towns from the 1950s until the Regime Change', *Belvedere* 28(2), 109–119.

Lara-Valencia, F., & Dołzbłasz, S. (2018) 'Border-city Pairs in Europe and North America: Spatial Dimensions of Integration and Separation', in Garrard, J. and Mikhailova, E. (ed.), '*Twin Cities: Urban Communities, Borders and Relationships over Time*', Abingdon: Routledge.

Molnár, I. and Szarka. L. (ed.) (2010) 'Memories and Reflections of the Dispossessed. A Collection of Memoirs for the 60th Anniversary of the Czechoslovak-Hungarian Population Exchange', Komárom: Hungarian Academy of Science Research Institute of Ethnic and National Minorities and László Kecskés Society.

Shmuel, S. and Wigoder, G. (ed.) (2003) '*The Encyclopedia of Jewish Life before and During the Holocaust I-III*', New York: New York University Press.

Sikos, T. T. and Tiner, T. (2008) '*One Town - Two Countries Komárom-Komárno*', Komárom: Research Institute of J. Selye University.

Slavik, V. (2000) 'New Trends in Settlement Development and Settlement Planning of the Slovak Republic in the Process of Transformations', *European Spatial Research and Policy* 7(2), 35–49.

Szarka, L. (2013) 'Hungarian National Minority Organizations and the Role of Elites between the Two World Wars: Addenda to the History of Minority Nationalism in Central and Eastern Europe', *Hungarian Historical Review* 2, 413–448.

14 Central European cross-border towns

An overview[1]

Jarosław Jańczak

With this chapter, we move far more clearly into a cross-border world where integrative positivity is expected, indeed where twins can become 'laboratories' for testing this. We are still mostly within the EU, but now exploring the border between the new and the old EU, and thus the world of recent accession states. Utilizing 14 case studies, the author focused on the tensions created by that accession: between integration and disintegration, deboundarization and defrontierization, and we catch sight of the interplay of pressures generated by the EU, national governments and local communities.

Against the background of European integration, cross-border cooperation involving a wide spectrum of actors has been a common borderland experience across the European continent in recent decades. Fueled by liberalized flows of people and goods, alongside eroding internal EU borders, various forms of local integration are detectable. Processes initiated in Western Europe spilled into its central and eastern parts along with EU enlargements in 2004/2007 and Schengen zone expansion. These processes are especially visible in border-twin towns.

   The aim here is to provide an overview of border integration and disintegration processes in Central Europe, taking border twin towns as examples. I argue that cross-border integration is not always linear, as suggested by the non-functional logics (Jańczak 2013) of the European integration theory (which assumes that integration initiatives are usually successful, leading to further steps). Its pace and dynamism vary according to different categories of borders in the region. The deboundarization observable in most cases after 1989 was motivated by continental integration involving national and local authorities as well as local communities in border-twin towns. This did not apply to the border that later became the external EU boundary. However, in the post-enlargement period, the process halted on some borders, even being replaced by defrontierization attempts, usually driven by central authorities and opposed by local actors. The chapter draws on secondary sources, complemented by semi-structured interviews with key actors responsible for local cross-border policy, mainly politicians and officials from border twin towns. They were conducted in Frankfurt (Oder), Giurgiu, Gmünd, Gornja

Radgona, Görlitz, Komárom, Komárno, Nova Gorica, Ruse, Słubice, Štúrovo and Zgorzelec between February 10, 2011, and July 12, 2013.

Further analysis requires definition of the categories used here. First, Central Europe is understood in the context of EU integration: the area of the European Union's 2004/2007 enlargement. Ignoring the term's different spatial-historical interpretations, I consider this region as a space where economic and political transformation has followed the collapse of communist regimes effectively enough to meet the Copenhagen criteria[2] and be accepted into western European structures. Consequently, the term denotes the area between the 'old Union' in the West (wealthy, democratic and integrating for decades) and the post-Soviet space in the East (with unsuccessful transformations and unstable structures), additionally characterized by similar political and economic processes since 1989.

Second, border twin towns, following Helga Schulz (Schulz, Stokłosa and Jajeśniak-Quast 2002), are seen as 'towns separated by an international state border', and directly neighboring each other. There is no terminological consensus among writers and such towns are also called 'twin cities' (Joenniemi and Segunin 2009, 231f), 'border double towns' (Szalbot 2011) or 'binational cities' (Ehlers, Buursink and Boekema 2001). In border studies, they are seen as anomalies (Lundén 2004, 1–2), undermining the border's separative logic, especially in the Westphalian sense. In the context of continental integration, specific processes seem to concentrate there, explaining why they are often seen as laboratories of European integration (Gasparini 1999–2000, 1). Their origins lie in the divisions of previously single towns resulting from border shifts, the duplication of a town on the other side of a border, or the connection of two towns separated by a natural barrier through new infrastructure (Buursink 2001, 8). In Europe, they appeared in several waves (starting in the west, with the most recent ones in the east): some remained from the feudal-medieval territorial-political order, and others appeared as a result of the Napoleonic wars, the First and Second World Wars and, finally, the collapse of communism (Jańczak 2013, 36–71). In Central Europe, almost all cases come from the last three waves, unlike Western Europe where pairs related to previous waves dominate.

On this basis, 14 sets of border twin towns are identifiable in Central Europe (Map 1), ignoring villages and other non-urban settlements as well as towns not located directly on the border. Two are located on the EU's outer border, where Central Europe abuts the post-Soviet space: Terespol–Brest (No. 3) and Narva–Ivangorod (No. 1); six on the region's 'internal' borders: Calafat–Vidin (No. 13), Giurgiu–Ruse (No. 14), Komárno–Komárom (No. 9), Štúrovo –Esztergom (No. 10), Cieszyn–Český Těšín (No. 7) and Valga-Valka (No. 2) and, finally, another six sit along the border between the 'old Union' and Central Europe: Gorizia–Nova Gorica (No. 11), Bad Radkersburg–Gornja Radgona (No. 12), České Velenice–Gmünd (No. 8), Frankfurt (Oder)–Słubice (No. 4), Guben–Gubin (No. 5) and Görlitz–Zgorzelec (No. 6) (Figure 14.1).

LEGEND

- Central Europe (CE)
- Eastern border of CE
- internal borders of CE
- Western border of CE

Border twin towns
- with conflict legacy
- with elements of conflict legacy
- without conflict legacy

1 - Narva - Ivangorod
2 - Valga - Valka
3 - Terespol - Brest
4 - Frankfurt (Oder) - Słubice
5 - Guben - Gubin
6 - Görlitz - Zgorzelec
7 - Český Těšín - Cieszyn
8 - České Velenice - Gmünd
9 - Komárno - Komárom
10 - Štúrovo - Esztergom
11 - Gorizia - Nova Gorica
12 - Bad Radkersburg - Gornja Radgona
13 - Calafat - Vidin
14 - Giurgiu - Ruse

Figure 14.1 Map of border twin towns in Central Europe. Map content by Jaroslaw Jańczak, map design by Ekaterina Mikhailova.

Finally, we need a short debate on the concept of a border. Considering borders as tools for distinguishing 'us' from 'non-us' seems to be widely accepted, but does not really take in the nature and dynamism of borders. Consequently, following Ladis Kristof, it seems useful to distinguish *frontiers* and *boundaries*. In the first case, borders are open and influence (and are also influenced by) both neighboring states; in the second, borders mark the limits of state territories unified by their respective centers (Kristof 1959). Frontiers can be understood as spaces of contact between various political, cultural and social orders. Boundaries, emerging in Europe following the Westphalian understanding of territoriality and sovereignty, are lines of exclusive control, separating political-territorial entities (Evans and Newnham 1998, 185).

As border organization changes over time, the European debate has in the last three decades concentrated on debordering and rebordering. This shows the dynamic nature of borders and the ongoing process of their creation and elimination, their hardening and softening (Scott 2016, 14–15). However, employing the concepts of boundaries and frontiers, it is more interesting to approach European border processes because they reveal the more complex nature of border-related changes. In their case, we can also apply the viewpoint of boundarization and frontierization (with the prefixes de- and re-) determining the direction of changes in border organization.

Boundarization in turn means the processes of boundary creation. Historically, this was connected with the appearance of nation-states in Europe, exercising exclusive sovereignty within precisely marked territories. Frontierization is linked to the reestablishment of frontiers and goes together with the erosion of the nation-state. European integration has contributed to this latter process on its internal borders by, among others, cross-border cooperation programs and the elimination of border controls. But boundarization and frontierization are not linear processes (following the logics of functional spill-over). They can accelerate, but also stop or even retreat. Consequently, periods of deboundarization can be followed by reboundarization and frontierization by defrontierization.

The following sections will empirically examine deboundarization and refrontierization processes in the three categories of borders in Central Europe.

Central Europe's eastern border

This border, being also the EU's and the Schengen zone's eastern border, has experienced constant and systematic reboundarization. Alongside the accession of states in the region to western European structures, the introduction of an evermore restrictive border regime has been experienced, strongly affecting towns located there, despite numerous attempts to overcome the border's dividing role. This has produced the ongoing separation of towns from each other. Debordering inside the Schengen zone has been followed by rebordering on external borders. Two cases represent this process: Terespol–Brest and Narva–Ivangorod.

The Terespol–Brest pair (No. 3), on the Polish–Belarusian border, straddles the Bug River. Originally side by side, they were relocated (Brest was moved 2 km to the east, and Terespol to the west) in the nineteenth century due to fortifications being constructed near the river. In 1945, the Polish–Soviet border was moved to the Bug line (Kisielowska-Lipman 2002, 137–139), separating both towns politically and functionally. And it remained practically closed until communism's collapse in Poland and the rise of an independent Belarus. The initially liberal visa regime hardened as Poland prepared to enter the Schengen area. This limited social contacts across the border again, additionally impaired by very poor border infrastructure: producing a permanent traffic jam, an inconvenient railway connection and a lack of facilities for pedestrian cross-border traffic (Kindler and Matejko 2008, 25–26, 31, 55). All this significantly reduced mutual contacts between the residents and the authorities of both towns (Kisielowska-Lipman 2002, 148).

The national and language structure of both towns reflects post-Second World War changes. Terespol residents almost unanimously declare their Polish nationality during censuses (Sadowski 2002, 293). In Brest, a Belarusian/Russian ethnic and cultural character dominates (Eberhardt 1994, 125, 150–151, 169). Poles and Jews formed the majority of the pre-war population; the latter group was exterminated by Nazi German troops, and the former was repatriated to Poland, or assimilated after intensive Russification (Książek 2009).

Everyday contacts between city authorities and between inhabitants are limited. Crossing the border is not a daily practice, except for smuggling groups earning their living from the border. Other groups maintain mainly commercial contacts, if any (Jańczak 2011, 47–48). Both the visa regime and poor infrastructure led most inhabitants to declare that they have never visited the city across the border (Kolanowski 2004). This situation, observed at the time of Poland's accession to the European Union in 2004, was maintained by further border-regime hardening, together with the Schengen zone enlargement in 2007. Non-governmental organizations rarely collaborate with partners across the border as reported by local actors (Terespol City Administration 2011, 28). This is reinforced by the political regime and political culture dominating Belarus, with weak civil society and a high level of control over local initiatives by the central authorities (Jańczak 2011, 47–48). The towns' authorities signed a cooperation agreement in 2002, but its outcomes so far are limited (mainly a Border Days festival and a cross-border marathon). There are limited EU-funded cross-border projects, but there were talks about opening the historical fortresses on both sides of the border to both countries' tourists and building a pedestrian bridge or passage linking both towns. The latter was initiated in 2011 but never materialized. In the meanwhile, sports and cultural cooperation between both towns accelerated. In 2017, Brest's local authority lobbied in Minsk, the Belarus capital, to introduce local visa-free facilities for Poles, which could enhance human flows and local cooperation (Burda 2017). Eventually, these were introduced on January 1, 2018, as part of a wider process of Belarus opening up for short-term border tourists initiated in 2017.

The case of Narva–Ivangorod (No. 1) on the Estonian–Russian border represents a story of numerous divisions and unification, with the last separation in 1991, alongside the USSR's collapse and the emergence of independent Estonia. In the middle ages, the city was a border-point between Denmark, Sweden and Russia, still symbolically visible in the two castles facing each other across the Narva River. As part of the Russian Empire, it became an Estonian city in the interwar period. After Estonia's incorporation into the USSR, the internal border, separating two socialist republics within the Soviet Union, was shifted, and Ivangorod reappeared as a separate town. Both Narva and Ivangorod were filled with Russian speakers replacing almost all Estonians (Kaiser and Nikiforova 2008, 545–546); the border was not noticeable in everyday life. After Estonia regained independence in 1991, each neighboring state claimed the town across the border (Burch and Smith 2007, 923). The state boundary was set and border controls introduced (Pihlak 2008, 28). Together with Estonian aspirations to join NATO and the EU, the border and border regime hardened, significantly reducing its permeability. Further separation eroded social ties (Boman and Berg 2007, 206) as well as official and infrastructural links (Joenniemi and Sergunin 2009, 18). Eventually, as a part of EU policies, projects aiming at reducing the separating character of the external Schengen border were initiated, especially around tourism, with special focus on the castles. Both were revitalized and linked for tourists between 2011 and 2014. These have contrasted, however, with both the 'Estonization' and 'Europeanization' of the space in Narva (Nyyssönen and Jańczak 2015) and Russian speakers being perceived as Russian Federation agents (Makarychev and Yatsyk 2016). The result is that any sense of commonality 'is quite rare in Narva–Ivangorod' as indicated by some researchers (Stokłosa 2017, 312). At the same time, others notice that, due to their ethnic and linguistic proximity, 'friendly and professional ties between the residents of both sides of the border are quite strong' and institutionalization of cooperation at the level of local authorities is high, especially in the context of the availability of European funds (Mikhailova 2018). This happens in opposition to high politics at the level of relations between NATO, the EU, Estonia and Russia (Tambi 2016).

Central European 'internal' borders

The cases located on the region's 'internal' borders reveal two, sometimes contradictory, tendencies. On the one hand, deboundarization is observable, strongly supported by local actors. On the other hand, central authorities, after initially supporting this process, sometimes tend to support the opposite process of de-frontierization, even reboundarization. Consequently, these internal pairs have also experienced two opposing tendencies. Because of the Copenhagen criteria and the 'European integration spirit', central authorities have had to establish friendly relations with their neighbors, often requiring reconciliation, openness, cooperation, etc., and leading locally to deboundarization. Following the 2004 EU enlargement and the 2007 Schengen expansion, the national centers of most states in the region achieved their political aims. Central authorities then often lost interest in further cross-border cooperation since it sometimes undermined traditional sovereignty in the border areas, especially where national minorities were strongly present and borders had a legacy of conflict. At the same time, however, local authorities and local communities are often interested in further integration for both politico-cultural and economic reasons. This sometimes leads to clashes with the national centers.

The emergence of the Giurgiu–Ruse twin towns (No. 14) on the Romanian–Bulgarian border is connected with the construction of the Danube Bridge in 1954 (ROBULNA 2012, 28) and provides an example of connected towns. They were established in the Middle Ages on both sides of the Danube, and existed – in practice – independently due to the river's size constituting a natural barrier. The current border was set in 1978, alongside the independence of Romania and Bulgaria. Until 1954, when the bridge linking the towns was first built (for several decades, this border's only bridge), the Danube isolated states, nations and cultures. Later not much changed, due to the border regime under the communist authorities. As with other borders in the region, EU accession sparked border relations between the national and EU authorities, inducing numerous projects softening this border's dividing character. In higher education, the Bulgarian–Romanian Interuniversity Europe Center was created (Assenmacher 2005, 102; Kormazheva 2004), alongside many other initiatives. They have, however, not significantly increased local communities' interactions (Interview 9, Ruse). This is partly because border-regime liberalization has never been completed. Romania and Bulgaria have never entered the Schengen zone, so border controls still exist on the bridge and crossing is additionally conditioned by a toll. Local authorities, lobbying to exempt local inhabitants from this fee, have never been positively answered in Bucharest and Sofia, and feel marginalized by the national centers (Interview 8, Giurgiu). The second pair, Calafat–Vidin (13), was connected only in 2013, when a second bridge was completed. Evaluating this case is impossible due to the recent character of the integrative processes involved.

Komárno–Komárom (No. 9) and Štúrovo–Esztergom (No. 10) are located on the Slovak–Hungarian border on the Danube. When the First World War ended, they were Hungarian towns. Both Austro-Hungarian disintegration and Czechoslovakia's creation in 1920 produced a new border splitting the towns along the river, and leaving a Hungarian minority on the Czech side (Hardi 2017). Both Komárno and Štúrovo were the centers of this minority. After the Second World War, the border was poorly permeable, and the bridge between Štúrovo and Esztergom was not rebuilt until 2001. Together with the collapse of communism and creation of an independent Slovakia, border liberalization was implemented, restoring ties between the two sides (Interview 6, Komárom). Again, the prospects of joining western structures pushed both states into creating a friendly environment, manifested, among other things, in cross-border cooperation, national minorities being granted rights and privileges. Both local communities and local authorities were deeply involved in this process, often stressing the cultural and national proximity of the neighboring towns (see Chapter 13 by Tamáska, this volume). The symbolic role of both pairs was visible during EU enlargement eastwards, when the Hungarian Selye Janos University was established in Slovak Komárno, the aforementioned bridge was reconstructed in Štúrovo–Esztergom and the European Square was constructed in Komárno (Schultz, Stokłosa and Jajeśniak-Quast 2002, 51).

The 2004 enlargement meant that Slovak and Hungarian political goals were achieved. Meanwhile, especially in Slovakia, there were vivid opinions about the re-Hungarization of the Slovak side of the borderland, as well as about Budapest's strong influence there (Interview 5, Komárno; Interview 7, Štúrovo). Consequently, Bratislava wanted to regain the former level of control over its territory. Symbolically, this was represented by Hungarian President László Sólyon being stopped on the border bridge in 2009 upon trying to enter Slovakian Komárno to unveil a monument to the Hungarian king, Saint Stephen. The towns still seek to deepen cross-border cooperation – especially since the ties are powerful with regard to market forces and identities (Balogh and Pete 2017). However, the central authorities think that it generates problems.

Cieszyn and Český Těšín (No. 7) are located on the Polish–Czech border. Until 1918, it was a single town controlled by the Habsburgs, divided alongside the Olsa River. The twin towns emerged during the rebirth of the Polish state and the rise of Czechoslovakia in the years 1918/19. A significant Polish minority remained on the southern side, illustrating the legacies of conflict there. After 1945, the border was relatively open, yet cooperation was framed by 'socialist friendship' principles, which means that it was moderated and organized by the communist authorities. The friendly international environment of eastern enlargement allowed both towns to become a spot where both real and symbolic forms of deboundarization were implemented, demonstrating good Polish–Czech relations. The context of the Visegrad Group was also useful here. After 2007, the interest of the central authorities started visibly decreasing, contrasting with local authorities' co-operative involvement. Additionally, legacies of conflict started manifesting in new forms, e.g. the erection of a monument in 2012 to General Josef Šnejdárk, the commander of the Czechoslovak troops who conquered Český Těšín in 1919; it was rapidly vandalized. Latest research shows there are still two mirror-town milieus, instead of a unified cross-border one in the pair. This is visible both in the duplicated spatial and functional structures (town centers, suburbs, etc.) and in the local communities' perception of those structures (see Chapter 13 by Tamáska, this volume).

Finally, Valga–Valka (No. 2) on the Estonian–Latvian border emerged after the end of the First World War, in 1920. The previously existing single town was divided by a state boundary, causing relocation for over 2000 Latvians to the southern part of the city. Together with the Baltic States' incorporation into the USSR in 1940, the border acquired administrative significance and again became invisible to the inhabitants. This changed further with Estonian and Latvian independence in 1991. Border controls were introduced, and both social and functional structures were disconnected. EU membership and, later, the Schengen zone again removed the border from the cities daily functioning (see Chapter 16 by Lundén, this volume). The year 2004, and then 2007 witnessed many events of European unity organized by both countries' governments. Local authorities also got involved in several joint projects (branded 'One Town, Two States'), mainly in spatial planning, education and culture. However, the local authorities often claim that the central authorities marginalize both local needs and cross-border problems.

Central Europe's western border

This border between the old and the new Union reveals ongoing political deboundarization, with subsequent refrontierization. This process has been strongly centrally inspired and fueled by political and historical reconciliation on historically problematic borders (Italian–Slovenian, Austrian–Slovenian, Austrian–Czech and German–Polish). Given the difficult legacy of border shifts, population expulsions, nationalization of space, etc., overcoming it was a precondition for new relations between the west and east in Europe's post-Cold War political environment. So, Germany, Italy and even Austria had to prove that they were peaceful and friendly actors on the border, whereas Poland, the Czech Republic and Slovenia had to establish their reliability as candidates for entry to western structures. Border twin towns after 1989 became interfaces between Old and New Europe.

Gorizia and Nova Gorica (No. 11) on the Italian–Slovenian border represent a town historically influenced by Italy, Austria and Slovenia and having strong borderland legacies. In the interwar period, Gorizia belonged to Italy but was inhabited mainly by Slovenians. In 1947, it was divided between Italy and Yugoslavia. As the latter obtained only the eastern suburbs plus the strategic railway station, Nova Gorica was constructed in 1952 (Jańczak 2009, 112). Alongside Slovenian independence, the EU's eastern enlargement, Schengen expansion and the Euro's introduction into both states, border relaxation started. This resulted in the towns' functional reunification, embodied in the creation of several joint projects and policies. The local communities have relatively high levels of social contact, especially because of Gorizia's Slovenian minority (Interview 11, Nova Gorica). The Italian and Slovenian central authorities used both towns to manifest friendly relations (Figure 14.2).

The pair of Bad Radkersburg and Gornja Radgona (No. 12) on the same border resulted from dividing Austrian Bad Radkersburg (Kurahs et al.

Figure 14.2 Gorizia–Nova Gorizia: a former border crossing point that cuts a 'normal' street, photo by Jaroslaw Jańczak (2013).

1997, 112–30) when the Austrian Republic and Kingdom of Serbs, Croats and Slovenes were established in the years 1918–20 (Just 2007, 21). The interwar period saw the Germanisation of Austria's Slavic population. The post-war years, after initial neighborly tensions, were marked by Austrian–Yugoslav relaxation, symbolically reflected also in this pair: among other things, stimulating the reconstruction of the city bridge on the Mur River border. After Slovenian independence and the accessions in 1995 and 2004, states and towns lay within the EU. Co-operation and mutual contacts were additionally strengthened by Schengen inclusion and the Euro. Intensive, functionally oriented cooperation, including education, culture, tourism, etc., was supplemented in 2009 by the reconstruction of the bridge, which became a meeting place for local inhabitants (Interview 10, Gornja Radgona).

České Velenice–Gmünd (No. 8) on the Czech–Austrian border was divided in 1919 by the Treaty of Saint Germain, partly along the Lainsitz River, partly by a land border. The strategic railway station was left on the Czech side, stimulating the creation of a new town there (Lohninger et al. 2005, 278–283). German speakers dominated both sides. After the Second World War, however, they were expelled to Austria. Consequently, two separate and relatively homogeneous German and Czech communities appeared on either side of the border. The EU's eastern enlargement, followed by the Schengen zone enlargement, led to functional reunification. Both towns cooperate, claiming to be a meeting place in the very center of Europe. The EU is considered a catalyzer of this process (Interview 12, Gmünd).

The emergence of three city pairs on the German–Polish border – Frankfurt on the Oder–Słubice (No. 4), Guben–Gubin (No. 5) and Görlitz–Zgorzelec (No. 6) – is connected with the division of German cities located along the Odra and Neisse River in 1945, when the German–Polish border was moved eastwards. The eastern suburbs (in the case of Guben's historical center) became Polish cities, and the German population fled or was expelled and replaced by Polish settlers. Most were refugees from Polish territories transferred to the USSR. The border was closed until the early 1970s, and then again in the early 1980s. Consequently, two alien and isolated communities were living adjacent to each other in the three pairs, without possibility of developing real contacts. German reunification and Poland's EU aspirations pushed both states into normalizing border relations and creating a new political environment. The pairs very quickly began evidencing these new grand relations, exhibiting the European project on a micro scale. Even before EU and Schengen zone enlargement, dozens of cultural, social and educational projects emerged, especially the European university Viadrina in Franfkurt (Oder) and Collegium Polonicum in Słubice. Similarly, infrastructural cooperation was very intensive there (Interview 2, Słubice) producing close cooperation in cross-border governance, tourism development (Interview 4, Zgorzelec), common usage of resources (Interview 1, Frankfurt/Oder) and attracting investors (Interview 3, Görlitz). Słubice and Frankfurt (Oder) promote themselves as a 'city without borders', forming a joint urban landscape (Opiłowska 2017, 340–341). A shift in the mutual perception of the neighbors in twin towns is visible in empirical research, changing from negative to positive (Dębicki and Makaro 2017, 366).

Conclusions

Central Europe's border twin towns are locations where border processes are intensively visible. Most are closely related to the dynamics of European integration, being a framework of debordering or rebordering in each pair. Continental integration and disintegration has rendered them 'laboratories of European integration'.

However, the transformations in the borders and border-twin towns have not been linear. After 1989, continental deboundarization affected the western and 'internal' borders of the region, whereas its eastern border experienced (re)boundarization. In the post-EU accession period, the former were affected by some elements of the reverse phenomenon. On Central Europe's western border, deboundarization and refrontierization were caused (alongside the efforts to access the EU) mainly by reconciliation between neighboring states and nations. This happened by 2004 and the dynamism of border processes slowed down. On the 'internal' borders, however, this trend was even reversed, especially in border-twin towns with strong conflict legacies. Here, centrally inspired trends towards defrontierization (fueled by sovereignty-based fears) have often been opposed by local authorities and communities.

At the region's eastern border, being at the same time the external EU's border, reboundarization has occurred, producing the ongoing disintegration of the border twin towns. This is usually fueled by central and local actors trying to prevent it and – more or less successfully – to soften the dividing nature of the boundary there.

Notes

1 This chapter is a substantially revised version of the text that was published in Polish in 2013 as Jańczak, J., Integracja i dezintegracja w Europie Środkowej. "Graniczne miasta bliźniacze" jako laboratoria współpracy transgranicznej. Rocznik Integracji Europejskiej, 7, 265–279.
2 EU rules established in 1993, determining the eligibility and preparedness of candidates joining this organization.

References

Assenmacher, B. (2005), 'Bulgarisch-Rumänisches Interuniversitäres Europa-Zentrum: Hochschulzusammenarbeit in der südosteuropäischen Grenzregion als Motor für regionale Entwicklung', in: Duda, G., and Wojciechowski, K., eds., *Trans-Uni. Herausforderungen des Managements bei der internationalen Hochschulzusammenarbeit in den Grenzregionen*. Bonn: HRK, 102–107.

Balogh, P., and Pete, M. (2017), 'Bridging the Gap: Cross-border Integration in the Slovak–Hungarian Borderland around Štúrovo–Esztergom', *Journal of Borderlands Studies*. https://doi.org/10.1080/08865655.2017.1294495.

'Biuletyn informacyjny miasta Terespol', (2008), No. 3, available: http://www.terespol.pl/podstrony/gazeta-biuletyn/biuletyn308.pdf (accessed January 8, 2018)

Boman, J., and Berg, E. (2007), 'Identity and Institutions Shaping Cross-Border Co-Operation at the Margins of the European Union', *Regional and Federal Studies,* 17, 195–215.

Burch, S., and Smith, D. J. (2007), 'Empty Spaces and the Value of Symbols: Estonia's "War Monuments' from Another Angle", *Europe-Asia Studies,* 59(6), 913–936.

Burda, K. (2017), 'Do Brześcia bez wizy? Chcą w ten sposób przyciągnąć Polaków', *Dziennik Wschodni*, 2 February. https://www.dziennikwschodni.pl/biala-podlaska/do-brzescia-bez-wizy-chca-w-ten-sposob-przyciagnac-polakow,n,1000193541.html (accessed January 3, 2018)

Buursink, J. (2001), 'The Binational Reality of Border-Crossing Cities', *GeoJournal,* 54, 7–19.

Dębicki, M., and Makaro, J. (2017), 'German neighbours in the eyes of the inhabitants of Gubin. An analysis of stereotypes in the Polish German twin town', in: Opiłowska, E., Kurcz, Z., and Roose, J., eds., *Advances in European Borderlands Studies.* Berlin: Nomos Verlag, 345–369.

Eberhardt, P. (1994), 'Przemiany narodowościowe na Białorusi', Warszawa: Editions Spotkania.

Ehlers, N., Buursink, J., and Boekema, F. (2001), 'Introduction. Binational Cities and Their Regions: From Diverging Cases to a Common Research Agenda', *GeoJournal,* 54, 1–5.

Evans, G., and Newnham, J. (1998), '*The Penguin Dictionary of International Relations*', London: Penguin Books.

Gasparini, A. (1999–2000), 'European Border Towns as Laboratories of Differentiated Integration', *ISIG Quarterly of International Sociology*, 4, 1–5.

Hardi, T. (2017), 'Asymmetries in the formation of the transnational borderland in the Slovak-Hungarian border region', in: Boesen, E., and Schnuer, G., eds., *European Borderlands. Living with Barriers and Bridges.* London and New York: Routledge, 175–192.

Książek, J. (2009), 'Historia "Karta" spisana', *Echa Polesia* 3 (23), 10–14. http://pdf.kamunikat.org/13197-1.pdf (accessed January 15, 2018)

Jańczak, J. (2009), 'Gorizia – Nova Gorica. Between unification and reunification', in: Jańczak, J., ed., *Conflict and Cooperation in Dicided Cities.* Berlin: Logos Verlag, 122–132.

Jańczak, J. (2011), 'Cross-border governance in central European border twin towns. Between de-bordering and re-bordering', in: Jańczak, J., ed., *De-Bordering, Re-Bordering and Symbols on the European Boundaries.* Berlin: Logos Verlag, 37–52.

Jańczak, J. (2013), '*Border Twin Towns in Europe. Cross-border Cooperation at a Local Level*', Berlin: Logos Verlag.

Joenniemi, P., and Sergunin, A. (2009), 'When Two Aspire to Become One: City-Twinning in Northern Europe', DIIS Working Paper, 21.

Just, F. (2007), '*Gornja Radgona*', Murska Sobota: MA-TISK.

Kaiser, R., and Nikiforova, E. (2008), 'The Performativity of Scale: The Social Construction of Scale Effects in Narva, Estonia', *Environment and Planning D: Society and Space,* 26, 537–562.

Kindler, M., and Matejko, E. (2008), 'Gateways to Europe. Checkpoints on the EU External Land Border. Monitoring Report', Warsaw: Stefan Batory Foundation.

Kisielowska-Lipman, M. (2002), 'Poland's Eastern Borderlands: Political Transition and the Ethnic Question', *Regional & Federal Studies*, 12, 133–154.

Kolanowski, K. (2004), 'Brest und Terespol – Eine Doppelstadt am Bug, Internationale Tagung "Stadt – Grenze - Fluss"', 29–30 April, Collegium Polonicum, Slubice.

Kormazheva, M. (2004), *'Risks and Opportunities of Trans-Frontier Co-operation in Higher Education. The Case of Bulgarian-Romanian Interuniversity Center'*, Rousse: University of Rousse Proceedings.

Kristof, L. K. D. (1959), 'The Nature of Frontiers and Boundaries', *Annals of the Association of American Geographers*, 49(3), 269–282.

Kurahs, H., Reidenger, E., Szedonja, S., and Weiser, J. (1997), *'Bad Radkersburg. Naturraum und Bevölkerung. Geschichte. Stadtanlage. Architektur'*, Wolfsberg: Stadtgemeinde Bad Radkersburg.

Lohninger, D., Hermann, M., Neunteufel, V., and Winkler, V. (2005), *'Gmünd. Chronik einer Stadt'*, Gmünd: Verlag Bibliothek der Provinz.

Lundén, T. (2004), 'European Twin Cities: Models, Examples and Problems of Formal and Informal Co-Operation', *ISIG Quarterly of International Sociology*, 3–4, 1–17.

Lundén, T. (2018), 'Border Twin Cities in the Baltic Area – Anomalies or Nexuses of Mutual Benefit?' in: Garrard, J., and Mikhailova, E. eds., *Twin Cities: Urban Communities, Borders and Relationships over Time*, Abingdon: Routledge.

Makarychev, A., and Yatsyk, A. (2016), 'Russia-EU Borderlands after the Ukraine Crisis: The Case of Narva', in: Besier, G., and Stokłosa, K., eds., *Neighbourhood Perceptions of the Ukraine Crisis. From the Soviet Union into Euroasia?* Farnham and Burlington: Ashgate, 105–120.

Mikhailova, E. (2018), 'Goroda-bliznetsy v rossiyskom pogranich'e: opyt sotrudnichestva na mestnom urovne', in: Kolosov, V., ed., *Rossiyskoe pogranich'e: vyzovy sosedstva.* Moscow: Matushkin's Publishing House, 278–305.

Nyyssönen, H., and Jańczak, J. (2015), 'Conflit et coopération dans l'espace public. (Re)négociation symbolique de la frontière historique et contemporaine entre la Scandinavie et la Russie', *Revue d'histoire nordique,* 19, 151–177.

Opiłowska, E. (2017), 'Creating transborderness in the public spaces of the divided cities, The case of Frankfurt(Oder) and Słubice', in: Opiłowska, E., Kurcz, Z., and Roose, J., eds., *Advances in European Borderlands Studies.* Berlin: Nomos Verlag, 317–343.

Pihlak, M. (2008), 'Estonian country report', in: Pääbo, H., and Pihlak, M., eds., *Monitoring European Border-Crossing Points on the EU's Eastern Border: Estonian and Russian Case Studies.* Tartu: University of Tartu, 22–75.

'ROBULNA. Romanian-Bulgarian Neighbourhood Area', (2012), Ruse: BRIE. http://robulna.eu/brains/wp-content/uploads/2012/01/Leaflet-ENG1.pdf (accessed January 21, 2018)

Sadowski, A. (2002), 'Polska polityka wschodnia a pogranicze polsko-białoruskie', in: Polskie pogranicza a polityka zagranicza u progu XXI wieku,(red.) R. *Stemplowski, A. Żelazo, Warszawa*: Polski Instytut Spraw Międzynarodowych, 281–298.

Schulz, H., Stokłosa, K., and Jaeśniak-Quast, D. (2002), 'Twin Towns on the Border as Laboratories of European Integration', FIT Discussion Paper, 4.

Scott, J. (2016), 'Rebordering Central Europe: Observations on cohesion and cross-border cooperation', *Cross-Border Review*, European Institute of Cross-Border Studies - Central European Service for Cross-border Initiatives (CESCI), Budapest, 9–28. ISSN 2064-6704.

Stemplowski, R., and Żelazo, A., eds., '*Polskie pogranicza a polityka zagraniczna u progu XXI wieku*', Warszawa: Polski Instytut Spraw Międzynarodowych, 281–297.

Stokłosa, K. (2017), 'Twin towns in Eastern Europe: on the way from bordering to the cross-border identity?' in: Opiłowska, E., Kurcz, Z., and Roose, J., eds., *Advances in European Borderlands Studies*. Berlin: Nomos Verlag, 307–315.

Szalbot, M. (2011), 'Społeczno-kulturowa specyfika przygranicznych miast podwójnych Europy jako problem badawczy', in: Bukowska-Floreńska, I., and Odoja, G., eds., *Studia etnologiczne i antropologiczne. Tom 11. Etnologia na granicy*. Katowice: Wydawnictwo Uniwersytetu Śląskiego, 141–152.

Tamáska, M. D. (2018), 'Spatial portrait of twin towns on the Danube', in: Garrard, J., and Mikhailova, E., eds., *Twin Cities: Urban Communities, Borders and Relationships over Time*, Abingdon: Routledge.

Tambi, S. (2016), 'The Contemporary Development of the Concept of the Twin Cities: The Case of Cross-Border Cooperation between Narva and Ivangorod', Master thesis. Tartu.

Terespol City Administration. (2011), 'Strategy of cooperation of Terespol with Non-Governmental Organizations in Years 2011–2016', Terespol.

15 Comines and Wervik

On the three-way divide of two historically integral towns

Peter Martyn

Due to historically friendly Franco-Belgian relations, nothing has been formalised between what, as we shall see, are the two parts of Comines and Wervik allocated on both sides of the naturally eroding border. However, an unconnected Belgian decision to federalise itself to try to cope with the Flemish–Walloon divide has greatly complicated relationships between these treaty-created twins – thereby emphasising another enduring characteristic of border-twin life – tensions between national, regional and local authorities/communities and the impact of sometimes unconnected decisions taken further up the political systems of one or other of the countries concerned.

The exceptional triadic (eventually fourfold) relationship between two once-complete urban centres, situated well under an hour's walk of each other, arose from the seventeenth-century Southern Netherlands' territorial truncation, as well as comparatively recent adoption of a federal system of government in the modern-age geopolitical successor: Belgium. Northern Brabant was thus lost to the Dutch Republic and a vital part of Flanders surrendered to the French Kingdom a century-and-a-half before Belgian independence. Devolution, from the 1960s, while giving spoken Flemish equal status with French, separated the so-called 'Flemish' (North) from the so-called 'Walloon' (South) autonomous regions. This crude divide has compromised historic distinctions between the constituent Belgian provinces of West and East Flanders, Hainaut, Brabant, the Namurois, Liègeois, Limburg and Luxembourg.[1] The once highly developed and politically self-determining, bilingual county of Flanders (Fr.: *Compté de Flandres*; Flem.: *Graafschap Vlaanderen*) has been banished to history and replaced by a territorially augmented, monolingual region. This expansion may be typically perceived by innumerable Flemings as amply compensating the exclusion of primarily Francophone[2] areas, even while historically Flemish cities fully integrated within pre-eighteenth-century Flanders have been lost to Belgian Hainaut (Tournai), the French-administered Nord *département* (Lille, Douai), as well as Dutch-administered Zeeuws Vlaanderen (Sluis).

The main squares of the Flemish towns of Comines (Flem.: Komen) and Wervik (Fr.: Wervicq) lie barely 3 km apart. Straddling the navigable

River Lys, the key determinant of local topography and spatial layout, the two urban centres occupy sites continuously inhabited for at least 2000 years. Insignificant in the context of globalised urbanisation, both towns boast prodigious histories and place-specific architecture with unique monuments of ecclesiastical and secular architecture. Microcosms of the most densely populated and intensively urbanised region of Northern Europe, their deeply entrenched local identities evolved from the social coordination required to restore life, labour and built environment following numerous localised fires, floods and conflicts. Foreign invasion and occupation from the later 1500s onwards caused increasing ruination that, in a country known as the cockpit of Europe, culminated in the catastrophe of World War I. Today, only northern, left-bank Wervik lies in contemporary Flanders: the reciprocal north-bank township of Comines belongs to an enclave of Hainaut created in the 1970s; right-bank Comines and Wervik have remained nominally integral parts of France for over 300 years.

The two towns' *de jure* separation from their cross-river suburbs goes back to the Treaty of Utrecht (1713–1715) where, following Louis XIV's 50-year military campaigns against the Spanish and Holy Roman Empires, a section of the Lys between Armentières and Menen became, by default, part of France's northern border. When the former left-bank suburb of Comines and right-bank extension at Wervik received their own municipal authorities, the two original urban entities effectively became four:

1   French-administered Comines;
2   cross-river, mainly Francophone Belgian Comines;
3   predominantly Flemish-speaking, Belgian-administered Wervik;
4   a previous south-bank *faubourg* with the Frenchified name of Wervicq-Sud.

No officially sanctioned relationship has ever operated between any of the towns. However, with regard to town–city matching, and the wider context of a continental '*Europe sans frontières*', the twofold trans-border relationships between French with Belgian Comines and Wervik with Wervicq-Sud appear to possess greater potential to operate as unofficial, cross-river 'urban twins'. Nevertheless, the third urban correlation, between Comines and Wervik themselves, remains highly relevant. This, the oldest link, traceable through much of their history, is of special importance on the Belgian side of the Lys, where localised identities and language have not been undermined to the degree imposed across the river by the centralised French state's culturally assimilative policies.

To grasp how the four closely interlinked municipalities arose and subsequently evolved, some in-depth account must be taken of their past (Figure 15.1).

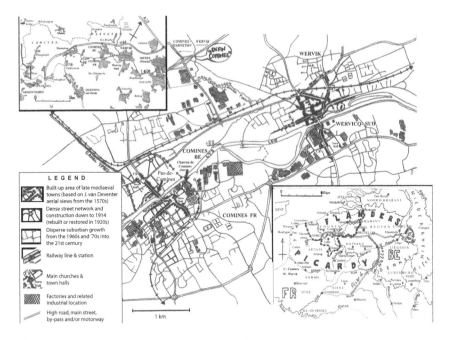

Figure 15.1 Map of Comines and Wervik by Peter Martin.

Origins to the mid-seventeenth century

Unfamiliar outside Belgium and the French Nord, Comines and Wervik possess closely interwoven, parallel histories going back many centuries. Wervik was a Celtic village predating Julius Caesar's invasions, next to which a late Roman *vicus* was founded on the ancient road from Cologne to Boulogne. The origins of Comines may lie in the etymology of its place name; cf. a Roman general's *nom de guerre* (Cominus), or the Flemish verb *komen*. A crucial function determining their survival was as fording points over the Lys: Wervik emerging on the left bank and Comines on the right. These locations presumably shaped Wervik's future cultural and linguistic orientation towards Ypres, while Comines entered the orbit of Lille.

First recorded in 1196, St. Pierre 'de Cumine' parish church was destroyed and rebuilt many times over the ages (*Le Patrimoine ...* 2001). The raising of three churches indicates that early mediaeval Wervik became a centre of pilgrimage, as well as trade, with a hospice, monastery and numerous chapels. Right-bank ribbon development here (Bouwen 1991) was echoed by a left-bank suburb along the highway northwards at Comines (Duvosquel and Lemoine-Isabeau 1980). The respective Lys bridges, conveying vital north–south trade routes, were protected by small forts that became fortified residences. As seigniorial towns, Comines and Wervik secured urban charters in the thirteenth century, but neither was permitted to construct defensive walls, making do with reinforced gateways that controlled movement of goods at main entry points, and to hinder attacking armies. At Comines, chief town of the Ferrain 'Quarter' of Lille Castellany, nominally French aristocratic families pledging allegiance to the rebellious Flemish counts successively built up a monumental castle which, heavily bombarded in 1667–8 and surrendered in 1674, was dismantled by the invading French.[3]

These two modest towns' remarkable activity and wealth between the thirteenth and fifteenth centuries attests to the tremendous socio-economic, civic and cultural apogee of Flanders as mediaeval transalpine Europe's most highly developed province. Comines-manufactured cloth was traded at the fair in Novgorod, while Wervik's fabrics were so sought after that, when the massive fire of 1397 halted production, the Tuscan Degli Alberti family lost their fortune. St Medardus' church, enlarged and redesigned in Brabantine Gothic (c.1414–1433), became Wervik's most prominent building, while at Comines a cloth hall was raised whose truncated tower became the *beffroi*.[4] Notable among the numerous pre-industrial buildings demolished in the late nineteenth or early twentieth centuries, the Market Square's *maison Renaissance* (Ferrant 2008), a hospice and Greyfriars' convent (*Le Patrimoine* ... 1978) deserve mentioning; as do Wervik's preserved "*t-Kapittel*' merchant's house, town hall and courts (*Inventaris* ... 2017).

At their peak, both towns each accommodated 5,000–8,000 inhabitants. Highly typical of Southern Netherlandish urban topography, the street plan and sites of major architectural monuments, faithfully recorded in bird's eye views by Jacob van Deventer (1560s) and Antonius Sanderus (1640s), remained little altered for 500 years (van Deventer 1545; Sanderus 1641). However, like most Flemish towns west of the River Dender, prosperity waned from the 1380s. The fanatical Spanish backlash after 1568 against the Reformation's Calvinist phase, and the ensuing flight of Protestant Flemings to the United Provinces, caused the region's severe decline, and socio-economic ruin for the two towns on the Lys.

French invasions and reverberations (1667–1715–1794)

Louis XIV's geopolitical ambitions and exaggerated dynastic claims produced a prolonged international conflict during the second half of the seventeenth century. Stretching from the Swiss Alps to the North Sea, the main theatre of war changed the fortunes of key continental powers. Of these, the defeat of Spain, the rising star of Habsburg Austria – acquiring the truncated Southern Netherlands – and Britain's 'Personal Union' with the Dutch Republic each profoundly affected the situation in Flanders.

French bombardment, moreover, of Brussels (1695) and Heidelberg (1693) revealed through such populous cities' vulnerability their loss of status as independent urban variables. Preponderance of state interests over urban freedoms had already decided the localised fates of Comines and Wervik. Complying with the new realities of early–modern warfare, French military engineer and architect, the Maréchal de Vauban, conceived a vast system of interlinked fortified towns. Fortunately, even while much of south-bank Wervik and outlying parts of northern Comines had suffered grievously during the invasions, neither town was turned into a military fortress.

The period between the Treaty of Utrecht and 'liberation' of the Belgian lands following the French Revolution marks an intriguing transition for both cross-border towns, during which, regardless of the frontier's formal existence along the Lys, integrated 'communal administrations' continued to function: the cross-river *fauxbourgs* did not as yet become distinct settlements. In the Pas-de-Comines, an island neighbourhood uniting north- and south-bank districts, urban life revived from around 1719 through the introduction of ribbon manufacture (Segard 2007). Adapting older houses or building anew on the rue de Moulins, the ribbon-makers funded the St. Nicholas's riverside church. Wervik's recovery, meanwhile, was stimulated by tobacco cultivation, reaching peak production in c.1710–50.

Formal separation, Belgian independence and industrialisation (1794–1830s–1914)

Typical of French centralisation, the two former cross-river district-suburbs only became autonomous urban entities, when Paris again administered both sides of the Lys. Extending Napoleonic regional government to the newly incorporated Belgian territories, municipal administrations were established in left-bank Comines (placed in the Lys *département*) and right-bank Wervik, officially renamed Wervicq-Sud (in the Nord *département*). Even though the border remained in force after 1815, thanks primarily to the traditionally integrated communities on both banks, and deeply ingrained Franco-Belgian cordiality, the trans-border municipalities were never to be isolated from each other.

On 25 August 1830, influenced by the July Revolution in Paris, an uprising spontaneously broke out in Brussels that, backed enthusiastically by the working classes, erupted into full-scale revolution against post-Napoleonic Dutch hegemony. The British political establishment, eager to minimise French influence, convened the first of two London Treaties guaranteeing Belgium's statehood.

Industrialisation turned Belgium into one of the world's most advanced economies. With coal and iron production established in the Meuse and Sambre valleys from the late 1700s, the southern provinces boomed. In Flanders, however, the traditionally home-based textile production and other handcrafts production declined, causing famine in the late 1840s. Populous, semi-rural and fervently Roman Catholic, the Flemings, while finding employment in the country's textile industries, proved a vital labour force in the Nord *département*. Partly reliant on the business acumen of Belgian industrialists, shop-floor production focussed on Roubaix and Tourcoing, just across the border.[5] In Comines, ribbon-making expanded from a cottage to factory-floor industry and the staple occupation. On the French side of the Lys, large-scale industrial location from the mid-1800s, including wool-based textile plants at Wervicq-Sud, attracted an immense Belgian workforce used to routinely crossing the bridge from the opposite bank. Pre-1914 photographs reveal the proximity of buildings on both banks where the river narrowed at the two respective drawbridges. The only hindrance to the increased movement involved stringent customs checks to counteract food, alcohol and tobacco smuggling into France. Many Belgians from rural areas north and east of the Lys even 'emigrated' to urban areas across the French border.

Long preceding these developments, the traditional activities of cross-river family get togethers, church going and mass attendance of annual fairs, held most typically in French Comines and Belgian Wervik, were mainstays of social life in all four towns (Malbranke 1976) that outlasted the Industrial Revolution. Such intimate associations between Belgian- and French-controlled Flanders ensured the north- and south-bank halves of Comines and Wervik effectively remained single, integrated urban bodies.

By 1911 French Comines had around 8,700 inhabitants, exceeded by a longer expanding Belgian Comines with c.10,000 (1900). Wervik's growth stagnated from mid-century to the 1890s, accelerating thereafter (1910: c.12,000). The former cross-river quarter of Wervicq-Sud developed modestly and haphazardly (1911: 2294),[6] but factory production – here also more concentrated on the French bank and employing far more Belgian than French citizens – brought unprecedented growth. The physical appearance of each town underwent partial transformation. Once all-pervading single- or, less typically, two-storey houses were often rebuilt or replaced by new domestic and occasionally three- or four-storey warehouse construction. Wood- and timber-framing once crucial to urban architecture, while retained in courtyards or nearby farmyard structures, gradually disappeared. Factory complexes occupied amalgamated building plots or (particularly in French Comines and Wervicq-Sud) were located along the navigable river. Pools and inlets accommodated small ships, boats and increasingly barges. Many house frontages were plastered or whitewashed, but the dominant building material remained dark red (or otherwise traditional, orange) brick. Various architects' groups applied diverse architectural forms and décor, bourgeois town-houses and suburban villas tending to demonstrate updated classicist, eclectic or avant-garde styles, yet the overall architecture remained closely analogous in all four built-up areas. These broad characteristics, typifying the *fin-de-siècle* built landscape of numerous Flemish urban centres straddling the Franco-Belgian border, reasserted themselves in the post-1918 reconstruction.

The 'Great' War and post-1918 reconstruction

Occupied by the *Wehrmacht* from September 1914, Comines lay within 10 km of the Western Front. The catastrophic phase did not come until late 1917, the civilian population having been evacuated the previous May (Vander-marlière 2016), followed in June by the people of Wervik/Wervicq-Sud,[7] and the German offensive of April 1918. Most of Comines was virtually flattened and much of Wervik with Wervicq-Sud obliterated. St. Pierre's and the *beffroi* stood to the bitter end, the former's bell tower eventually collapsing into its burnt-out wreck, the latter, with its iconic bulbous cupola, dynamited by the retreating Germans weeks before the Armistice. A sketch of St. Nicolas' church in ruins was retrieved in May 1945 from Hitler's bunker in Berlin.[8]

In the reconstruction, a regionalised architecture, previously applied in restoring historic monuments, was adopted. Traversing the Franco-Belgian border, and known in Belgium as the *stil traditionel*, but in France as *le style flamand*, this conservationists' historicism, reliant upon modern construction methods, drew inspiration from the Flemish Gothic and Renaissance, or a Mannerist fusion of the two. If the Belgian Royal Commission for Architectural Monuments respected age-old urban traditions and thus locally inspired solutions, in France stringent planning laws were enforced, ostensibly to improve living conditions. A modernist aesthetic conformity was thereby imposed on the built landscape, as yet inconceivable in Belgium.

In French-administered Comines, no attempt was made to rebuild the mediaeval hall church or town hall and a drastic redefinition of the Grand'Place ensued. Eager to procure a monumental religious temple fit for the 'new age', the post-war mayor approached established Parisian architects to design a pseudo-Moorish brick colossus faced with ceramic tiles. Transforming the town's physiognomy, it was given a 50-m bell tower – where the town hall had stood. Replacement municipal offices were designed by the celebrated regionalist, L.M. Cordonnier, complete with a reconstituted copy of the lost *beffroi* (Martyn 2016). This rejection of historic spatial layout and aesthetics contrasted with rebuilding barely 100 m away in Belgian Comines, whose parish church was precisely reconstructed, using bricks recovered from the ruins (*Mémoires...*, vol. XXI, 1995). The restoration of Wervik's historic monuments was overseen by H. Hoste, the chief architect of war-devastated towns in the Westhoek (Verdonck 2012). While extensive use of modernist decorative arts broadly differentiated French from Belgian 'reconstruction' programmes, comprehensive transformation from white-washed houses of varying heights and proportions to more standardised bare red-brick construction prevailed in each town. At this crucial conjuncture in the four border towns' modern history, despite the overriding influence exerted by centralised legislation (above all on the French side), a vital role was played by localised communities returning to the ruins, frequently rebuilding destroyed homes on the same foundations.

Contemporary circumstances and the growing threat to urban identity

Demographic increase down to the late 1990s ranged from a meagre 10% and 20% (Belgian and French Comines) to more substantial 35% (Wervik) and 50% (Wervicq-Sud) of the population recorded in the 1910 (France) or 1911 (Belgium) censuses, the last-mentioned town containing barely 2000 people until the late 1950s (*Mémoires...*, vol. V, 1975). Growth in the twenty-first century, on the other hand, reflects a burgeoning new generation of home buyers from Lille, Roubaix–Tourcoing, Mouscron and Courtrai attracted to healthier, semi-rural surroundings from where they commute to work.[9]

Comines and Wervik enjoyed a mediaeval prosperity known far and wide beyond Flanders. Substantial nineteenth-century growth, despite the trans-border split, made all four towns important centres of regional socio-economic and related changes. Transformed thereafter by the first of two cataclysmic international conflagrations, the four adjoining communities – further undermined by worldwide economic disaster in the 1930s – endured mass closures of production plants from the 1970s that ushered in the post-industrial age.

If intensive post-1918 rebuilding (Comines Fr/Be) and extensive restoration (Wervik/Wervicq-Sud) still largely accounts for each town's present-day exterior appearance, extensive low-density construction in the suburbs reveals that a far-reaching social change is currently under way. Indicative of a new age that is not merely post-industrial (or 'post-modernist') but effectively post-urban (the most recent stage in the towns' related histories seeming a travesty of the two mediaeval core-towns' buoyant mediaeval life and labour or subsequent four trans-border centres' industrial-age development), a point – as elsewhere in the crisis-ridden 'developed world' – has been reached tantamount to post-history.[10] The self-evident contrast between this late twentieth/early twenty-first-century demise of urban identity and communal life is strikingly reflected in public ceremonies, marked by (ex-)religious pageants, secularised parades involving retinues of giants, fairs, etc., held throughout the year (Maurent 1979). Although mere echoes of mediaeval street entertainment, often reintroduced from the 1800s, these well-attended events are offshoots of place-specific customs and rituals, providing a fleeting sense of community that current realities no longer engender.

Administrative changes accompanying Belgian federalisation have further undermined regional identity. In 1963, local government reforms created a Francophone administrative unit along the Lys, from Le Bizet (outside Armentières) to Wervik's western edges. Attached to Mouscron *arrondissement*, this enclave was re-incorporated from West Flanders into Hainaut. Comines and Warneton urban boroughs and Ploegsteert, Le Bizet, Bas-Warneton and Houthem village/hamlets constituting this territorial pocket were amalgamated into a single district area and electoral constituency. In 1977, Comines was united with Warneton, some 5 km upstream, into a single urban body and a joint-hyphenated name generated for the former group of district-areas.[11]

On the Flemish-speaking side of this new linguistic boundary, the district area (*gemeente*) of Wervik was combined with Geluwe to form a single canton. Whereas Belgian Comines was reduced to a '*section*' (*sic!*) of Comines–Warneton, Wervik retained its status as a self-governing community, although Geluwe was demoted to a 'sub-'district-area (*deelgemeente*). Viewed in the context of the closely related histories of Comines and Wervik, this late twentieth-century redefinition of the regional map effectively sounded the death knell for the historic relationship. Inflated Belgian demographic figures, compared with those of French Comines and Wervicq-Sud, demonstrate statistical inflexibility, which prioritises the district-area over the urban centre at its core. The terms *ville* and *stad* are being eroded as statutory institutions in Belgium's administrative system (Koppen 2011). Replacing Warneton town hall and Comines police station underlines how borough fusion facilitates local government expenditure and service cuts. Ominously referred to as *pacification communautaire*, this '(district-)community pacification' (*sic!*) has so far been avoided in France, where the town-and-country distinction still survives. Right-bank Comines and Wervicq-Sud retain their age-old district-areas within Quesnoy-sur-Deule canton. While consigning the inhabitants of Comines–Warneton to the *Région wallonne* (avoiding their probable '*Flamandisation*'), the Belgian changes have further provoked segregation of Flemings within 'their' respective 'region' from Francophone citizens.

This ongoing process drastically compromises traditional intermingling within a population once unconcerned by ethnic identity and used to communicating in localised dialects of West Flemish *Nederlands* and Picardian French, whence the intermarriage of communities transcending the political frontier. At Comines, cross-river contacts between residents remain intimate, going far beyond formal relations involving joint-municipal ceremony, or 'initiatives' of common planning, architectural projects, etc. (Figure 15.2). In financing a college, sports and cultural centre with main library and numerous museums, it is apparent that the Warneton–Comines enclave has also gained from its 'special status'.[12] With a historians' and genealogists' society based in the district history museum,[13] and the French right-bank funding its own heritage centre and museum, Comines has become a prodigious research centre of poignantly nostalgic local history.[14]

In Wervik and Wervicq-Sud, on the other hand, the once far-reaching association across the Lys has undergone imposition of French on the right bank, and resurgent Flemish self-determinism on the left. Seen in this broad context, the recent partnership agreement between the two municipalities merely accentuates Wervicq-Sud as yet another French town, while reinforcing the Flemish ethnic-cum-linguistic region's very own international border.[15] The age-old cross-river connection has become dependent on the respective town halls and related interests on either side of this re-enhanced frontier, including the enormous revenue Flemish tourism guarantees the Nord. Recent replacement of the post-1945 road bridge by an ultra-modern

Figure 15.2 Poster advertising the Spoon Festival (Fête des louches) in French and Belgian Comines.

construction (at the site of the mediaeval crossing) to facilitate trans-border movement, thereby augmenting the density of transit traffic through the city centre and Wervicq-Sud, is indicative of the respective councils' short-sighted planning policies. Wervik, although possessing its own heritage centre, apparently suffers from reduced investment in community cultural and research institutions.[16] Furthermore, the spectre of demolition widely employed in numerous Flemish towns has reared its ugly head with the partial destruction of Hobbe's avantgarde *Volksbelang* – an early 1920s complex with concrete-faced frontages illustrative of the evolution in Belgium of *Art déco* – to make way for 'state-of-the-art' apartments.[17] The apparent decline of self-determining in favour of bedsit communities, combined with tax-reduced 'luxury' housing for the aged, epitomises the threat small towns like Wervik face in the current age.

Picardy: the missing link

Comprehending trans-border relations within the pre-modern county of Flanders requires appreciation of the intricate complexities of Belgian regional history. The zone in transition between standard French and the language traditionally spoken in Picardy begins just beyond the Paris metropolitan region. Comprising the Oise, Somme and much of the Aisne *départements*, this southern border region marks the furthest reaches, where Picardian French was traditionally spoken. Distinguishing Walloon-French from Picardian-French-speaking regions is crucially important (Hardelin 2014). *Picard*, entering gradual decline from the fourteenth century, survived as a regional tongue spoken around Mons, Tournai, Lille and Douai no less than in rural areas near Amiens, Beauvais and Laon.[18] Frenchification, through schooling, work-place, military conscription and a broad intolerance in social mores towards any form of spoken language other than standardised French, weakened unique regional characteristics from 1789 onwards.

The ancient lands of the Belgae tribes stretched northwards from Compiègne, Soissons and Rheims, across almost ubiquitous flatlands as far as the maritime area where, extending from the Scheldt estuary beyond Thérouanne (Terwaan; Picard–Flemish: Terenburg) to Boulogne (Bonen), Flanders emerged from the fifth century.[19] Since Gallia Belgica, inhabited by Celtic *and* Germanic tribes, formed part of Gaul, the remarkable degree of continuity in Belgium's internal complexities is echoed in its close relations with France. Conversely, the Flemings' and Picardians' mediaeval struggle with French hegemony reveals the relationship between a Belgium committed to propagating regionalism and a (once) progressive France moving towards a uniform nation-state, to be multi-faceted.

The extent to which the fusing of Romance with Germanic tongues evolved a place-specific *patois* is probably best demonstrated in the one city where both localised variations of French and *Nederlands* still exist, the Brussels metropolitan area, famed for its *Bruxellois* dialect.[20] In Ostend, where communities of Francophone Flemings and Belgians have persisted, as well as in Belgian Comines itself, where Flemish speakers still live and work, something has also survived of a Francophone-influenced West Flemish dialect. Such urban-based vernacular is most likely a residue of the bilingualism practised by merchants, traders and craftsmen in 'bygone' times. The regionalised variations of place names for Comines and Wervik, other than standardised French and *Nederlands*, demonstrate the extent to which the Romance and Germanic linguistic twain truly did meet and intermingle; cf.: *Comen – Wervi* (Picardian); *Koomn/Coomn – Werueke/Wervike* (West Flemish); *Komen* (Flemish-*Nederlands*); *Cômene – Wervicque* (Walloon) (various sources); cf. additionally: *Verovino* (vernacular Latin) and a sixteenth-/seventeenth-century Anglicisation: Werwick.

Conclusions

Seen from a historical perspective, Belgium's current internalised ethnic-linguistic border is a reverberation of what, from the early Middle Ages, comprised a localised belt of convergence between predominantly Francophone areas to the south and primarily Flemish/Dutch-speaking areas to the north. It was not until the early 1900s that this informal fusion, as reflected in the case explored here between communities inhabiting the Flemish towns of Comines and Wervik astride the River Lys, began to break down. Placing the modern-age offshoots of left- and right-bank Comines, Wervik and Wervicq-Sud in the regional context of pre-modern Picardy and the County of Flanders are of incomparably greater relevance to analysing the two historically integral towns' millennial association than basing it on today's superficially defined Flemish and 'Walloon' regions. Picardian–Flemish Comines and West Flemish Wervik lie at the heart of the Belgian question; not merely in terms of the present internal divide but altogether wider issue of appreciating the country's historic cultural reach well beyond its post-1839 borders.

Assuming the predominant expression of regional identity to be place related, most locals apparently continued regarding themselves as Flemish (Guillaume and Moke 1860). On the right bank, the historic continuity of social and cultural relations was undermined by state policies aimed at creating a modern-age French society. On the Belgian side, however, issues concerning language and ethnicity remained firmly in the background until Flemish regional identity became increasingly linked to an ethnically orientated movement focussed on the Teutonic language of *Nederlands* (to the effective exclusion of the Francophone population).[21] Although the struggle to reintroduce a standardised version of Flemish as an officially operative language predated World War I, it noticeably intensified through the politically induced nurturing of Flemish nationalism under German occupation. Prior to 1914, localised perceptions of identity and inter-communal toleration (in complete contradiction with twentieth-century, race-orientated ideologies and broadly alien to nineteenth-century ethnographic theory) had persisted among rural and small-town populations across much of Europe.

A geographically and historically integrated Flanders remains divided between the Belgian provinces of East and West Flanders and shadow Flemish districts subordinated to the march-land Nord *département* and Belgian Hainaut.[22] With *Nederlands* speakers prevailing in Belgian *Vlaanderen* while French speakers predominant in *Flandre wallonne* (*sic!*) and traditionally Flemish-speaking *Flandre maritime*, Francophone inhabitants of Belgian West Flanders have been alienated from their historic association within the *Compté de Flandres*. This negative process is symptomatic of Belgium's failure to establish a Francophone Flemish area, composed of Tournai and Mouscron *arrondissements* with left-bank Comines and neighbouring communities (contrary to their ahistorical incorporation into Hainaut), to bridge today's Flemish region and French *'Flandre'*. In 'constructing' a cross-border region affiliating Tournai, Courtrai (Kortrijk) and Ypres (Ieper) with the Lille conurbation (Baert et al., 2004), recent EU-related policies are apparently championing reintegration of Flemish-speaking with misnamed 'Walloon' and French-controlled Flanders.

Are relationships between settlements bridging an international border of secondary importance to the nonetheless unique qualities of an interrelationship between two neighbouring towns straddling what has, in part, become a linguistic border? Whatever the opinion may be, the waning *rapport* between left-bank Comines and Wervik offers a sad reflection of the virtually moribund state of 'Flemish'–'Walloon' relations in the two-states-in-one solution that is today's Belgium. The rapid demise of localised cultural and linguistic diversity ironically appears to have been accelerated by the politicised ethnic-linguistic boundary traversing the entire country from East to West between the Francophone and *Nederlands*-speaking regions, which has crystallised over the past half century. Where hope remains for community identity, and historic architecture, it is in the ability of places like Comines and Wervik to continue functioning as urban (or otherwise rural) communities and thus nurture something of their traditions; dependent although they might already be on a localised kind of tourism. The institutionalised 'twinning' of Wervik with Wervicq-Sud may actually undermine genuine bonds between inhabitants from either side of the Franco-Belgian border. However, at Comines (irrespective of Belgian administrative 'reforms' into the twenty-first century), a vestige of communal relations does still hold the cross-river halves together as a transnational, *urban* body, thanks to living social, cultural and educational (as well as economic) ties apparently transcending policies sanctioned by bureaucrats and politicians in distant Paris, Brussels or Namur.

Notes

1 As opposed to the independent state of Luxembourg, separated from Belgium by the 1839 Treaty of London, reference here is to the Belgian province.
2 A term differentiating the Picardian and Walloon *langues d'oïl* from standardised French. The German minority (c.0.5% of the Belgian population) and French national community possess no autonomous territory.
3 For early history: www.youtube.com/v=VFN3pzJ6RYM (20/04/2017).
4 UNESCO World Heritage Committee, Belfries of Belgium and France, http://whc.unesco.org/en/list/943 (20/04/2017).
5 Benoit-Cattin, R. (ed.), 2005, *Roubaix-Tourcoing et les villes lainières d'Europe*, Septentrion-Lille, 31–5. Zola, E., 1885, *Germinale*; https://www.free-ebooks.net/ebook/Germinal (20/04/2017).
6 Cf.: demographic statistics at https://fr.wikipedia.org/wiki/Comines_(Nord)#D%C3%A9mographie; https://fr.wikipedia.org/wiki/Comines-Warneton#D%C3%A9mographie; https://fr.wikipedia.org/wiki/Wervicq#D%C3%A9mographie; https://fr.wikipedia.org/wiki/Wervicq-Sud#D%C3%A9mographie (15/04/2017).
7 www.wervik.be/de-groote-oorlog-in-wervik (02/10/2014).

230   *Peter Martyn*

8 http://www.hitlerpages.com/pagina12.html, 1914-15-16-17-18, Comines-Wervik-Wervicq-Sud (16/04/2017).
9 www.linternaute.com/ville/comines/ville-59152/demographie (20/02/2018).
10 For example, Horrocks, C., and Jevtic, Z., 1996, *Introducing Baudrillard*, Icon Books-Cambridge, 171–4.
11 Local Belgian administrative units, composed of *arrondissements*, divided into *cantons*, subdivided into *communes*, have since been replaced by a total of 589 *communes* (Flem.: *gemeente*).
12 *Guide d'architecture moderne et contemporaine Tournai et Wallonie Picarde*, Mardaga-Bruxelles, 182–7.
13 https://amisdecomines.blogspot.com/ (16/03/2017).
14 www.ville-comines/maison-patrimoine.html (25/03/2017).
15 https://nl.wikipedia.org/wiki/Partnerstad (05/05/2017).
16 www.wervik.be/vrije-tijd/cultuur/erfgoed/erfgoedhuis-ren-defrancq (10/04/2017).
17 Unnoted, as yet, by Flemish architectural heritage: https://inventaris.onroeren-derfgoed.be/erfgoedobjecten/32817 (04/05/2017).
18 www.languepicarde.fr (14/02/2017): '*Nou picard, nou future*'.
19 For abundance of West-Flemish place names in the Pas-de-Calais, cf., for example, https://archive.org/details/dictionnairetopo00mencuoft (20/02/2018).
20 L. Quievreux, *Dictionnaire du dialecte Bruxellois*. Libro-Sciences Bruxelles 1985.
21 A linguistic union between Flanders and the Dutch Netherlands with its ex-colonies, signed in Brussels in the *Verdrag tussen het Koninkrijk België en het Koninkrijk der Nederlanden inzake de Nederlandse Taalunie* (09.09.1980), never evolved into political amalgamation.
22 As though responding to recent moves by local communities and groups to re-suscitate regional traditions and identity, the French authorities have renamed this predominantly low-lying region '*Les Hauts de France*' (*sic!*).

References

Baert, Th. et al. (2004), 'Architectural Guide to the Lille Metropolitan Area, Lille Métropole – Courtrai – Tournai – Ypres', transl. Ch. Penwarden, Le Passage-Lille, 11–15.
'*Bouwen door de eeuwen heen in Vlaanderen: Provincie West-Vlaanderen, Arrondissement Ieper, Kantons Mesen – Wervik – Zonnebeke*', (1991), ed. Delepiere, A.-M., Huys, M., Brepols-Turnhout, 108–72.
van Deventer, J. (1545), 'Stadsplattegronden van de zuidelijke Nederlanden', available at Biblioteca Nacional de España: http://bdh-rd.bne.es/viewer.vm?id=0000015403 (accessed March 17, 2018).
Duvosquel, J.-M., Lemoine-Isabeau, Cl. (1980), 'La région de Comines-Warneton, sept siècles de documents cartographiques et iconographiques', 2011, Crédit communal...-Bruxelles, 1–151.
Ferrant, P.D. (2008), '*Comines, La Grande Place*', Les Amis de Comines, 1, *passim*. Guillaume, H., Moke, Ph. (1860), 'La Belgique ancienne et ses origines gauloises, germaniques et franques', Lebrun-Devigne-Bruxelles, 107–9, passim.
Hardelin, J.L. (2014), '*Les Chtimis, sont-ils des Belges?*' Broché-Lille Valenciennes, 25–30, *passim*.
Inventaris Onroerend Erfgoed (2017), https://inventaris.onroerenderfgoed.be/erfgoe-dobjecten (04/05/2017).
Koppen, J. (2011), '*Belgische politiek voor Dummies*', Pearson-Amsterdam, 161–3.

Malbranke, Cl. (ed.) (1976), '*Guide de Flandre et Artois mystérieux*', Princesse-Paris, 218–20, 466.

Martyn, P. (2016), 'Comparable and divergent Methods Applied in the Rebuilding and Restoration of Towns and Cities in Belgium and the Nord Region of France after the First World War', in: *Reconstructions and Modernizations of Historic Towns in Europe during the First Half of the 20th Century*, ed. I. Brańska, and M. Górzynski, TPNK-Kalisz, 62–5.

Maurent, R. (1979), '*Géants processionnels et de cortège en Europe, en Belgique, en Wallonie*', Editions Vey-Tielt, 426–51.

'*Mémoires de la societé de l'historie de Comines-Warneton et de la région*', vols I-XLVII (1971–2017), ed. J.-M. Duvosquel, Comines-Warneton.

'*Le Patrimoine des Communes du Nord*', (2001), ed. Despature, P., Flohic-Paris, vols. II, 1369–78.

'*Le patrimoine monumental de la Belgique: Province de Hainaut, Arrondissement de Tournai, Arrondissement de Mouscron*', (1978), Solédi-Liège, 856.

Sanderus, A., '*Flandria illustrate* (1641–4)', Amsterdam, 648.

Segard, J. (2007), '*Comines au fil du temps*', Les Amis de Comines.

Vandermarlière, M. (ed.) (2016), '*Ombres et Ténèbres. Crépuscule d'une époque: Comines 1914–1918*', Les Amis de Comines, 35, 165–77.

Verdonck, A. (2012), '*Huib Hoste, pioneer van het modernisme bij het heropbouw van de Westhoek*', Cahier: 1914–1918 in Wervik en Geluwe, 3, 77–85.

16 Border twin cities in the Baltic Area

Anomalies or nexuses of mutual benefit?

Thomas Lundén

In this more general chapter, we shift focus northwards and return to the world of EU accession states, though now where three sets of twins are portrayed against a longer period of historical change. Immediately, we see how border twins and their inhabitants become subject to forces very different from those facing governments at the national centres. We also start coming across the notion of international borders not so much as points of integration or disintegration, but as 'resources' and points of opportunity. Thomas Lundén attempts to classify the varying results.

Introduction: borderland twin cities

Twin cities – defined as two urban settlements divided by an administrative boundary – represent nexuses of conflict and cooperation. If divided by international boundaries, they represent rather unique combinations of local and international relations: perhaps causing problems unforeseen in often-distant capitals, meanwhile providing opportunities for their inhabitants, assuming permeable boundaries and differing regulations. Border twins are rarely equal; there is often differentiation caused both by the area's specific history and influences from each territorial state.

Where states are juxtaposed, differences between jurisdictions of specific hierarchical levels can produce asymmetries inducing what one could term misfits. Asymmetries have negatively influenced authoritative organizations' attempts to create local cross-border communities. Issues one state treats at one level may be handled at a different level by its neighbour. Consequently, they cannot be locally negotiated by similar level officials, but rather referred upwards, thereby losing priority (Lundén 2009a, 135).

Border twins are roughly of three types:

- Towns post-dating boundaries and developing on each side, often with one town acquiring a trans-border suburb that later becomes a full-fledged town.
- Towns pre-dating borders where only one town existed before the boundary, the border cutting through the urban settlement, often due to ethnic delimitation.

• Special cases when a new border is drawn between two existing, separate, but proximate, urban settlements that later merge.

This chapter provides examples of the two main types from the Baltic area.

Central place theory sees towns as market places serving hinterlands. Single border towns, in this theory, are truncated, each losing half its supply-area and purchasing power, and needing a 'twin' across the border (Christaller 1933). This has truth, but most towns result not just from market functions but political decisions, often born of military strategy: after all, in many European languages, the word for town derives from *castrum, fortress,* places often originally on the periphery of political territories. Since then, most European boundaries have changed, with fortress functions becoming obsolete. However, in peaceful times, suburbs may develop on the former enemy side.

The Baltic Sea Region (the catchment area of all rivers leading into the Baltic), has undergone fundamental geopolitical changes since territorial states emerged in Northern Europe c. 1000 years ago. Some towns emerged during the Hanseatic League period, at sites favourable to harbouring and trade, while others appeared at existing state borders protecting the interior. Subsequent territorial changes and developing military and transport technology made earlier locations obsolete, but most old towns survive, albeit with changing or decaying functions; some have evolved at present-day state borders. Governance affecting the respective sides of the borders has changed from authoritarian to democratic, language policies playing important roles in the new, or re-born 'nationalizing states' of Finland, Estonia and Latvia. This chapter analyses three such towns: Tornio–Haparanda, Narva–Ivangorod and Valka–Valga.

Tornio–Haparanda: a Finnish mother-town and its ethnically mixed trans-border 'suburb'

Haparanda and Tornio, until recently separated by a strip of grass and wetland, are divided by the only inhabited Finland-Sweden boundary in the Torne River's southern section. This borderland is mainly lowland, forest and farmland divided by the Torne and two tributaries. Both populations were mainly Finnish speaking, with a Sámi minority in the north. Torneå, near the rivermouth, emerged in the Swedish province of Västerbotten in 1621. Until the 1808–9 war with Russia, Finland belonged to Sweden. Under the peace treaty ceding Finland to Russia, Torneå (Finnish Tornio), situated west of the Torne, became part of the new Grand Duchy of Finland, which kept the old Swedish constitution but with the Czar as Grand Duke. This sudden suzerainty-change ruptured a hitherto fairly homogeneous river valley, populated mainly by Finnish-speaking peasants, except for Tornio – a bilingual (Finnish/Swedish) trading town. Losing the river valley's central place, Sweden built a new town, Karl Johans Stad, some distance from the boundary. But, 'voting with their feet', the new town dwellers appeared in Haparanda, a small suburban village outside Tornio, immediately south-west of the urban and state boundary. Karl Johans Stad's charters were transferred to Haparanda in 1842.

Haparanda's emergence attracted many merchants and administrators from elsewhere in Sweden while many of Tornio's Swedish-speaking burghers left or merged into the Finnish-speaking surroundings. On the Swedish side, preaching and teaching, in the Lutheran tradition, occurred in the area's native language. However, with increasing state nationalism from the 1880s, administration and teaching became exclusively Swedish.

Serving mainly market, trade and educational functions, Haparanda in neutral Sweden was important during both world wars for war-prisoner exchange. Post-war, the boundaries re-opened, but Sweden long remained richer than its neighbours, unhurt by wartime destruction. Wealth and regulatory differences incited smuggling, especially across the Finland–Sweden river boundary (Prokkola 2008). Except for necessary regulation of the riverine borders, formal local cooperation remained limited.

Local, regional and military authorities even viewed friendly crossover connections as suspicious. Finnish women moving across to compensate for young women moving south helped keep Finnish local dialect alive, a general decline of the local Finnish language notwithstanding. The media were mostly strictly 'national' in language and coverage. Swedish state radio's introduction of a Finnish program for its borderland met with suspicion but soon became popular.

The situation of a richer Sweden surrounded by poorer neighbours gradually changed in the 1980s, as Finnish industry, education and living standards progressed. Some industrial workers, who earlier migrated to Sweden's metal and textile industries, now returned, but many northern Finns settled in Haparanda to enjoy lower prices and housing costs, certain social benefits and linguistic comprehension, while still remaining near their birthplaces (Lundén & Zalamans 2001).

Haparanda and Tornio have about 10,000 and 23,000 citizens, respectively, including the rural hinterland. Haparanda's *western* municipal border coincides with the old Finnish–Swedish linguistic boundary. Culturally, the trans-border population comprises four groups (Lundén & Zalamans 2001), though the lack of an ethnic census in Sweden, plus declining ethnic identity, mean that there are no useable figures:

Finland Finns in Tornio and east of the river, speaking modern Finnish with some regional dialect. Except for higher officials and people with boundary-related jobs or interests, they have little Swedish language or experience, only using Sweden for some supplies: groceries being cheaper, and berry-picking fields better available there, due to Swedish public-access rights. With Finnish-speakers on the Swedish side, they can often be served in their own language.

Sweden Swedes, mainly in Haparanda municipality's urban area, are either descended from urban settlers from Southern Sweden or have migrated to serve as officials, teachers etc. They generally know little Finnish, and see little reason to cross. However, increasing knowledge of English on both sides has facilitated contacts.

Torne Valley inhabitants from rural areas on the river's Swedish side speak Swedish quite fluently, even though Finnish is the region's original language. These usually have kinship and friendship contacts on both sides. Due to long neglect of teaching Finnish in schools, only actively-interested or family-connected people are fluent in standard Finnish, while others speak a 'wild dialect' Finnish (*Meänkieli*), now a Swedish minority language alongside standard Finnish (Lundén 2011b).

Sweden Finns in urban Haparanda, mostly immigrants from Northern Finland settling in Haparanda after living in Southern Sweden as labourers. Many speak or read little Swedish, having moved/stayed there apparently to retain social benefits and lower housing costs. Living just a few hundred metres from the boundary, they place great emphasis in preserving their Finnishness and Finnish contacts.

Compared to other Swedish municipalities, Haparanda thus has an unusually high foreign-born population (c.40%), almost exclusively Finnish. Assessing the population's ethnic and linguistic characteristics is difficult, but Swedish–Finnish bilingualism is widespread and promoted. While language divides the two unilingual groups, Protestantism is shared. With many having lost their ancestors' Finnish, bilingualism has become advantageous.

There are conspicuous differences in how Haparanda and Tornio's inhabitants use the media. Each town is dominated by newspapers issued for the province in either Finland or Sweden, and in the majority language of each state. Sweden Finns quite frequently read the local Tornio paper. They also often access local stations on the Finland side, and watch Finnish television channels. While around 25% of respondents in Haparanda and Tornio participate in civil organizations, they mostly do it in their own town.

The inhabitants perceive each other somewhat differently. In Haparanda, 36% saw differences between inhabitants in Haparanda and Tornio and a 'we and they' mentality, against 23% in Tornio. Generally, Tornio people are more positive about their neighbours than Haparandans. Haparandans most negative about their neighbours are Swedes speaking little Finnish. Studies of teenagers' future plans in both towns indicate low willingness to remain or move to the other town, but vague dreams of futures in their respective capitals, or – the only thing uniting the two groups – in some international metropolis (Jukarainen 2001).

In 2000, Tornio and Haparanda presented a plan, implying common development of the boundary zone into a commercial and administrative centre, the boundary dissecting the centre. Subjected to a referendum in Haparanda alongside the September 2002 general elections, the proposal was narrowly rejected on a low turnout. In low-turnout districts, yes-votes dominated and in high-participation areas no-votes dominated – indicating that Finnish residents, with a low tendency to vote, supported the project, whereas less apathetic Swedes mostly reacted negatively (Election results, Haparanda municipality).

Finland–Sweden cooperation derives from structural and physical similarity along the rivers and Bay of Bothnia. Tornio–Haparanda's relative remoteness and smallness has incentivised common efforts to make the agglomeration a joint centre, utilizing legal and cultural differences and combating hindrances. Natural conditions and lack of border controls favour amalgamation. Inter-municipal cooperation began in the 1960s with decisions to jointly use Haparanda's public baths. In 1971, a joint sewage purification plant opened. While these decisions were *ad hoc*, they highlighted the advantages of joint ventures and suggested that border obstacles could become a resource. Municipal sporting and cultural cooperation started in the late 1970s with joint-use of sports facilities and joint cultural festivities. Libraries also cooperated: a Tornio library bus covers both sides, serving Finnish speakers on the Swedish side.

Provincia Bothniensis HaparandaTornio emerged in 1987 as an organization dedicated to developing, deepening and increasing cooperation. Formally, it is a 'government' with five politicians from each municipality preparing suggestions for the two units' decision-making bodies. Their local governments meet half-yearly to coordinate. Many public services are used jointly, sometimes defying national legislation. Provincia Bothniensis' establishment formalized cooperation. Much is done by working groups of municipal officers covering tourism, industry, education, physical planning, technical services, fire services, sports and recreation, social welfare and health, culture and youth, plus a special group for the *On the Border* project (see below). Directly and indirectly, Provincia Bothniensis has helped implement joint projects linking the two municipalities. Haparanda–Tornio's community planning is characterized by close cross-border cooperation on all possible projects, reducing costs and enhancing efficiency via common usage of resources.

The twin towns co-operate widely over education, including a joint primary/secondary school taking pupils from both native languages and preparing them for bilingualism (plus a foreign language). For over 20 years, local students can choose which side of the border they attend school.

The most visible project is *On the Border,* aimed at amalgamating the towns. This started in 2005 with a new municipally planned street, part-financed by the EU's Interreg program. IKEA's new furniture supermarket in 2005, probably inspired by borderland access, has produced further investments making Tornio–Haparanda a regionally important commercial centre. In autumn 2014, a joint bus-terminal opened on the Swedish side, almost on the border, after resolving legal problems (e.g. Finnish military appearing on Swedish territory). The Finnish power company, *Tornion Voima,* provides district heating. Tornio processes Haparanda's waste. There is joint-co-operation during fires and accidents.

Civil organizations from both sides, many with cross-border member-ships, often jointly organize cultural events. In 2007, a joint youth policy started. Sports are often nationally divided, but in the Provincia area there is much cooperation. Finland's best bandy team plays home games on a Swedish rink, and a golf course transcends the boundary, even covering two time zones in one stroke.

Not all jurisdictions match because of hierarchical asymmetry (Lundén 2009a, 135–136). While County Councils control aspects of Swedish health-care, in Finland municipalities administer them. Nevertheless, both bodies co-operate in utilizing expensive equipment, specialists and ambulances. A project providing free choice of medical care irrespective of the border began in 2008.

Overall, Haparanda–Tornio twin-town cooperation is one of the most successful examples of trans-state local governance in Europe: effectively establishing a cross-border polity. However, peaceful co-existence was re-cently challenged by the Middle-Eastern refugee influx coming through Sweden and entering Tornio in October 2015, causing pro- and counter-demonstrations on the Finnish side, even demands to reintroduce border controls.

Narva–Ivangorod: a fortress town united and divided over centuries

Narva, strategically located on the Narva River, was built as a Danish for-tress in 1256 and sold to the German Order in 1346, becoming an impor-tant market centre. The Russian Kreml at Ivangorod on the opposite bank was built by Muscovy in 1492 (Figure 16.1). The 1595 Peace of Teusina secured Narva to Sweden but left Ivangorod to Russia. Until the Peace of Stolbova in 1619, Narva was a Russian border town, while after 1619 both sides of the river were Swedish and after 1703 (effectively) Russian. The town and its Russian suburb have thus undergone several different geopolitical changes: sometimes united, sometimes divided by state bor-der. The 1920 Tartu Peace drew Russia's state boundary 20 km east of the river, leaving Ivangorod in independent Estonia. After the Soviet and German occupations and the Estonian Soviet Republic's re-establishment in 1944, its boundary was drawn in the river (Mälksoo 2005, 144–149). Long after the USSR's break-up, Estonia regarded the border as a *de facto* limit, albeit strictly controlled by each side. In February 2014, Russia and Estonia signed a border treaty but, at the point of writing, neither country has ratified it (http://news.err.ee/v/6c90aec3-8263-4c8c-a84c-ed1427b36a33 [02/05/2016]) .

Figure 16.1 Narva Castle (Estonia) and Ivangorod Kreml (Russia) and the border bridge by Thomas Lundén (2005).

Today's Narva has approximately 70,000 inhabitants located on the Narva River's western bank in north-eastern Estonia. On the Eastern bank, in the Russian Federation, lies Ivangorod with 10,000.

After World War II, Soviet Narva was designated an important centre for textile production, furniture and metal industry, and oil-shale extraction for the Leningrad market. Russians and other Soviet ethnic groups replaced the original population, producing near-total Russification of language-use in Narva and elsewhere in Ida-Virumaa province. After the collapse of the USSR's command economy and Russia's economic decline, Ivangorod lost almost all its industry and Narva was also severely affected (Jauhiainen & Pikner 2009, 6–8).

Narva and its environs are replete with symbols referencing the city's many upheavals. Monuments, graveyards and buildings in Stalin-era style remind inhabitants and visitors of different, often contradictory, interpretations of the past (Lundén 2011a, 15–16). With the area's ethnic character, Russian Orthodoxy has re-appeared, alongside a new old-style church; the 1906 cathedral occupies an important location; the Protestant church, largely ruined by the 1944 bombings, has been totally renovated, notwithstanding small congregations. Ivangorod's dilapidated Orthodox Church, built in 1875 by the industrialist Stiglitz family to celebrate converting to Orthodoxy, has been likewise renovated. However, while supplying symbolic landmarks, religion is a minor issue, superceded by ethnicity and allegiance.

Narva's *citizens* divide into three roughly equal categories. One-third has acquired Estonian citizenship; another third is Russian; the rest comprise those considered aliens. This last group need special permits which, until recently, also functioned as valid entry-permits at the boundary. Interviews suggest that many Narvaites see little advantage in Estonian citizenship. Russian citizens can get free basic medical care in Ivangorod and, until recently, could easily cross the boundary. Now that Estonia conforms to EU standards, visas are required. Some multi-visit visas were distributed free to some residents, but access required valid passports. However, being 'an alien' is also difficult. Aliens must apply for work- and residence-permits, a slow and tedious process. Their action-radius is circumscribed: it being difficult to travel abroad, invite relatives to Estonia, do business and acquire land in border-areas, including Narva.

Under the Soviets, released criminals and deportees often had to settle outside their home *oblast'*. Narva, just beyond the Leningrad oblast', naturally attracted persons from that region. Statistics show disproportionately high violent crime in north-eastern Estonia. This is unsurprising in an area with border traffic, high unemployment and a working-class culture characterized by visible crime, like theft, robbery, drug-addiction and smuggling.

Nearly, all schools taught Russian as first language. Some have begun teaching in Estonian, but few Estonian-speaking teachers are available notwithstanding vigorous municipal and state recruitment campaigns. Narva schools do not cooperate with Ivangorod's. In principle, Narva's population can access Estonian- and Russian-language media from Estonia and Russia. Most watch Russian TV, including Russian news. Local Narva radio stations broadcast mostly in Russian, trying also to cover Ivangorod news. Narva has some local newspapers, all in Russian, but younger people rarely read newspapers (Lundén & Zalamans 2002, 188–189).

In analysing Russian–Estonian border relations, Kolosov and Borodulina (2006) mention three factors necessitating at least minimal co-operation:

1   territorial proximity of urban settlements;
2   common interest in maintaining transport infrastructure;
3   prevention of possible natural and technological disasters caused by harnessing water from the Peipus Lake and Narva River, bordering both states.

Though the Narva River separated the two Soviet Republics, most local infrastructure was managed jointly. In 1991, after the USSR dissolved, most such services were discontinued, starting with public transport in 1992 and telephone services in 1994. In 1996, the state oil-powered *Baltic Electricity* plant near Narva stopped providing Ivangorod with heat. Urban planning started in Narva in compliance with Estonian legislation. The two towns do not co-operate on this issue, partly because Ivangorod lacks organized city planning. Narva is supplied with water and sewage treatment from a local plant that originally also supplied Ivangorod, but the local water supplier terminated supplies in November 1998, after sending repeated reminders of overdue payments and after Ivangorod was granted EU and Estonian development funding. Eventually, even sewage treatment and purification was discontinued, with effluents flowing into the river. This situation has been eased somewhat by the sad fact that most industries on the Russia side now lie idle. A new plant has been built, with Danish development assistance (Jauhiainen & Pikner 2009). On the Estonian side, the largely industrial Ida-Virumaa region experienced redundancies, with the large Kreenholm textile plant closing; recently, the biggest shale-oil producer *Viru Keemia Grupp* announced that, with falling oil prices and over-high prices for oil shale from Eesti Energia, it must create hundreds of redundancies (http://news.err.ee/v/business/economy/186cdff1-2246-493c-a988-67f0ef282bf0/eesti-energia-to-lay-off-150-due-to-low-energy-prices [25/03/2016]).

Narva's hospital has no contact with its Russian counterpart. On the Russian side, hospitals treat many Russian Estonians, because they have free-care rights in Russia, whereas Estonia's system requires that, as aliens, they pay substantial sums for health-care insurance. This, alongside other real and imagined benefits, explains why many Estonian Russians retain Russian citizenship. To Ivangorod this poses financial problems, further enhancing cross-boundary tension (Lundén & Zalamans 2002, 188f; Kolosov & Borodulina 2006).

Narva politicians fully accept that the city belongs to Estonia, but complain that central government ignores Narva and that Estonian Russians are seen as an 'alien mass'. Narva's Russian-speaking intellectuals often express bitterness over Estonia's language legislation, which they (wrongly) consider violates Council of Europe National Minority Rights (Lundén 2009b, 93). Even if Narva and Ivangorod form a historical unit, necessarily cooperating on some technical issues (water, cross-border traffic), their coexistence receives no official encouragement. This stems partly from hierarchical asymmetries and partly from frosty inter-state relations. Initiatives have sometimes been taken to develop the Narva–Ivangorod area for tourism (including the Narva-Jõesuu seaside resort); the area is extremely rich in historical monuments and has breath-taking scenery. Most recent has been a project, *Development of the unique Narva–Ivangorod trans-border fortresses ensemble* as a single cultural and tourist object, financed by both local authorities, Estonian and EU funds. Obstacles are bureaucratic border controls and ethnic Estonian resentment towards the eastern borderland, reciprocated by local Russians' scepticism about the Tallinn government.

Valga/Valka: a town split into twins

Valga–Valka lies in a shallow valley, where the small river Pedeli/Pedele flows from Latvia into Estonia, crossing the boundary almost at right angles. The town is mostly south-east of the river. A brook, under a metre across, flowing into the river, defines the boundary in one central part of town.

The place called Pedele was first mentioned in 1286, belonging to the Livlandian Confederation, on the highway between Dorpat (Tartu) and Riga. It was included in the Polish–Lithuanian Commonwealth in 1561 and chartered as the town of Walk in 1584. While from 1627 to 1721, it belonged to the Swedish Realm, the Russian Guberniya of Livland then annexed it as an administrative centre. In 1886, Walk became a railway junction on the important Pskov–Riga line, with a branch to the university town of Dorpat (Tartu), and later a narrow-gauge line to Pernau (Pärnu). Population growth brought not only ethnic Latvians and Estonians, but also Germans and Russians. During World War I and its aftermath, Walk was under German, Latvian and finally Estonian control. However, in 1919, both new republics claimed it. A 1920 Arbitration Convention decided the matter on ethnic grounds: Estonia received the main part, including the centre and railway station; Latvia acquired a suburban area mainly of wooden houses (*Arbitration Convention* 1920, 188–189). Of over 10,000 people, around 2,500 Latvians moved into Latvia, where a new railway station was built (Kant 1932).

In 1932 the Estonian geographer, Edgar Kant, described the resultant situation. Estonian Valga had 13,289 inhabitants in 1925, while Latvian Valka counted 3,339. Valga, and probably also Valka, declined, and Kant noted that the part of the old urban centre now adjacent to the boundary lost much of its commerce and shops. In the 1920s, Valga's principal occupations were industry and transport, nearly all railway connected. It became an important export centre for lighter and more expensive products, while heavy goods were exported through Tallinn or Pärnu. Valka meanwhile got a new railway station with connections to Valga and the junction, and Riga. Kant concluded that, among Estonian towns, these boundary towns were the most modest in terms of population and commerce, both having lost much of their hinterlands (Kant 1932).

When the USSR annexed Estonia and Latvia in 1940, the state boundary became the delimitation between two Soviet republics, thereby losing much significance. After the Nazi-German occupation, these Soviet republics were reinstated. Valga–Valka regained its position as a junction on the Pskov–Riga line, and Valka station was closed. Within the Soviet system, the inter-republic border had little significance; many infrastructural arrangements were located to render them useable by both populations. The town was extended, especially on Valka's side, along Soviet planning lines, with a park, small parade square, some soviet-style neo-classical buildings and a common bus station. The mandatory Soviet war memorial was placed very close to the Estonian Soviet Republic, and the graveyard used by both sides.

Since the USSR fragmented, Valga (Estonia) and Valka (Latvia) are legally separate. However, morphologically, and to some degree functionally, they constitute one unit divided by a state boundary. In the Soviet era, Valga/Valka functioned as one city over health-care, planning and infrastructure development. During the first years of independence, this situation reversed: all co-operation ceased. Today, health-care access is strictly determined by the country of residence. Some years after Estonia's independence, Valga opened its own hospital. Valka's hospital is now over-dimensioned, severely problematic for an already stressed economy. Both towns have had problems with drinking water quality and each has constructed its own sewage-treatment plant, either being sufficient for both towns. Each town has also expressed an impossible wish to supply the other with clean water.

Although Valga/Valka is a twin town on the border, its media are strictly Estonian and Latvian. Today some Latvian pupils use Valga's sports arena and swimming pool. Apart from visa-requirement difficulties (see below) and language differences, waiting time at the boundary has been highly problematic, especially for non-citizens. When Estonia and Latvia joined the EU, particularly after joining the Schengen Union, most problems disappeared. Paradoxically, Valga/Valka is today the 'open' city it was under the Soviets, at least outwardly. But each part is necessarily subject to laws, regulations and planning systems of its own state, and many infrastructural initiatives taken separately after independence in 1991 still shape inhabitants' daily lives.

One small street shows how painful the consequences can become. In the USSR, deported citizens were not allowed to return home. On returning from Siberia, some Valga Estonians therefore settled in Latvia, on a street that was actually an extension of the Põhja road, near the town centre. Savienība is just a small unkempt road flanked by a few tidy one-family houses, but has played an important role in serious but fruitless Latvian–Estonian negotiations. Whoever moved to Savienība would have known there was an old boundary running through, but could not have imagined its importance becoming apparent so soon. Until the Schengen regime was implemented in late 2007, Savienība's inhabitants had to choose between three rather distant official checkpoints when wanting to cross the border, causing great inconvenience. The Valga/Valka authorities therefore issued special permits to cross that part of the border running through the street. They were available to people with permanent addresses on Savienība. The problem was, however, that several inhabitants had permanent addresses elsewhere on the Estonian side, even while actually living in Savienība. Thus, they could not get the permit. There are several reasons why they wanted permanent Estonian addresses, especially insurance and labour issues. Ten years after Latvian and Estonian independence, some 35 people lived in Savienība, in nine houses. Of these, 26 were Estonian citizens, 7 Latvians and 2 so-called aliens. Of the seven Latvians, two were of Estonian origin, who took Latvian citizenship to make life easier, but still considered themselves Estonians. Savienība people's average age is relatively high. Savienība Estonians face other problems, like inadequate electricity supply, telephone service, road maintenance and emergency health-care. Electricity originally came from the Estonian side, but understandably the Latvian authority wants to take this over. Estonian electricity is much cheaper than Latvian, so Savienība's inhabitants do not want this. Road maintenance is conspicuously absent. Neither Estonian nor Latvian authorities maintain it. The Estonian side sees no reason to cross the state border to do so, and the Latvian side is obviously uninterested. Any Estonian on Savienība needing emergency health-care must cross the border into Valga, risking fatal border delays. A Valga doctor making a house-call in Valka encounters the same problem. Land ownership also causes uncertainty. It is unclear who owns the land and houses in Savienība. Legally, it belongs to a Latvian citizen who left Valka long ago. If he wants to sell his land, another problem arises. Under Latvian law, no foreigner may own land within a 2 km zone along the Latvian border. This means that Estonians in Savienība can never own the land they reside on, only their houses. Some Estonians on Savienība have, reluctantly, considered selling their houses and moving to Valga, but Savienība is on Valka's 'periphery', a 'problem area' to which apparently no Latvians wish to move. The area covered by Savienība, a mere 2.6 acres, apparently represents an extremely difficult, even insoluble, problem for municipalities, counties and states. A territorial exchange has been discussed, and seems reasonable, but this is against Latvia's laws. The two presidents met in the twin towns as recently as October 2012, promising a solution: so far nothing has happened.

Savienība aside, the two titular nations are well separated by the border. However, the Russian-speaking population lives on both sides; rarely speaking Estonian or Latvian, they mostly disregarded the Soviet republic boundary in their settlement and contacts. Independence in 1991 left them severely hampered by the border. For many years, there were passport requirements, particularly affecting aliens. The Schengen regime and change to the Euro resolved many problems: young people especially are assimilating into the language and culture of each state (Lundén & Zalamans 2002, 196). Compared to most other European border twin towns, Valga and Valka are unique in that their two majority populations clearly differ in (ethnic) nationality, while both have large ethnic-Russian minorities. Advised by the Haparanda/Tornio administrations, Valga and Valka have entered a program of contacts and communication to improve living conditions around the boundary. Several difficulties have been reported, however: mostly where local intentions clash with state legislation and each country's 'estonification' or 'latvification' policy. Most inhabitants do not see the border as important, perceiving few opportunities on the other side. Ethnic Russians and other Russian-speakers appear to be the only group with cross-boundary contact networks.

Unlike Haparanda–Tornio and Narva/Ivangorod, Valga and Valka's populations represent different religious denominations, with Russians predominantly Orthodox, Estonians and Latvians belonging to separate national protestant churches plus several minority confessions.

Cooperation between border twins: incentives for personal contacts?

The three cases have things in common: historically, they have all been part of a Baltic Swedish-Russian zone of geopolitical influence, obviously with highly different outcomes. In Soviet time–space, the twin cities were joined by infrastructural arrangements now mainly abandoned, while for Haparanda–Tornio the situation has moved from almost total separation to profound integration. After the USSR's demise, Sweden and Finland, mainly through *Provincia Bothniensis,* have actively promoted cross-border co-operation, especially in the Estonian–Latvian case. All sites have overlapping ethnicities: in the Haparanda–Tornio case Finnish, in Narva–Ivangorod Russian, whereas Valga/Valka differs in that the Russian speakers on both sides represent ethnic groups not autochthonous to the area. Ethnic and linguistic affinity may not only facilitate co-operation, but also create fears of *irredenta* in the affected capitals.

Even if two neighbouring but state-separated municipalities wish to cooperate and can legally do so, they often find that they differ so much regarding legal competences that other hierarchy levels must get involved. This is so for both Estonian–Latvian and Finland–Sweden relations, though the countries are culturally similar and all EU members. While the Nordic countries have long histories of co-operation, Estonia and Latvia are 'nationalizing states' (Brubaker 1996), trying to consolidate internal allegiances. For Narva and Ivangorod, inter-state tensions hamper co-operation, which is also negatively influenced by a double imbalance between the small, EU and NATO-aligned Estonia and the Russian Federation.

Cross-boundary political cooperation cannot itself increase individual interaction, but can facilitate it. Nevertheless, many initiatives are occurring regardless of formal legalities and financing. On individual levels, border dwellers act according to perceived incentives: costs, and benefits of crossing or not crossing the border. Apart from actual landscapes of opportunities, there are also information landscapes, where foreign neighbourhoods may be blank spots mentally. Through family networks, going to school, finding jobs, partners or searching for recreation, individuals acquire domains of knowledge. Language is important here, both as symbol, and means of communication between individuals and the territorial state. Many boundary-dwellers are enmeshed in the formalized social life of their 'nation state', or as with cross-border immigrants, with their kin-state through membership in sports clubs, professional and interest organizations, in turn creating social contacts, friendships and family formation. Driving through customs to buy meat or liquor at a hypermarket will not increase social interaction or cultural understanding. Immigrants from and residents of the kin-state ethnicity may isolate themselves from the surrounding 'alien' nation state. People of Finnish origin in Haparanda and ethnic Russians in Narva may thus live physically in one state but cognitively in the bordering state, the border on the ground being less influential than the border in the mind.

References

Arbitration Convention between the Esthonian and Latvian Governments, signed at Walk, March 22, 1920, League of Nations Treaty Series vol. 2, 1920, 188–89. Online www.forost.ungarisches-institut.de/pdf/19200322-1.pdf [02/05/2016]

Brubaker, R. (1996), *'Nationalism Reframed. Nationhood and the National Question in the New Europe'*, Cambridge: Cambridge University Press.

Christaller, W. (1933), *'Die zentralen Orte in Süddeutschland'*, Jena: Gustav Fischer. [*Central Places in Southern Germany*. Englewood Cliffs, NJ: Prentice Hall, 1966].

Jauhiainen, J., and T. Pikner (2009), 'Narva–Ivangorod: Integrating and Disintegrating Transboundary Water Networks and Infrastructure', *Journal of Baltic Studies* 40:3, 415–436.

Jukarainen, P. (2001), *'Rauhan ja raudan rajoilla: nuorten maailmanjäsennyksiä Suomen ja Venäjän sekä Ruotsin ja Suomen rajojen tuntumassa'*, Helsinki: Helsingin Yliopisto.

Kant, E. (1932), *'Valga. Geograafiline ja majanduse ülevade'*, Tartu: Tartu Ülikool Majandusgeograafia Seminari Üllitised.

Kolosov, V., and N. Borodulina (2006), 'Rossiysko-estonskaya granitsa: bar'ery vospriyatiya i prigranichnoe sotrudnichestvo', *Pskovskiy regionologicheskiy zhurnal* 1, 145–157.

Lundén, T. (2009a), 'Valga-Valka, Narva-Ivangorod Estonia's divided border cities – cooperation and conflict within and beyond the EU', in J. Jańczak (ed.). *Conflict and Cooperation in Divided Towns and Cities*, Berlin: Logos Verlag, 133–149.

Lundén, T. (2009b), 'Language landscapes and static geographies in the Baltic Sea Area', in M. Andrén, T. Lindqvist, I. Söhrman, and K. Vajta (eds.). *Cultural Identities and National Borders*. Göteborg: Center for European Research, Göteborg University, 85–102.

Lundén, T. (2011a), 'Religious symbols as boundary markers in physical landscapes. An aspect of human geography', in J. Jańczak (ed.). *De-Bordering, Re-Bordering and Symbols on the European Boundaries*. Berlin: Logos Verlag, 9–19.

Lundén, T. (2011b), 'The Creation of a Dying Language', *Folia Scandinavica Posnaniensia 12*, Poznań: Wydawnictwo Naukowe UAM, 143–154.

Lundén, T., and D. Zalamans (2001), 'Local Co-Operation, Ethnic Diversity and State Territoriality – The Case of Haparanda and Tornio on The Sweden-Finland Border', *Geojournal* 54, 33–42.

Lundén, T., and D. Zalamans (2002), 'National Allegiance' and Spatial Behaviour in Baltic Boundary TwinTowns', *Journal of Baltic Studies* XXXIII/2, 177–198.

Mälksoo, L. (2005), 'Which Continuity: The Tartu Peace Treaty of 2 February 1920, the Estonian–Russian Border Treaties of 18 May 2005, and the Legal Debate about Estonia's Status in International Law', *Juridica International* 10, 144–149.

Prokkola, E.-K. (2009), 'Unfixing Berderland Identity: Border Performance and Narratives in the Construction of Self', *Journal of Borderland Studies* 24, 21–38.

17 City-twinning as local foreign policy

The case of Kirkenes–Nikel

Pertti Joenniemi

In this chapter, we shift focus to twins on the external border of the Schengen zone, the one between Norway and Russia, this time 54 km apart and thus of the fourth more distant, engineered sort outlined in the introduction. Again, we hit tensions between central governments and events on the twin-city ground (and here the pressures towards cross-border intimacy are coming from the centre rather than the locality). Pertti Joenniemi raises the possibility of twins becoming informal venues for the making of foreign policy – rendering explicit a theme implicit in previous chapters.

Introduction

City-twinning has some puzzling features. Entities like cities generally originate from moves of differentiation and by expressing what they are not. Twin cities do the opposite: they unsettle the ordinary by emphasizing virtues of similarity, of being alike.

The exceptional and in a sense quite defiant mode of constitution underlying city-twinning becomes even more challenging when involving cities engaged in twinning across a shared border as with Kirkenes and Nikel, a city-pair located at the Norwegian–Russian border in the North as core actors part of a broader cross-border arrangement between Sør-Varanger municipality and Pechenga Rayon. In entailing a denial and radical downgrading of the divisive and difference-producing impact of national borders, their engagement in twinning unavoidably problematizes the nature and alters the functioning of the border. If there is sufficient similarity present across the border for Kirkenes and Nikel to construct themselves as twins, what is left of the divisive impact of the Norwegian–Russian border?

Twinning as a top-down initiative

Their bonding thus invites a probing of quite profound and far-reaching questions, particularly since twinning did not emerge in the standard bottom-up fashion as an initiative from the cities themselves. It was instead launched by the Norwegian and Russian foreign ministers in 2008, with Kirkenes and

Nikel being requested to contribute through twinning to cross-border cooperation and the unfolding of shared Norwegian–Russian space.

Twinning obviously landed on the agendas of the various actors part of the Kirkenes–Nikel constellation in a promising manner. The two foreign ministers tabled a joint proposal and the two cities responded by pledging to engage in twinning as expected. The Norwegian Foreign Minister, Jonas Gahr Støre, initiated matters through a letter sent to the mayors in Sør-Varanger municipality and Pechenga Rayon, instructing them to jointly develop a scheme on twinning between Kirkenes and Nikel. For Norway and Russia to be able to initiate the so-called Pomor zone, originally developed and proposed by Norway in 2006, city-twinning could signify a starting point.

The initiative apparently testifies to significant changes in how the two states see themselves, define their border and exercise power in their borderlands. The relationship between identity, sovereignty, territoriality and borders in both countries has traditionally been quite tight. Border-drawing, with borders seen as barriers rather than frontiers, has appeared crucial not only in singling them out as independent states, but also in presenting them as entities opposite each other. In fact, encouraging city-twinning as one aspect of broader Norwegian–Russian cross-border cooperation conflicts sharply with traditional techniques of sovereign power.

The twinning between Kirkenes and Nikel also apparently offers insight into broader issues, particularly since Norway and Russia have invited Kirkenes and Nikel to contribute to conducting their foreign policies. Local concerns are fused with state-related interests, thereby undermining the states' traditional prerogative and contributing to a decentralization of foreign affairs.

Kirkenes–Nikel's pairing did not start from scratch since they had already signed a friendship agreement in 1973 during the Cold War. Their position as an integral part of broader plans for cooperation and cross-border regionalization between Norway and Russia provides them with considerable potential agency in an international context.

However, perhaps the two cities are less than enthusiastic about contributing as local actors to cross-border cooperation. Switching from depicting themselves as 'friends' to 'twins', that is, riding on far-reaching similarity and mutual sympathy, appears a formidable step since friendship entails a close relationship between two distinct entities, whereas twinning rests on notions of unity and being alike. Riding along would imply that Kirkenes and Nikel both accept that their previous 'soft' idealist and politically loaded endeavours of friendship are provided with and traded for new, more extensive and quite instrumental contents.

The initial step

In responding to the initiative, the Sør-Varanger municipality and Pechenga Rayon announced their preparedness to apply the powers they had been

given, promising to turn their existing cooperation into a twin-city relationship focusing on trade and commerce, social questions, civil society interaction, environmental security and tourism.

Overall, much pointed to steady progress. Norway and Russia seemed
to be on their way to establishing a closely co-operative relationship in the
North; Kirkenes and Nikel took steps to position themselves within that
constellation not only to utilize the opening but also to spearhead the emergence of a shared borderland.

Yet only modest progress has apparently occurred. Several seminars
were organized in 2009 to provide twinning with concrete substance
(Haugseth, 2013); action plans have been developed. However, the resultant dialogue has not produced the improved cooperation in sectors such
as industry, logistics, trade and commerce the foreign ministers requested.
Kirkenes and Nikel have been unwilling and unable to deliver on their
promise of deepening their relationship through various forms of functional
cooperation. Although smaller steps have been taken, Kirkenes and Nikel
appear generally not to have spearheaded cross-border integration.

Togetherness remains thin and cooperation has not significantly increased. Friendship apparently worked well during the Cold War, and
inter-city closeness might continue since it contains no claim of being fully
alike as twinning does. The problematic aspect of twinning thus consists
of containing too far-reaching claims to similarity, therefore obstructing
rather than enhancing close and concrete cooperation. What worked during
the Cold War period as an expression of deviance and protest does not seem
to contribute to the aims of close local and functional cross-border cooperation border set by the Norwegian and Russian states.

Overall, the cities have not apparently joined forces to conduct local foreign policies. They have not utilized options opened up by the top-down initiatives to obtain various advantages, and to strengthen their economic and
functional cooperation for this to yield a shared borderland (Figenschou,
2011, 14, 16). Kirkenes and Nikel are increasingly in contact and the border
has lost much of its divisive impact, but cross-border projects between
them pertain mainly to 'soft' cooperation between libraries, kindergar-
tens, schools, cultural entities and sport clubs and have not really embraced
'hard' areas like those of industry and commerce (Sergunin and Joenniemi,
2013, 253).

Signs of increased familiarity

These disappointing outcomes need accounting for. What prevents twinning from yielding tangible results in terms of conducting a kind of local
foreign policy and significantly down-grading the Norwegian–Russian
border through concrete and instrumental cooperation?

Obviously, the encounter between Kirkenes and Nikel consists of meeting
the other as different from oneself – despite the term 'twins' presupposing

far-reaching similitude. Their difference can contribute either to intensifying the relationship or to problematizing the encounter. Differences may create curiosity, fascination and nostalgia, thereby prompting cross-border interaction; alternatively, they can seem too outstanding and hence bring aversion, resentment and avoidance.

Both options are generally available as noted by Bas Spierings and Martin van der Velde (2008, 2013) in their studies on cross-border interaction. They argue that differences can induce feelings of familiarity or unfamiliarity and conclude that the prevalence of positive unfamiliarity, that is, unfamiliarity cleansed of its negative and threatening aspects tends to produce moves of inclusion. It hence also contributes to the formation of porous borders, whereas forms of unfamiliarity felt to be uncomfortable invite exclusion and preservation of divisive borders.

These findings then raise questions about the impact of emphasizing likeness rather than the difference embedded in familiarity. Arguably the likeness integral to twinning is even more conducive to cooperation and interaction than familiarity since difference has been discursively downplayed to the extreme in the context of twinning. However, perhaps, it is precisely the difference encountered in others and not similarity, which makes cooperation and reaching beyond borders interesting. The results of Spiering and Van der Velde suggest this. Surely, the similarity and likeness integral to twinning does away with the threatening aspects of difference, although likeness can also seem quite negative, cleansing all the differences and borderlines allowing an entity to know where it starts and ends, what it is and is not. Thus, twinning becomes problematic, a discourse about removing difference, an encounter between entities arguably similar to each other. One entity is merged or swallowed by another with both parties surrendering their previous, difference-based identities. The overall outcome may be a loss of identity, there being no longer any difference left in the relationship allowing identity to be defined through border-drawing in relation to the significant other and in terms of who one is not. The consequent lack of a constitutive outside may produce profound anxiety as collective identities are always premised on difference, mostly in the form of otherness (cf. Abizadeh, 2005).

Thus, engagement in twinning understood as a form of familiarity rather than something embedded in non-bordered similitude can produce positive outcomes, functioning as an unconventional strategy related to the construction of identities. In Kirkenes and Nikel, switching from relations based on the previous concept of friendship to that of twinning – and twinning induced from above – is felt rather problematic, undermining the difference still part of friendship and eliminating in conceptual and symbolic terms the option of meeting the adjacent city as different in interesting and positive ways.

It is yet to be noted that engagement in societal relations and cooperation among citizens across the Norwegian–Russian border does not seem similarly hampered by switching from friendship to twinning. Difference has not been traded for similarity and feelings of being alike. Instead, difference prevails as the dominant notion, with friendly as well as less friendly forms of difference present in relationships between the cities' inhabitants. The region's past contains considerable elements of familiarity that can be drawn upon in grounding a cooperative relationship. Modern bordering, part of extending state-formation into the north, occurred relatively late and was presided over by a 'common land' epoch. Interaction occurring during that period paid scant attention to state-related borders and efforts at bordering (Viken, Granås and Nyseth, 2008). In particular, a Sámi population moved flexibly across borders and also Finnish-speakers were strongly present in the region straddling various borders. The region's borders emerged gradually during the Swedish–Norwegian Union (1826–1905) and were finally drawn in 1826, although demarcation and delimitation proceeded quite slowly with borders actually preserving their nature of frontiers up to the early twentieth century (Niemi, 2005).

However, the period of openness was then followed by one of strict closure, especially because of the Russian Revolution in 1917. With Finland gaining independence the same year, most of the Norwegian–Russian border became a Finnish–Norwegian border-line up to 1944. This implied that Nikel (Nikkeli) and the surrounding area of Pechenga (Petsamo) were part of Finland up to the end of World War II when the border returned to being a Norwegian–Soviet/Russian one.

The various historical turns implied that the Sør-Varanger municipality long remained quite multicultural, containing Norwegian, Kven (Finnish-speakers), Sámi alongside Russian elements. Interactions were locally premised largely on familiarity and feelings of togetherness, whereas the policies of the Norwegian state rested on 'Norwegianization', that is, the introduction of threatening unfamiliarity as a key constitutive narrative (Rogova, 2008, 11; Viken, Granås and Nyseth, 2008, 27).

The Second World War implied that the position of security as a formative influence was further strengthened. There was much resistance to German occupation, particularly in Northern Norway with Norwegian partisans cooperating to some extent with Soviet forces (Niemi, 2005). Towards the end of the war, the Red Army liberated north-eastern areas of Norway, including Kirkenes, from German occupation. Although the fighting destroyed most dwellings in the region and caused profound destruction, the Soviet forces have remained rather positive as indicated among other things by that a statue devoted to Soviet soldiers still stands on a hill-top in Kirkenes (Figure 17.1).

However, the outbreak of the Cold War and Norway joining NATO implied that threatening unfamiliarity and exclusion rather than familiarity and inclusion became dominant. A garrison-mentality prevailed on both the Norwegian and Soviet sides. The rather securitized national discourse implied that the border remained almost entirely closed, being long

Figure 17.1 Soviet Liberation Monument in Kirkenes by Ekaterina Mikhailova (2016).

comprehended as 'a symbolic end of the world for people living on the two sides of it' as noted by Anastasia Rogova (2009, 33).

Yet notions of familiarity did not vanish entirely as indicated by the sporadic cross-border contacts taking place even during the Cold War. They occurred, for example, in the form of port visits by Russian fishing vessels and some tourism. More significantly, Pechenga Rayon and Sør-Varanger municipality signed a friendship agreement in 1973. This stood out as a symbolic gesture signalling political dissatisfaction with the mood and policies pursued during the Cold War, although it also produced some intermittent contacts across the border (Brednikova and Voronkov, 1999). For example, the two municipalities have celebrated the end of occupation and commemorated jointly Finnmark's liberation (Figenschou, 2011, 23).

In general, increasing space has opened up for the (re-)application of notions premised on familiarity since the end of the Cold War. Consequently, the border has gained flexibility, allowing various transactions to unfold. Increased room has also been provided for stories depicting the Norwegian–Russian borderlands as multicultural and loosely bordered. Some previously rather dominant and divisive tales have been rewritten, as also evident from the proliferation of the term *Pomor* (a term referring to ancient coastal trade and feelings of togetherness between Russians merchants and the broader local population (Niemi, 1992)).

Overall, the legacy of conflicts and profound unfamiliarity has apparently lost standing, whereas narratives indicating a common past, with familiarity as a key departure, have increased in weight. The application of these narratives imply that the future is provided with features of a recreated past, and projected as less framed by interstate relations and strict bordering.

From Kirkenes to 'Kirik'

Actually, much points to the emergence of a shared borderland, although this change seems to have happened without much contribution from Kirkenes and Nikel through their policies of twinning. Particularly, Kirkenes has become a major meeting point for Russian–Norwegian contacts since the 1990s on various levels and in different forms. The town's multicultural character has become even more pronounced because, alongside a Norwegian majority, it contains a Sámi population and many Finnish speakers; also increasing numbers of Russians and Russian speakers. This latter group in 2008 comprised over 10% of Kirkenes' population (Rogova, 2008, 29) and has steadily grown.

Notably, Russians visiting Kirkenes do not seem to have the feeling of being abroad. There is considerable commonality present as also indicated by Kirkenes being named 'Kirsanovka' or 'Kirik' with connotations of a small local and nearby entity/village in the language used in the Murmansk region. Another sign of a decline in the unfamiliarity felt to the threatening alongside increased feelings of familiarity consists of the region occasionally being called 'little Murmansk' (Figenschou, 2011, 10). In other words, it is comprehended as being in-between with the familiar kind of difference part of the town, implying that it attracts significant interest and curiosity.

Nikel: end of closure

Nikel has been less touched than Kirkenes by changes in the external environment. The local economy has been based on heavy industry and that remains the case despite relative decline in industrial production. The town has clearly not opened up and changed as rapidly and profoundly as Kirkenes.

Yet Nikel also is far less closed and defined by a closed border than heretofore, and the constitutive discourses of relevance for the part of Nikel show signs of de-securitization. This is most clearly evidenced by the town no longer having official status – as it predominantly had prior to 2008 – as a closed border zone despite still hosting some smaller military units.

Compared to Kirkenes, Nikel has also been apparently less able to utilize and take advantage of the border's changing and more porous nature. Actually, the town seems to have declined rather than increased in importance due to factors like reduced support from the central government, cuts in military personnel as well as declining production at the Norilsk Nikel plant. The diminished standing is reflected by the fact that its population (around 12.500 in 2012) has dropped by a third since the Cold War ended.

However, alongside increasing numbers of inhabitants visiting the Norwegian side, the new and more flexible border-regime has added significantly to the number of Norwegians visiting Nikel (Haugseth, 2013). Nickel has changed and opened up towards the exterior instead of remaining 'a city squeezed between extreme peripherality and a closed border' as characterized by a person interviewed by Haugseth (2013).

Much of the interaction has unfolded around initiatives from individual people. However, Nikel's city administration has also contributed to increasing interaction. For several years, it has organized annual Norwegian–Russian events focusing on cross-border cooperation.

The city administration seems generally to favour twinning and has tried to improve the record by producing an unofficial assessment of the achievements, obstacles and failures in the conduct of twinning (Figenschou, 2011, 87–89). On the positive side, it notes there is increased interest among the inhabitants in participating in various activities. Moreover, mutual confidence as well as understanding vis-á-vis the Norwegian neighbours has apparently grown. The list of problems foregrounds, in turn, the inadequacy of Russian legislation for not offering the clarity needed by local actors interested in twinning. Insufficient coordination between different authorities and levels of decision-making alongside inadequate funding are also raised as issues to be tackled. Anne Figenschou (2011, 86) views the assessment as a positive sign testifying to Nikel's serious interest in twinning.

Signs of a standstill

More generally, bottom-up contributions premised on twinning to the development of Norwegian–Russian relations have remained modest. The option of engaging in pairing has not been much utilized, and the cities have, for example, not engaged in joint image building and self-promotion. Overall, Kirkenes and Nikel have not been willing or able to break with their previous tradition of getting together as friendship cities and move to constructing a relationship premised on far-reaching similitude and therefore also extensive functional cooperation.

However, balanced judgment must also take account of top-down input from Norway and Russia about the emergence of a cooperative cross-border relationship and the construction of an integrated borderland. Overall, their record appears mixed. A major step towards reducing the border's divisive impact clearly consists in establishing a visa-free zone. Agreement on a visa-facilitation regime between Norway and Russia was reached in November 2010, coming into force in May 2012. Those living within the 30 km border area on the Norwegian and Russian sides around Kirkenes and Nikel can acquire three-year identity cards allowing holders to cross the border visa-free and stay on the other side up to 15 days each time. Border-traffic increased significantly with over 250,000 crossings in 2012. Notably, while the city administration in Sør-Varanger municipality decided in autumn 2015 to establish a fence at the site of border-crossing to shield

Norway at least symbolically from refugees using Russia as a transit route, local inhabitants on the Norwegian side strongly opposed this apparent restoration of features of the border as a barrier.

The visa-free arrangement is clearly important, although further top-down contributions are required for Kirkenes and Nikel to break with their domestic being and reach out into the sphere of the foreign without this conflicting with rules and regulations generally applied in conducting foreign relations. The two cities need a mandate plus various resources for broadening and de-centralizing of foreign policy to become reality.

While there have been modest restrictions on the availability of financial means to engage in far-reaching cross-border cooperation, the question of mandate has apparently posed more significant issues. This goes particularly for Nikel. Twinning for Russia has become an important aspect of a rather heated debate regarding the relative treaty-making powers of the federal centre, regions and municipalities. Despite Moscow's resistance since the early 1990s, quite a number of Russian border municipalities have concluded agreements with their international partners (Alexeev, 2000). Eventually, compromise between the centre and local actors emerged; it was decided that such agreements should not have the status of full-fledged international treaties (seen as a federal prerogative), and should be prepared with assistance from the Russian Foreign Ministry. Local actors' power to engage in treaty-making is thus limited in principle, although they have in reality been able to establish cooperative relations across national borders and strengthen their international standing. Hence, Nikel also has effective, if not fully crystallized, power to engage in twinning.

Overall, the constitutive discourses impacting the unfolding political landscape relate increasingly to various joint projects, particularly those dealing with cooperation around oil, gas and shipping. The more future-oriented narratives pertain to the northern areas' growing importance for Russian and Norway's national economies.

Yet, there has also been some stagnation around Norwegian–Russian cooperation. Success has not always been as quick and complete as expected; some plans have failed and others postponed. One disappointment has been over the Shtokman gas and oil field. With declining world oil and gas prices, the project has lost some urgency and attraction. Consequently, Norwegian Statoil has relinquished its share in the field and Russian companies have put their development plans on hold. The various backlashes in cooperation generally have then affected many of the specific Norwegian and Russian plans. Among others, the building of a metallurgy plant on Pechenga Bay and the construction of a liquid–gas plant have not materialized.

However, the northern areas still remain quite dynamic, among other reasons because of the impact of climate change and the opening up of the north-eastern sea route connecting Europe and Asia. Kirkenes particularly, but to some extent also Nikel, remain important as part of areas that are clearly increasingly important in terms both of national development and of international cooperation.

Conclusions

Unfriendliness as a form of othering has clearly lost significance in determining relations between Kirkenes and Nikel. It does not affect how they unfold nor the border's significance as forcefully as in the past. It has been overlain by departures depicting the border as something to be transgressed, even rendering it a unifying element between the cities as well as Norway and Russia generally.

This links with more general trends in post-Cold War Europe. Furthermore, Norway and Russia have, alongside many other states, abandoned some of their previous nature as Hobbesian territorial states and turned increasingly into Lockean competition states, although the trend is far from linear as indicated by that security has more recently re-gained some of its constitutive weight.

Twinning seems to be the exception here. In the context of the Pomor plan, Norway and Russia invited and encouraged Kirkenes and Nikel to engage in city-twinning – this would necessarily alter the nature of the Norwegian–Russian border. It would, if extensively implemented, introduce significant likeness into the intercity relationship. It would also turn their border *vis-à-vis* the exterior into a constitutive one – the difference crucial to their being residing in being Norwegian and Russian on special terms, that is, cities at the border and unified by the border rather than border cities. The difference would undoubtedly be benign, but twinning as a constitutive category can nonetheless alter the border-related delineation of both Norway and Russia. The two countries would become less sharply separated around this point of the border, not exactly one nation or the other because cities located at the border can define themselves more flexibly, even use twinning, and the similarity that concept assumes, as their point of take-off. The outer edges defining what is Norwegian and what is Russian get blurred – allowing for new combinations to emerge, perhaps permitting increasing numbers to think of themselves as both Norwegian and Russian, even neither Norwegian nor Russian.

Already the visa-free arrangement represents a step in this direction. It singles out the two adjacent towns as belonging to a category of their own by abolishing restrictions upon Kirkenes and Nickel inhabitants making quite frequent cross-border trips while normal rules and regulations still apply at their external borders. Crucially, being able to skip visas implies that they form an internally undifferentiated entity with difference delineated by their external border. The arrangement not only allows Kirkenes and Nikel, in eliminating internal distinction between the two cities, to be simultaneously Norwegian as well as Russian, but also furnishes them with the option of figuring as a third, being neither Norwegian nor Russia and residing in a distinct category of their own. This option emerges if they prefer to articulate their similarity as something entirely different from the exterior and play their internal similitude sharply against the difference part of the exterior.

Particularly, the latter option suggests that twinning can be quite demanding for the cities themselves. The step from previous friendship into twinning might outwardly seem short and easy. However, this is not the case: friendship is in several ways quite different from those of twinning with the latter allowing far more radical choices about identities and the more general unfolding of political space.

Policies of friendship were present during the Cold War and allowed cities to express a deviant opinion. Cities came together as locals and non-securitized actors mainly to symbolically demonstrate that alternatives existed to the rather antagonistic and securitized policies pursued by states like Norway and the USSR. Through engaging in cross-border, they aspired to show ways out of the polarized, strictly bordered and dangerous situation created by the states.

Present-day twinning is obviously different. Whereas the friendship concept allows the preservation of the difference between the parties involved, twinning foregrounds similarity and stands out as something rather drastic in undermining notions of difference, challenging various moves of bordering and adopting different integrative and cooperative endeavours. Twinning has ordinarily been confined to relations between the cities and has not been launched as part of broader schemes integral to some broader international aspirations. However, Kirkenes and Nikel deviate from this pattern as the initiative here has been taken by actors at state level.

It is hence unsurprising that city-twinning has been difficult for Kirkenes and Nikel. Indeed, their policies appear more like friendship than twinning. As such, there seems a considerable dose of familiarity between Kirkenes and Nikel. They remain far from similar but their difference is no longer viewed as threatening unfamiliarity. It connects rather than isolates with difference being conducive to curiosity and interest in exploring the not-fully-familiar entity across the border. Thus, the change in approaches and perceptions of difference allows significant increases in trans-border interaction; this in turn blurs previous constellations by internationalizing the domestic and domesticating the international. Yet, it is difference rather than similarity, which is conducive to growing interactions and contacts.

Increasing familiarity has even produced some degree of commonality and has been conducive to the emergence of shared mental space as indicated, for example, by the use of images like 'little Murmansk' on the Russian side. The familiarity present testifies to the Norwegian–Russian border being seen less as a barrier, more as a frontier. This is positive but hardly to the credit of the local administrations and their twinning efforts: they seem to have allowed this to unfold rather than actively contributing to it. In addition, twinning between Kirkenes and Nikel has been hampered by a certain competitive logic. For example, disputes over where investments should go have militated against a pooling of resources. Whereas Finnmark and Kirkenes on the Norwegian side apparently have a constructive, co-operative and largely frictionless relationship, this has been less true for Nickel and the Murmansk Oblast. The paralysis detectable in Nickel seems to have originated largely from the rather competitive policies pursued by Murmansk. The presence of competitive aspirations is unsurprising, and these may well be present in the future, affecting all relevant levels: states, regions as well as the cities themselves. We should note that the emergence of competition and other difficulties around cooperation and cross-border contacts nonetheless tell us that the Norwegian–Russian border is significantly changing in essence.

References

Abizadeh, A. (2005) 'Does Collective Identity Presuppose an Other? On the Alleged Incoherence of Global Solidarity'. *American Political Science Review*, 99(1), 45–60.

Alexeev, M. (2000), Russia's Periphery in the Global Arena: Do Regions Matter in Kremlin's Foreign Policy? *Ponars Policy Memo* No. 158. San Diego State University.

Brednikova, O. and Voronkov, V. (1999). 'Granitsy i restrukturirovaniye sotsialnogo prostranstva [Borders and the restructuring of a social space]', in Brednikova, O. and Voronkov, V. (eds.), *Kochuyushchieye granitsy* [Migrant borders]. Saint Petersburg: CIRP, 19–25.

Figenschou, A. (2011), 'The Twin Cities Petchenga Rayon and Sor-Varanger Municipality'. Master thesis. The Faculty of Humanities. University of Oslo.

Haugseth, P. (2013), 'Tvillingebysamarbeid i den norsk-ryssiske grensesonen', in Viken, A. and Fors, B.S. (eds.), *Grenseliv*. Tromsø: Orkana forlag.

Niemi, E. (1992), *Pomor. Nord-Norge og Nord-Ryssland genom 100 år*. Oslo: Gyldendal.

Niemi, E. (2005), 'Border Minorities between State and Culture', in Jackson, T.N. and Nielsen, J.P. (eds.), *Russia-Norway. Physical and Symbolic Borders.* History Department of the University of Tromsø and Institute of World History, Russian Academy of Sciences. Moscow and Tromsø, 69-79.

Rogova, A. (2008). *From rejection to re-embracement. Language and identity of the Russian speaking minority in Kirkenes, Norway*. Kirkenes: Barents Institute.

Rogova, A. (2009). 'Chicken Is Not a Bird – Kirkenes Is Not Abroad. Borders and Territories in Perception of the Population in a Russia-Norwegian Borderland'. *Journal of Northern Studies*, 1, 31–42.

Sergunin, A. and Joenniemi, P. (2013), 'Another Face of Clocalization. Cities Going International (The Case of North-West Russia)', in Makarychev, A. and Mommen, A. (eds.), *Russia's Changing Economic and Political Regimes. The Putin Years and Afterwards*. London and New York: Routledge, 229–259.

Spierings, B. and van der Velde, M. (2008), 'Shopping, Border and Unfamiliarity: Consumer Mobility in Europe'. *Tijdschrift voor economische en sociase geografi*, 99(4), 497–505.

Spierings, B. and van der Velde, M. (2013), 'Cross-Border Differences and Unfamiliarity: Shopping Mobility in the Dutch-German Rhine-Waal Euroregion'. *European Planning Studies*, 21(1), 5–23.

Viken, A., Granås, B. and Nyseth, T. (2008), 'Kirkenes: An Industrial Site Reinvented as a Border Town'. *Acta Borealia*, 25(1), 22–44.

18 The Finnish–Russian border as a developmental resource

The case of Imatra and Svetogorsk

Matti Fritsch, Sarolta Németh and Heikki Eskelinen

The next two chapters focus on a twin-city pair – Imatra and Svetogorsk – on the border not just between two nation states but between the EU as a whole and Russia. The difficulties of such a location have been briefly explored in the context of Narva–Ivangorod and Kirkenes–Nikel (Chapters 14, 16 and 17). However, the experience of these border twins is distinct due to mixed collective memories, resulting from their recent past with population transfer and territorial redistribution. Again we come across tensions between local authorities, local inhabitants and decidedly distant central and state governments. Chapter 18 views the relationship from an elite perspective and analyses it through the lens of the benefits the cities have tried to derive from their adjacency.

Introduction

Collaborative activities between Imatra and Svetogorsk have attracted significant interest in debates about border-twin cities since around 2005 (e.g. Eskelinen and Kotilainen 2005; Kosonen et al. 2008; Joenniemi and Sergunin 2011). The two cities have often been compared to others in Europe and beyond. However, these contributions have generally focused on the symbolic meaning and value of the twin-city label (and ensuing collaborative activities) rather than on the border's diverse functions and mobilisations and the opportunities it provides for mutual development. The key defining aspect in border-twin relationships is the border itself. Borders are more than lines on a map: they can serve different and overlapping functions. O'Dowd (2002, 14) identifies four: as barriers, bridges, resources and symbols of identity. These functions and their interconnections are contextual, depending on particular border-settings. Hence, borders change their functions over time. Employing an actor-centred approach, it can be argued that the Finnish–Russian border has been re-conceptualised, at least among local elites, from a dividing line and impermeable barrier to a useable resource following the momentous changes from the early 1990s onwards.

Christophe Sohn (2014a) has recently enriched the academic debate on cross-border integration and using the border-as-a-resource. He distinguishes two contrasting models of cross-border integration. The *geo-economic model* mainly rests on mobilising the border as 'a differential benefit', aiming 'to generate value from asymmetric cross-border interactions'. By contrast, the *territorial project model* is based on a common identity or at least shared understanding of the prerequisites for joint development and convergence through 'hybridisation/innovation or via the territorial and symbolic recognition that borders entail' (Sohn 2014a, 587). We must understand that these two models represent ideal types and are not mutually exclusive. In practice, the two are often combined in different variants. We can argue that the territorial model comes close to the underlying aim of many international town-twinning initiatives, that is, 'creating – to varying degrees – communality and joint space, thereby providing the ground for the usage of the concept of a "twin city"' (Joenniemi and Sergunin 2011, 240). Over time, changing circumstances or framework conditions – and the resultant fluctuations in terms of intensity and degree of collaboration – can make one model more pertinent to a twin-city relationship than the other.

With regard to the resource function of an increasingly permeable border, Sohn (2014a, 594–596, based on Sohn 2014b) distinguishes five advantages or benefits available to regional and local actors: (1) *positional* benefits emphasise the importance of border-proximity and closeness to the adjacent region, on the other side, facilitating, for example, tourism or gateway functions; (2) *transaction* benefits arise from 'competences to cross or to bridge the border despite its barrier effects' resulting from knowledge of the adjacent economic, cultural, political, legal and administrative system, involving, for example, border-dependent business ventures or smuggling; (3) *differential* benefits involve exploiting 'factor-cost differentials (labour, land, currency or water) or differences in tax regimes and regulations', along with differentials in residential costs and qualities plus shopping experiences. Borders can also represent (4) *loci of hybridisation* when processes of mutual learning and exchange occur producing 'new ways of doing and thinking' in fields like planning, civil society (inter)action or cross-border governance, and (5) borders can become recognised as *symbolic resources* in place-making in order to 'reinforce the international character of a region in its strategy of territorial marketing' and highlight the specific advantages of a border-setting.

In this chapter, we hypothesise that, in relation to Imatra and Svetogorsk, regional and local actors have identified the border as a potential *resource* for local development, applying various initiatives/approaches to utilise it, learning from the results and experiences. These initiatives have occurred in, and been conditioned by, changing geopolitical and socio-economic conditions in this EU–Russian borderland. To evaluate and enrich this hypothesis, we will analyse how preconditions for and interaction across the border have developed and how local actors have adapted their co-operation and twinning activities around these conditions. We will also explore how priorities set by local actors as regards cross-border interaction and co-operation have evolved in terms of the two models of cross-border integration identified above. This policy- and local government-oriented approach complements the examination provided in the next chapter focusing on how such policy initiatives and actions are experienced 'on the ground', that is, by residents in Imatra and Svetogorsk.

The context

Imatra–Svetogorsk is an anomaly on the Finnish–Russian border, which, since 1995, has also constituted the boundary between the European Union and the Russian Federation. Nowhere else on this 1340-km long and sparsely populated border are two urban centres from the different countries in such close proximity. The reasons are historical events following the Second World War when the Finland–USSR border was redrawn. As part of this, the USSR claimed the industrial community of Enso, part of a relatively dense network of Finnish settlements, most likely because of the existence of a large pulp and paper plant. The settlements were effectively 'partitioned' (Joenniemi and Sergunin 2011, 237, based on Buursink 2001, 8) and the now-Russian and Finnish townships and industrial facilities started developing separately within their new territorial and political contexts. Consequently, Imatra–Svetogorsk exhibits features of a 'duplicated city' (Joenniemi and Sergunin 2011, p. 235, based on Buursink 2001). After the Second World War, Enso (re-named Svetogorsk) was re-settled with population from elsewhere in the USSR and the paper mill continued production within the Soviet system. On the Finnish side, the remaining three settlements were re-organised into the municipality/town of Imatra, which has since become an industrial centre in its own right. During Soviet times, the Svetogorsk–Imatra border was completely closed for almost 30 years.

In the early 1970s, a state-orchestrated, large-scale development project involving reconstructing the pulp and paper plant in Svetogorsk by a consortium of Finnish construction companies as part of bilateral trade relations between Finland and the USSR resulted in establishing a border crossing-point (Lilja et al. 1994). However, only persons involved in the construction project and with special permits were allowed to cross until 6 July 1990 when local Imatra citizens had their first opportunity to visit their neighbours. Since then, border crossings have increased rapidly reaching around 400,000 in 1994, but proved sensitive to changing regulations about required documents and imports. The closest international crossing-point was approximately 40 km south in Nuijamaa.

The two city centres are very close (approximately 7 km) and the border's character changed from total impermeability in the 1940s, 1950s and 1960s towards strictly selective permeability in the 1970s and 1980s and, finally, relative openness within the confines of the current EU–Russian border regime (see Figure 18.1). Thus, it is unsurprising that the opportunities and potentials of cross-border interaction and co-operation attracted much attention among the two sets of the local and regional actors from the early 1990s. This was clearly indicated by the signing of the 1993 town-twinning agreement.

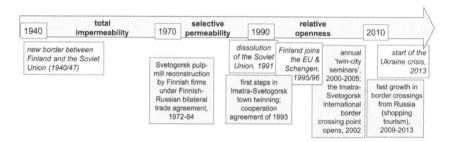

Figure 18.1 Conditioning factors, the nature of the border and the milestones in
local interaction between Imatra and Svetogorsk, 1940–2013.

Establishing an international border crossing-point, providing a long-term
and predictable border-crossing regime between Imatra and Svetogorsk,
was seen as crucial to realising this potential. In practice, however, this took
a long time – only opening in 2002, utilising EU cross-border co-operation
and national funding.[1]

The resource perspective on cross-border co-operation and interaction in Imatra–Svetogorsk

We can now start identifying and analysing the different processes of
cross-border interaction and co-operation in Imatra–Svetogorsk, setting
them against Sohn's typology of border-induced benefits. Because of their
close proximity, utilising the differences and potentials inherent on each
side of the border arrived on the agenda very early. The clear-cut constraint
to utilising these benefits was the unpredictable nature of border rules and
regulations at the (initially) temporary border crossing-point.

Positional benefit

The positional benefit Imatra–Svetogorsk can exploit from the border is obvious. It is on one of the three road connections between the St. Petersburg
metropolitan region (with approximately 6,000,000 inhabitants) and
Finland. Public-sector investments (Finnish government, EU programmes)
have created preconditions for a more permeable border by providing a better border crossing-point and transport infrastructures.

   As mentioned earlier, the opening of the international crossing-point in
2002 greatly intensified interaction between the towns. Border-crossings
grew rapidly from 400,000 in 2003 to 2,500,000 in 2013, as did its share of
total border-passenger traffic (Figure 18.2). More recently, the Ukraine crisis
and the rouble's resultant plummeting value has turned this trend downwards. Until 2015, Imatra–Svetogorsk's share of total border-crossings was
also growing: Russians constituting the vast majority of border-crossers.

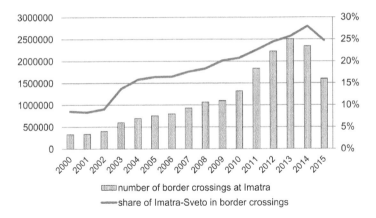

Figure 18.2 Changes in cross-border passenger traffic at Imatra–Svetogorsk, 2000–15 (data source: Finnish Border Guard).

The growth in cross-border traffic until 2014 largely resulted from rising affluence among people on the Russian side and their growing demand for shopping and leisure facilities. In 2013, the number of Russian visits to Imatra exceeded 900,000. Of these, 66,000 were made by Svetogorsk residents in 2014, meaning four trips a year per person on average. This is also enhanced and vividly illustrated by the fact that cross-border cycling has become popular among local people: cyclists being able to 'jump' the frequent car queues along the 'local gateway'.

Total spending by Russian tourists in Imatra was around 82,000,000 euro in 2013 and 65,000,000 in 2014 (TAK OY 2015). Consequently, Russian demand for shopping and leisure facilities has produced significant private investments in these sectors in Imatra. Importantly, this economic impact derives not only from Imatra's locational benefits in relation to Russian customers (i.e. proximity), but also from differential benefits attached to shopping tourism, particularly price differences.

Imatra's second positional benefit is its potential role as a transport gateway. This is particularly important because it is the only Finnish–Russian border-crossing, where road and rail crossings coincide (Joenniemi and Sergunin 2011). The Imatra–Svetogorsk road-crossing-point has become a main gateway for Finnish–Russian trade and Russian global imports, that is, transit trade. However, rail potentials have not been realised. Rail cargo's share of total cross-border carriages was 23% in 2014 comprising Russian imports entirely, that is, no exports and no transit traffic to Russia (Finnish Customs 2015).

Joint interest in exploiting this potential by raising the status of the Imatra–Svetogorsk crossing-point received much attention in the early phase of town twinning. More recently, the gateway function for rail traffic has been embraced by Imatra's municipally owned local development agency (KEHY), which co-operated with Russian and Finnish partners in an EU-funded collaboration project.[2] However, there is no cross-border passenger rail-traffic link despite extensive talks to open one between St. Petersburg and Imatra, thereby capitalising on historic tourism links. From Svetogorsk's viewpoint, its gateway position in road transport admittedly produces some disadvantages for local citizens, who complain about heavy lorry traffic through the town. However, border-proximity provides them with relatively easy access to Finland, and this positional benefit is actively utilised.

Transaction benefit

Exploiting positional and differential (see below) benefits generally requires familiarity with the situation and practices over the border and knowledge of the neighbour's cultural, economic, political, legal and administrative system. Particularly for individual citizens, as opposed to institutionalised co-operation between public and semi-public actors, language skills, cultural knowledge and personal contacts are important for effectively taking advantage of the border's potential as a resource.

Particularly relevant are individuals who can perform bridging functions between activities and organise their lives on both sides of the border, exhibiting what could be called 'cross-border competences'. Although Finnish–Russian migration is modest, the number of Russian-speakers in Imatra has tripled since 2000 to over 1000 (Statistics Finland). The Imatra area has also been popular amongst wealthy Russians seeking to purchase holiday homes, facilitating frequent movement between permanent Russian and temporary Finnish homes (Pitkänen 2011). It is however difficult to gauge how many Russians actually live more or less permanently in Finland.

At least on the Finnish side, there has also been a proactive drive to develop transactional skills. Imatra, like other Finnish border cities, has actively promoted Russian-language education in schools. Particularly important has been the Finnish–Russian School of Eastern Finland (with three active branches), established in 1997 to advance the so-called 'Russia know-how' in Finland nationally, while locally supporting the development of border regions in Eastern Finland (Hannula 2008). The School initially offered grade 5–9 education, but later developed into a comprehensive school, providing education across the full secondary level. In fact, the Finnish-Russian School of Eastern Finland in Imatra has more students than the other branches.

The development of transactional skills is also important for local entrepreneurs and business people, supported by the local development company and Chamber of Commerce. An interesting example of using transactional benefits is entrepreneurs and investors who, through bi-cultural competence or knowledge of Russian preferences, establish businesses on the Finnish side of the border, capitalising on Russian customers. Thus, supermarkets and smaller shops, almost entirely Russian focused, opened a few kilometres from the border to meet and generate Russian visitor demand.

Differential benefit

Utilising the border's potential differential benefits received particular attention in Imatra–Svetogorsk, even during Soviet times. During the later phases of reconstructing the pulp and paper mill in Svetogorsk by Finnish companies, a business park (named Imsveto) was planned. This initiative continued and became a key element in launching town-twinning activities (and signing the 1993 agreement). In 1998, Imatra's Mayor argued that Svetogorsk provided ample opportunities to access the immense Russian market, work behind the customs wall and take advantage of Russian cost-levels regarding production factors. In Imatra, the 'Key East' project was started to facilitate such ventures by providing bonded-warehouse services; in Svetogorsk, an industrial production village named DeCoSve OOO was planned; a car-assembly plant was even envisaged. However, these received no Russian central government support and came to nothing, despite other Finnish companies having production facilities in Russian border towns, for example, Kostomuksha (Zimin 2013). These projects sought to exploit the border's differential benefits, intrinsic differentials and asymmetries. An interesting example of utilising differential (and positional) benefits is evident in the operations of Svetogorsk's dominant employer, the American-owned International Paper pulp and paper factory.[3] This company has its production facilities on the Russian side (utilising lower wages, energy and land costs) while its higher management personnel (plant director and approximately 15 other employees) live and enjoy better living conditions in Imatra, commuting daily to Svetogorsk. This has caused some irritations in Svetogorsk since the plant director is Imatra's biggest individual taxpayer.

Plans for industrial production facilities taking advantage of differential benefits have not been on the agenda for many years. Instead, the differential benefit has come to play a major role in cross-border shopping, which over the past five years has become an important form of cross-border interaction (notwithstanding declining cross-border tourism after 2014). Here, the exploitation of differentials by Russians shopping in Finland is based on 'emotional' (different shopping experiences and some general sense of difference and 'foreignness') and 'rational' (e.g. price, quality) considerations (Spierings and van der Velde 2008). Prices of quality goods are often very competitive on the Finnish side and the availability of tax-free shopping adds another advantage to buying things in the neighbouring country. Russians also tend to trust Finnish-bought products more, often believing that many products sold in Russia are fake or of lower quality. Emotional reasons also play their part – the different shopping experience and products on offer certainly appeal to Russian customers. Russians often see shopping in Finland as a relaxed affair in a safe and clean environment, with shopping trips often combined with visits to the abundant spas and other tourist attractions.

Unsurprisingly, strong efforts to attract and increase Russian visitors and provide a functioning border and suitable shopping facilities/experiences have been made. This includes significant investments in Russian-oriented Finnish shopping infrastructure. For example, three major department stores have opened (with the telling names of *Laplandia*, *Rajamarket* [border shop], and *Scandinavian Market*) very close to the border-crossing station, also indicating the utilisation of positional benefit. Investments in tourism facilities have also been made, for example, a large spa just outside Imatra. However, the rouble's recent drop in value produced a significant decline in Russian border-crossers after 2013 (Figure 18.2). Simultaneously, Finns increasingly take advantage of price differentials by shopping for fuel, alcohol and cigarettes on the Russian side consequent upon the rouble's declining value.

Shopping tourism's importance for the local and regional economy in Imatra and south-eastern Finland is clearly considerable, and border-resource thinking has been strongest within this sector. Indeed, employment growth in the service sector and infrastructure investments around shopping tourism have somewhat cushioned the adverse effects of restructuring traditional industries, mainly in the forest and paper-related sector. Consequently, the local debate on and attitude towards Russians in Imatra has changed from marked suspicion towards seeing them as good for business and industrial activity.

Locus of hybridisation

Hybridisation points towards processes of mutual learning and exchange leading to (cross-border) innovations in various policy and governance fields, alongside civil society interaction. Facilitated by initially local and later national and EU funding opportunities, hybridisation has become crucial to co-operation both between Imatra and Svetogorsk and across the Finnish–Russian border generally since the mid-1990s, for example, producing teacher-exchange programmes and air-quality monitoring systems (Eskelinen and Kotilainen 2005). Communities of practice in several fields of civil society (sports, music, culture, education) and administration (business and the economy, local governance, planning and infrastructure development, environmental protection) have engaged in knowledge exchange and transfer, often with Finnish or EU financial support (NAC, Interreg/TACIS and ENPI[4]). Decades-long traditions of such institutional and people-to-people collaborations among relatively narrow circles of influential actors support mutual learning and the establishment of new and shared practices between the two towns. Thus, the local governments have co-operated in technical fields like water and energy supply, mainly by improving the situation and practices on the Russian side. Other relevant fields of activity are border-guarding and customs operations, emergency services, developments in education and sports-training methods, youth and social care (Németh et al. 2015). Teacher exchange and co-operation between secondary schools were further boosted by establishing the School of Eastern Finland (1997). There is a long tradition of multi-lateral sports collaboration between the cities of Imatra, Lappeenranta, Vyborg and Svetogorsk: the joint project *Step Up – Cross-Border City in Action* (financed by the EU's ENPI CBC instrument) aimed at building cross-border networks among sports clubs not only sharing infrastructures but also facilitating exchanges and learning. There is also an international bicycle ride called 'City Twins' between Imatra and Svetogorsk, carried out annually between 2006 and 2015 as one of the regular events organised within the framework of the co-operation agreements between the two towns – also helping present this area as a common trans-border space.

Integral to twin-city activities was the organisation of annual 'twin-city seminars' (2000–2005) designed to set an agenda and flesh out twin-city co-operation. Judging by individual seminar topics, a shift in emphasis (away from utilising differential benefits towards processes of mutual learning and exchange) is visible. The topics were:

* Industries and cities on the border (2000)
* Cross-border entrepreneurship (2001)
* The border ajar – trade amidst upheavals (2002)
* Health and welfare on both sides of the border (2003)
* Joint seminar with the FP5 EXLINEA border research project (2004)
* Media, democracy and the perception of the border in the EU and Russia (2005)

Hybridisation thus appears to have been challenging in economic fields of co-operation and easier to achieve in civil society, people-to-people and administrative co-operation.

Symbolic resource: object of recognition, branding

At least initially, the twin-city label, or vision of a cross-border twin city, had an important symbolic value to the two cities and their collaboration. The label represented – in its changing orientations – a search for uniqueness and an aspiration to raise the place's profile in both national and international contexts. Initiated in a co-operation agreement in 1993, the twin-city modes of operation were elaborated via an EU-funded project producing a joint development strategy between the two cities in 2001. A key element in these intensified co-operation activities after 2000 was the above-mentioned annual 'twin city seminars' on specific themes.

Based on ad-hoc co-operation between twin-city pairs across Europe, a wider European dimension and platform for co-operation under the Twin-City label was established in December 2006 via the City Twins Association, as a result of the Interreg IIIC[5]-funded City Twins Cooperation Network project 2004–2006 (Joenniemi and Sergunin 2011). Imatra–Svetogorsk joined this association alongside six other European twin-city pairs in 2006, but the activities of the network have apparently have waned in recent years.

Political and institutional changes in Russia after 2000 have somewhat complicated, even undermined, collaborative efforts between the towns. Specific problems arose when Svetogorsk became a second-tier municipality within the Vyborg district as part of local administrative reforms in 2006 (Zimin 2013). In addition, in 2010, Svetogorsk merged with the adjacent Lesogorsky municipality, which, due to the increased population and associated tasks, reduced its organisational capacity and took resources away from CBC activities (Iliyna and Mikhailova 2015). Essentially, this entailed the municipality losing decision-making powers to the regional level in fields important to Imatra, like education, health and social services. This meant that the two cities could co-operate only in comparatively 'soft' fields like the contents of the regular 'twin-city days', bicycling and sports events. Meanwhile, strategic economic, business and infrastructure projects – which could also reinforce the twin-city brand and image – have given way to individual mobilities, like shopping tourism and related retail investments between 2009 and 2013, which have been general to the Finnish–Russian border, not particularly linked to Imatra–Svetogorsk.

Particularly in co-operation with Russia, personal contacts and working relationships are extremely important. In Imatra–Svetogorsk, relations on the more official level deteriorated after the mayor of Svetogorsk who served in the early 2000s was changed. Previous to his replacement, the Russian side had significant interest in proactively planning and initiating co-operation with the Finns. However, with his successor, twin-city co-operation started to decline perceptibly. Anecdotal evidence suggests that, not possessing an international passport, the new mayor of Svetogorsk could not even visit the Finnish side. Partly as a result, Imatra has started to focus more and more on co-operation with the regional level on the Russian side as well as with the City of St. Petersburg, often pooling resources with Finnish cities nearby, like Lappeenranta.

Conclusions

Definitions and interpretations of what 'twin cities', and what they represent, as evident in this book, are quite broad. Our central concern has been with how the defining factor in the relationship between the Finnish city of Imatra and the Russian town of Svetogorsk on the external border of the European Union (i.e. the border) is mobilised by local actors, to some extent but not exclusively under the label of 'twin-city co-operation', as a resource for development. This analysis provided an evidence base for position-ing Imatra–Svetogorsk in terms of Sohn's 'geo-economic' and 'territorial project' models of cross-border integration. As the two towns' relationship spans several decades, we have tried to employ a temporal approach by in-terpreting the dynamics and changing patterns of interactions between var-ious processes triggered by the city pair's border-location.

Owing to the increasingly permeable border from the early 1990s on-wards, the two border cities actively invoked the twin-city concept in order to draw attention to the fact that they were the only cities in close prox-imity on the Finnish–Russian border. This did not include an ambitious drive towards the development of a common identity and convergence. The geopolitically sensitive setting, a territory lost by the Finns and re-settled with people from the USSR, worked against this ambition. Nevertheless, by building on the historical connections established in the 1970s as part of the state-orchestrated, enclave-type of modernisation project of constructing the pulp and paper plant in Svetogorsk, local actors in the 1990s initiated several fields of joint learning and co-operation (including innovative ac-tions such as measuring air quality and teacher exchanges), and contacts between individual citizens gradually developed from increasing border crossings. Yet despite this initial impetus and multifaceted collaboration, there was no serious attempt to create a joint twin-city identity or advanced integration. Therefore, a territorial project was not on the cards.

Imatra, in particular, was initially interested in gaining from utilising differential and positional benefits, as the topics of several initial twin city seminars organised by the two cities show (focusing on potential differential benefits achieved through industrial co-operation and cross-border trade). Imatra's 'geo-economic' flagship project of establishing a cross-border industrial and business park in Svetogorsk never took off due to lack of national-level support. The cities subsequently reoriented their activi-ties from using the border for positional and differential benefits towards co-operation and strengthening interaction in the fields of culture, social aspects and health using financial support from EU-funded cross-border co-operation programmes. This provided the ground for developing trans-action skills. Selective migration from Russia also resulted in a formation of a significant Russian-speaking minority in Imatra able to use possibilities ef-fortlessly on either side of the border. From 2009 to 2013 (until the economic consequences of the Ukraine crisis), rapidly growing Russian shopping tourism on the Finnish side provided an increasingly 'geo-economic' ele-ment in the cross-border relationship between the two cities.

Notes

1 We should remember that the border-regime narrowly defined – i.e. its man-agement, procedures and guarding – has not changed dramatically since Soviet times. Russian and Finnish border authorities co-operate on the basis of a

stable set of border treaties signed between the USSR and Finland that were renewed after the USSR's collapse. Border-crossers must also organise a single or multi-entry visa in advance and, despite the short distances between the cities, there is no Local Border-Traffic (LBT) regime, even though LBTs exist, for example, at the Norwegian–Russian border.

2   *Two-way Railway Cargo Traffic via Imatra/Svetogorsk Border-crossing Point.*

3   The pulp and paper mill, reconstructed by Finnish companies in the 1970s/1980s, was privatised after the USSR collapsed. Eventually, it was bought by International Paper in 1998.

4   The Finnish–Russian agreement on Neighbouring Area Cooperation (NAC) was signed in 1992. The programme, financed entirely by the Finnish Ministry for Foreign Affairs, ran until 2012. The EU's Interreg programme (strand A, for promoting CBC at internal EU borders) operated from 1996. It was combined with the CBC-oriented part of the EU's TACIS programme (designed originally to offer support to partner states undergoing transition in Eastern Europe and Central Asia) in order to support EU-external CBC, including co-operation across the Finnish–Russian border. In 2007, ENPI (the European Neighbourhood and Partnership Instrument) was created to replace these two tools providing a single and more effective financial instrument funded jointly by the EU and the 'neighbours', including Russia, where it was implemented during 2009–13 (Németh et al. 2015).

5   INTERREG is a series of five programmes designed to strengthen economic and social cohesion in the European Union by stimulating cross-border (strand A), trans-national (strand B) and interregional (strand C) co-operation between European regions. It has been implemented since 1989, funded by the European Regional Development Fund. INTERREG IIIC was the 'interregional' strand of the third INTERREG programme implemented between 2000 and 2006.

References

Buursink, J. (2001). 'The Binational Reality of Border–Crossing Cities', *GeoJournal*, 54(1), 7–19.

Eskelinen, H. and Kotilainen, J. (2005). 'A Vision of A Twin City: Exploring the Only Case of Adjacent Urban Settlements at the Finnish-Russian Border', *Journal of Borderlands Studies*, 20(2), 31–46.

Finnish Customs (2015). Statistics on Finland's land border traffic, 5/2015 [online] Available at: www.tulli.fi/fi/suomen_tulli/julkaisut_ja_esitteet/Ulkomaankaupan_tilastojulkaisut/seurantaraportit/tiedostot/maarajojen_liikennetilasto.pdf [Accessed: 10/12/2015]

Hannula, M. (2008). 'Vuoroin kouluvierailuilla', In: Hannula, M. and Hämälainen-Abdessamad, M., eds., *Vuoksen varrella vieretysten. Imatra ja Svetogorsk yhteistyössä*, Report 8 of the South Karelian Institute at the Lappeenranta University of Technology (LUT), 53–58.

Iliyna, I. and Mikhailova, E. (2015). 'Drivers of Cross-Border Community Formation along Russia-Finland and Russia-China Borders', *Bulletin Of PNU*, 36(1), 151–160.

Joenniemi, P. and Sergunin, A. (2011). 'When Two Aspire to Become One: City-Twinning in Northern Europe', *Journal of Borderlands Studies*, 26(2), 231–242.

Kosonen, R., Feng, X. and Kettunen, E. (2008), 'Paired Border Towns or Twin Cities from Finland and China', *Chinese Journal of Population Resources and Environment*, 6(1), 3–13.

Lilja, K., Tainio, R. and Törnqvist, S. (1994), 'Adjusting to Macro-Economic Reforms in Russia. The Crisis Case of the Svetogorsk Mills', *Industrial and Environmental Crisis Quarterly*, 8(1), 55–70.

Németh, S., Fritsch, M. and Eskelinen, H. (2015), 'Cross-Border Cooperation and Interaction between Southeast Finland and Its Neighbouring Russian Regions of Leningrad Oblast and St. Petersburg', *Publications of the University of Eastern Finland. Case Study Report, Reports and Studies in Social Sciences and Business Studies No 7.* Joensuu. [online] Available at: http://epublications.uef.fi/pub/urn_isbn_978-952-61-2018-8/urn_isbn_978-952-61-2018-8.pdf [Accessed: 20/04/2017].

O'Dowd, L. (2002). 'The Changing Significance of European Borders', *Regional & Federal Studies*, 12:4, 13-36, DOI: 10.1080/714004774.

Pitkänen, K. (2011), 'The Changing Significance of European Borders'Contested Cottage Landscapes: Host Perspective to the Increase of Foreign Second-Home Ownership in Finland 1990−2008', *Fennia*, 189(1), 43–59.

Sohn, C. (2014a), 'Modelling Cross-Border Integration: The Role of Borders as Resource', *Geopolitics*, 19, 587–608.

Sohn, C. (2014b), 'The Border as a Resource in the Global Urban Space: A Contribution to the Cross-Border Metropolis Hypothesis', *International Journal of Urban and Regional Research*, 38(5), 1697–1711.

Spierings, B. and van der Velde, M. (2008). 'Shopping, Borders and Unfamiliarity: Consumer Mobility in Europe', *Tijdschrift voor economische en sociale geografie*, 99(4), 497–505.

TAK Oy. (2015). *TAK Rajatutkimus 2014. Tuloksia* (Border research 2014. Findings). [online] Available at: http://www.ekarjala.fi/liitto/wp-content/uploads/2015/04/TAK-Rajatutkimus-2014-esittelyaineisto-Etela-Karjala.pdf [Accessed: 03/07/2018].

Zimin, D. (2013), 'Company towns on the border: the post-Soviet transformation of Svetogorsk and Kostomuksha', in: Eskelinen, H., Liikanen, I. and Scott, J. W., eds., *The EU-Russia Borderland. New Contexts for Regional Co-operation*, London and New York: Routledge, 151–166.

19 Impacts of town-twinning on the communities of Imatra and Svetogorsk through different fields of cross-border cooperation

Ekaterina Mikhailova and Sarolta Németh

Chapter 19 emphasises the human dimension of twin city relationships and sees what twinning efforts look like at grassroots level. Presenting findings from surveys and extensive in-depth interviewing, the chapter exemplifies that the input twin-city study may receive through sociological research. By advancing a citizen-perception continuum model, the writers underline the volatility of twin city initiatives, and create a specific tool capable of measuring and comparing integrative progress by various twin-town examples.

Introduction

Imatra and Svetogorsk sit on the mighty Vuoksi River, Svetogorsk being 7 km downstream from Imatra. The Vuoksi runs through the northern-most part of the Karelian Isthmus and is known for its rapids (including the Imatra rapid, Finnish *Imatrankoski*). The river drops 60 meters in elevation over a length of 26 km from the source; no wonder four hydropower plants operate within this distance, two on either side of the border, including those in Imatra and in Svetogorsk (Figure 19.1).

As Chapter 18 noted, the cities emerged from the partition and duplication caused by redrawing the Soviet–Finnish border after the Second World War. They emerged in their present form after the settlement-cluster in and around the town of Enso, one of Finland's most modern forest-based industrial centres at that time, was split. The consequent development processes of the settlements on each side of the border occurred in mutually insulated ways even though directly linked to their shared industrial heritage. On the Soviet side, where the production units stayed, Enso was re-populated from distant parts of Russia and renamed Svetogorsk. Svetogorsk has currently about 16,000 residents, and the pulp and paper factory, after several phases of modernisation and ownership change, remains the dominant employer. On the Finnish side, Imatra with its modern boundaries emerged in 1948 as a result of three nearby settlements merging. Thereafter, a separate forest-based industrial profile started developing. The two production plants and one research centre of the global pulp and paper manufacturer, Stora Enso,

Figure 19.1 The dam of Imatra on the Vuoksi River by Alexander Zaslavsky (2017).

still operate here, employing about 1,000, although currently, due to this traditional industry being consolidated, the regional economy is seeking ways to diversify. Today, Imatra has nearly 28,000 inhabitants, and is a secondary centre serving the Finnish region of South Karelia following the nearby city of Lappeenranta.

As also noted, the Imatra–Svetogorsk border remained closed from 1970 until the major political changes of 1991 began gradually increasing interactions, producing the first cooperation agreement in 1993. This remarkable step encompassed diverse spheres, including trade, tourism, industry, infrastructure development, education, culture, sport, youth exchange and environmental protection. Thereafter, the two municipalities started signing annual co-operation protocols specifying planned activities for the coming year, providing the local political framework for various joint development projects co-financed by external (EU) funds. Furthermore, the period 2009–2013 saw remarkable rises in shopping and leisure tourism by Russian visitors in eastern Finland, to which local administrations and businesses including those in Imatra attached high hopes for increased profit and economic base (Németh et al. 2015).

In the light of the above, we can say that both historical continuities and discontinuities have imprinted on collective memories in and around Imatra and Svetogorsk. This heritage (though relatively distant and somewhat unequally shared) combined with the more recent common experience of heightened interaction and hopes for a reconnected future has shaped collective memories and popular awareness of commonality as well as difference. This chapter focuses on these factors, connecting their significance to a developing twin-city initiative. More concretely, we focus on residents' awareness and evaluation of the twin-city agreements and the diverse cross-border co-operation (CBC) initiatives, alongside their impacts on everyday life in the two communities. Has a sense of community developed between citizens in the two towns due to freer cross-border movement and interactions, facilitated by funds from CBC programmes since 1993? How far and in what respects have residents' perspectives been changed? Has the twinning process happened consistently and continuously since the early 1990s?

To answer these questions, we start by reviewing academic literature on the twin-cities theme to pinpoint some criteria of town-twinning, and then discuss Imatra–Svetogorsk fieldwork results. We analyse informants' data as follows: local knowledge and identification with the 'twinning' concept; frequency and reasons for visiting the 'other place'; actual participation in CBC activities contributing to tightening the bond between the two towns, and informants' perceptions of visible impacts of changes or activities resulting from the above. We collected data in two stages: a citizen survey, and interviews with local experts. We conclude by formulating a town-twinning-perception continuum model, wherein these and other cities may be positioned.

Cross-border town-twinning and residents' engagement

Latest developments in twin-cities theory suggest the plural nature of this phenomenon (Joenniemi 2014, 3). Every twinning experiment has different variables – geographical, historical, political, social and cultural – and identical features, which substantially affect interactions between the cities concerned. Thus, it is hard to provide one analytical framework defining how a twin-city pair should look. Since relevant scholars came to agree that geographical adjacency is just a prerequisite for cities to be 'real' twins (Buursink 2001; Ehlers 2007; Joenniemi and Sergunin 2011; Anishenko and Sergunin 2012), town-twinning conceptualisation witnessed several attempts to isolate the most significant traits required to single out twin towns from the mass of neighbouring settlements (see Table 19.1). However, none has been accepted as a 'universal toolkit' to verify twin-town status.

In contrast to this list of features which rather treat border-city pairs as 'frozen' and unchangeable, we prefer to see town twinning as a dynamic process (Mikhailova 2014, 4). We assume that the presence of each characteristic can vary over time influenced by, or independent of, each other. To grasp changing twin-town relationships, we explore how far people are involved in and aware of twinning initiatives and see residents' opinion as crucial to town-twinning formation. As often argued in border studies, a distinct type of relationship between residents of adjacent towns across a border can signpost mature town-twinning. Buursink talked of 'cordial relations' (2001, 15), while Ehlers referred to 'a sense of having common interests and belonging together' (2001, 2).

Table 19.1 Twin-city features

Schultz, Stokłosa and Jajesniak–Quast 2002	Joenniemi and Sergunin, 2011	Anishenko and Sergunin, 2012
Delimiting should have been traded for open borders		Aspiration to cooperate today and in future
Joint history (shared existence before separation by national borders)		Common past
A preferable case consists of cities where a river both separates and connects the cities facing each other		Immediate border location
Connecting factors and features conducive to cooperation (e.g. common ethnic minorities)		Ethnically mixed societies, with some command of the language of neighbours
A certain level of institutionalized cooperation between the twins in terms of unified administrative structures and common urban planning		There should be legal and institutional bases for twin cities (agreements and organizations coordinating twin-city activities)
	Breaking the spatial fixations about national borders	
	Trans- and international features of twin cities (formation of commonality reaching beyond national configurations)	

There have been many studies of borderlanders' opinions and feelings towards the integration process (Bucken-Knapp 2001; Figenschou 2011) and towards people across the border (Lundén and Zalamans 2001; Dürrschmidt 2006). Recently, the gap between CBC activities and local residents' awareness of them has attracted increased attention. In 2001, Bucken-Knapp pointed out that 'the top-down process of region-building can succeed only if internalized' by many local people (2001, 51). Schultz, Stokłosa and Jajesniak–Quast (2002, 9) gave residents a more active role, interpreting 'grass-roots networking' as an essential pillar of trans-frontier cooperation. Löfgren (2008, 204) noted that social and cultural feelings accompanying border-crossing are as important as the physical proximity of the border itself and the monetary gains available on the other side. Six years later Trillo-Santamaría named 'inhabitants participation in cross-border regions' matters' on a regular basis as pre-conditional for building 'a more democratic and integrated Europe' (2014, 266). With these arguments in mind, we conceive twinning as a four-layered process comprising top-down efforts to inform residents about twinning, bottom-up responses to twinning rhetoric, plus formal and informal cross-community interactions.

We identify four major themes we see as essential for successful town-twinning, and keywords used in interview questions and in the respondents' answers that correspond with these themes in order to code interviews:

A   town-twinning rhetoric;   C   CBC activities (projects);
B   general interactions;   D   impact on residents' lives.

The first thematic block – 'Town-twinning rhetoric' (A) – contains phrases explicitly connected with notions of familiarity and twin towns, local identification with the 'twinning' or 'twin-towns' concept. The second criterion of town-twinning evaluation – 'General interactions' (B) includes such keywords as interactions, attractions, access and barriers (affecting town-twinning and CBC), frequency of and reasons for visiting the 'other place', perception of 'them' vs 'us'. The third topic is devoted to CBC activities (C), and relates to residents' awareness and perception of the main fields of collaboration, its actors, as well as resources and motives for it. The final criterion assesses impact on residents' lives (D). Keywords here comprise 'visible results', 'effects on life', 'changes in the position and perception of towns due to CBC/twinning'.

Local perspectives on Imatra–Svetogorsk as 'twin towns' connected by CBC

Survey findings

We administered the survey in both towns during May, August and November 2013. Overall, 93 questionnaires were collected – 48 from Svetogorsk and 45 from Imatra (33 Finns and 12 Russians) (see Figure 19.2). These included younger and older citizens, business-owners, heads of NGOs, sports clubs, cultural institutions, social care and educational organisations, plus some local politicians and decision-makers. Compared to their respective populations (16,000 and 28,000) and considering the sample's ethnic composition, the survey represents more perceptions from Svetogorsk and from those of Russian ethnicity. This, however, still allows for drawing useful and indicative conclusions. What follows is an analysis of the survey results and in-depth discussions.

The survey questions directly targeted the first three themes (A, B and C above) and helped also gather opinion (perceptions) about the visible impact of various interactions between the two communities (D above).

We start with perceptions of town-twinning rhetoric. As indicated in Figure 19.3, 29 Svetogorsk respondents (60%) knew about the twin-city concept before filling in the questionnaire. In responses about their source(s) of information about town twinning, 16 mentioned television, 10 mentioned local newspapers, and 7 indicated the Internet. 31 (68%) Imatrans (26 Finns,

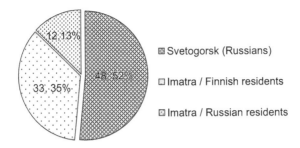

Figure 19.2 The distribution of the 93 survey respondents by place and ethnicity.

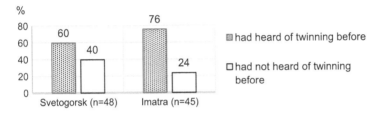

Figure 19.3 Awareness of the 'twin-city' concept among Imatra and Svetogorsk residents.

5 Russians) knew about the twin-city concept before completing the survey. Here, local newspapers were mentioned most frequently, indeed by a significant majority (17, 4), followed by speeches of municipal representatives,[1] mentioned six times (4, 2), and radio programmes, four times (2, 2).

Fourteen respondents from Svetogorsk and nine from Imatra (8 Finns and one Russian) applied the 'twin-city' concept to their towns. Nearly, a quarter of respondents from both towns (10 from each) could mention other examples of twinning. The twin city most frequently mentioned by Svetogorsk respondents was the relatively adjacent example of Narva–Ivangorod on the Estonian–Russian border (approximately 200 km south from our case; mentioned five times), while second place was shared by Blagoveshchensk–Heihe on the Russian–Chinese border, and Minneapolis–St. Paul in the USA (each mentioned twice). Among Imatrans, interestingly the Tornio–Haparanda case was best known, some 800 km to the north-west in the 'High North' at the Finnish–Swedish border (7 mentions), while Narva–Ivangorod was mentioned only twice, alongside another EU example, the German–Polish Frankfurt (Oder)–Słubice.

Thus, we apparently have rather similar awareness patterns among Svetogorsk and Imatran residents of the twinning phenomenon generally, and their own twinning experiment in particular. It is indicative of Imatra–Svetogorsk's currently relatively weak image as twin cities that, although two-thirds of respondents are familiar with the general twin-city or town-twinning concept, only a fifth of the total sample and only a third of those who know what 'twinning' means actually would apply it to Imatra–Svetogorsk. However, the circles of other known town-twinning examples are somewhat different for people answering in Svetogorsk and Imatra: Imatrans are familiar with border-crossing twins involving not just Finland but also elsewhere in Europe, but Svetogorsk respondents know about border twins only in Russia, the only exception being the USA's Minneapolis–St. Paul. This may indicate contrasting understandings of the phrase 'twin cities' in Finland and Russia: the national frame seems more important than the regional one in Russia (i.e. Russians know mainly Russian examples), while Finns are familiar with Finnish and other EU twins. European as well as American examples show how twinning is interpreted in the mass media: either as a European or as a global phenomenon[2]. Finally, within Imatra, there is a major difference between the Finnish and Russian communities, with Finns much better informed about 'twinning' than Russians.

Observing those responses that concern visits to the 'other place', it shows that over 84% (41) of Svetogorsk respondents have been to Imatra, whereas 58% (26) of Imatran respondents have visited Svetogorsk. The former also seem to have visited a neighbouring town more frequently (Figure 19.4). The survey also shows that, among Imatran respondents who have visited Svetogorsk, Russian residents were over-represented (25 of the 26), also among those visiting most frequently. This is understandable due to their strong links to Russia through family and friends.

The two sides are different in terms of visiting purpose (see Table 19.2). Svetogorsk people tend to visit Imatra for tourism (both leisure and shopping). By contrast, Imatrans visit Svetogorsk less as tourists than as people travelling on to other Russian destinations (42%) – for example, Vyborg or St. Petersburg, indicating that their visits are short (often only to buy petrol). Imatran Finns tend to travel to bigger, better-sustained and, thus, more attractive cities nearby like Vyborg or St. Petersburg, while Imatran

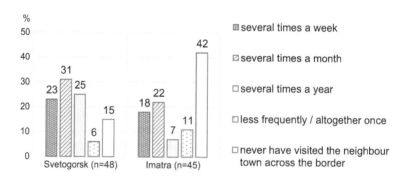

Figure 19.4 Frequency of visiting the other twin.

Table 19.2 Main reasons for visits from Svetogorsk to Imatra and from Imatra to Svetogorsk

	Tourism (leisure) % (no.) of respondents	Shopping % (no.) of respondents	Transit % (no.) of respondents
From Svetogorsk (n=41)	73% (30)	88% (36)	5% (2)
From Imatra (n=26)	19% (3+2)*	50% (8+5)*	33% (4+7)*

* Finnish and Russian respondents from Imatra, respectively.

Russians often go to visit family and friends in the Leningrad Oblast' or in the North-West Federal District of Russia.

Some of what underpins these differences emerges from information the survey provides on personal linkages across the border and perceptions of the other place (Figure 19.5). More Svetogorsk respondents claim personal links on the other side than *vice versa*, and Imatra's image among Svetogorsk people is generally much better (clean, hospitable, green, quiet, cultural, orderly, with good quality roads) than Svetogorsk's among Imatrans (especially among Russian Imatrans). Neutral descriptions of Svetogorsk (mainly by Finns in Imatra) included its proximity and low prices.

Overall, Svetogorsk residents are more active 'consumers' of the neighbouring city's space. Among Imatrans, there is a clear division between the Russian community, for whom crossing the border is easier (therefore every Russian informant has been to Svetogorsk), and the Finnish community, for whom visiting Russia requires a visa application and facing another culture (or even overcoming some prejudice and conflicts rooted in border history). The majority of informants on both sides of the border cross it several times monthly; however, some social groups cross many times a week. Svetogorsk residents form the most active element of these groups (one in

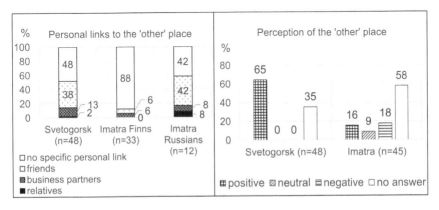

Figure 19.5 Personal links to and perceptions of the other twin.

four informants from there makes weekly crossings). It is also important to point out that, in general, Russians on both sides have more personal connections across the border (approximately one in three Svetogorsk or Imatran Russian residents has them compared to one in six Imatran Finns). Notwithstanding the small size of our sample, the difference is indicative between the two ethnic groups: the minority population in Imatra has an essential and active role in linking the two towns across the border on a more personal level. Imatra has a better image among Svetogorsk residents than Svetogorsk does among Imatran residents. Clearly, Imatran Russians have more negative opinions about Svetogorsk than Imatran Finns. Their critical remarks about Svetogorsk focused on the city's unappealing outlook, poor urban infrastructure and reputedly high air pollution.

We now turn to awareness of and participation in activities organised within the framework of CBC projects by local people in the twin towns. Finns in Imatra are more involved in, and aware of, CBC between Imatra and Svetogorsk than Russian residents. Among those surveyed, Finnish residents of Imatra and residents of Svetogorsk seem to demonstrate similar levels of interest in CBC events (approximately every fourth respondent visits such happenings). Svetogorsk respondents and Finnish Imatrans interpret such cooperation as a long process dating back several decades and mainly reliant on efforts by the public authorities, while Russian Imatrans see it as relatively new, and more citizens driven (Figures 19.6 and 7).

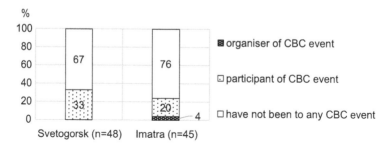

Figure 19.6 Participation in CBC events by Imatra and Svetogorsk residents.

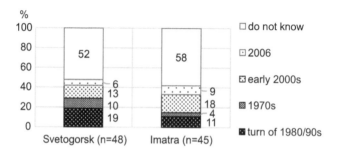

Figure 19.7 Awareness of CBC activities: perception of the start of CBC.

Opinions vary about the main 'engine' of CBC. In Svetogorsk, equal credit is given to efforts by public authorities and citizens (9-9 respondents); some (5 of 48) mention the builders of the pulp and paper factory, and business is mentioned twice. By comparison, Imatrans see public authorities as most important and more influential than do Svetogorsk residents (14 of the 45 respondents). Yet here too citizens are given importance (9), as are efforts by business and consultants (6 in total).

Finally, opinions differ somewhat about the most successful joint activities: for Svetogorsk respondents, trade comes first, closely followed by tourism and culture; for Imatrans, culture is clearly the most important CBC-field, followed by trade and other activities like education, sports and road-construction (a few mentions each). Perhaps significantly in both towns, only Russians see trade as a field where CBC has achieved much (maybe actually meaning private shopping tourism), while Imatran Finns mainly mention culture (9 of 10).

Findings from the interviews

In-depth interviews comprised 20 experts (10 from each town) involved in CBC. They are drawn from a range of public, private business, non-governmental and not-for-profit organisations. Interview transcripts were coded along the four interconnected criteria of town-twinning assessment identified above. This coding was not entirely straightforward, for example, some statements could be linked to more than one theme related to town-twinning, unsurprisingly since those themes are interlinked themselves. Nevertheless, based on the discourses from the interviews, we attempt to map and interpret the inter-relationship between the four thematic dimensions of town-twinning. The conceptual scheme for our conclusions from interviewees' assessments of Imatra–Svetogorsk as a twin-town is as follows (Figure 19.8):

The first line indicates mutual interactions between the observed phenomena in citizens' perceptions. Increased local identification with the twinning concept and engagement in cross-community activities (here, general cross-border interactions or more particularly, CBC) seems to lower barriers and induce more exchanges between citizens of the two towns. However, an increase in accessibility and general interaction (ability, willingness and tendency to cross the border) is seen by the interviewees as a factor that further raises the level of awareness of 'twinning' and promotes participation in CBC activities. Also, CBC projects can potentially strengthen awareness

Figure 19.8 Conceptual scheme for interview analysis.

of twinning between the communities, while internalising the idea of the twin-town project makes engagement in CBC projects seem more natural and interesting for the actors on both sides. More positive perceptions of the 'other', and increased interactions resulting from all these processes enhance mutual learning across the twin-town pair, and visible improvements in the way of life, such as tidier public spaces on the Russian side, livelier cultural life for residents of the two cities, borrowing cultural traditions from neighbours, for example, marching on the 'Day of the City' in Imatra, and creating new joint traditions (the annual women's run called 'Women of the Vuoksi River', the Annual International Art Festival 'Vuoksi' or annual meeting of the Joulupukki and the Father Frost – Finnish and Russian Santa Clauses).

Theme D – 'Impact on residents' lives' – was largely analysed by interview content analysis. Fourteen of the 20 experts made an assessment statement to summarise the chief results of 20 years of intercity cooperation. Six evaluations (five by Imatran Finns and one by a Svetogorsk expert) were rather positive, mentioning the practical, stable and regular character of CBC as its central advantage. For instance, Imatra's mayor of that time, Pertti Lintunen, stressed the habituality of intercity dialogue, suggesting that 'today cooperation is a normal thing'. Others remarked that it had shifted from authority-led formats to informal interactions among Svetogorsk–Imatra residents: 'cooperation is already among normal people, not administrations. It's slowly developing' (Imatra, business representative).

However, six interviewees (four Svetogorsk residents and two Imatran Finns) perceived a decrease in CBC intensity, range of directions and participants. Expressions like cooperation *'is not that active anymore'*, *'has become sluggish'* and *'effectless'* or *'used to be more interesting'* demonstrate dissatisfaction with CBC's current mode and scope and almost nostalgic perceptions of earlier years of interactions. Two interviewees even said that there is no cooperation between Imatra and Svetogorsk. Given the aforementioned frequency of border-crossing (Figure 19.4) and personal links to the other side (Figure 19.5), we assume that this zero assessment of cooperation refers to official dialogue between the two city halls. One reason for such slowdown in inter-authority communication perceived by interviewees is the mayoral change on the Russian side of the border and consequential alteration of Svetogorsk priorities.

General findings

With these interconnections and dynamics in mind, Figure 19.9 introduces the perception continuum comprising data from the Imatra–Svetogorsk surveys and interviews. As is evident, this model requires grading dispersed residents' opinions from 1 to 4. Thus, filling table cells creates a systematised overview of town twinning.

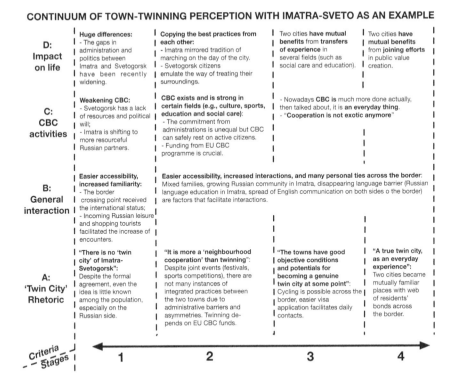

CONTINUUM OF TOWN-TWINNING PERCEPTION WITH IMATRA-SVETO AS AN EXAMPLE

Criteria / Stages	1	2	3	4
D: Impact on life	**Huge differences:** - The gaps in administration and politics between Imatra and Svetogorsk have been recently widening.	**Copying the best practices from each other:** - Imatra mirrored tradition of marching on the day of the city. - Svetogorsk citizens emulate the way of treating their surroundings.	Two cities have mutual benefits from transfers of experience in several fields (such as social care and education).	Two cities have mutual benefits from joining efforts in public value creation.
C: CBC activities	**Weakening CBC:** - Svetogorsk has a lack of resources and political will; - Imatra is shifting to more resourceful Russian partners.	**CBC exists and is strong in certain fields (e.g., culture, sports, education and social care):** - The commitment from administrations is unequal but CBC can safely rest on active citizens. - Funding from EU CBC programme is crucial.	- Nowadays CBC is much more done actually, then talked about, it is an everyday thing. - "Cooperation is not exotic anymore"	
B: General interaction	**Easier accessibility, increased familiarity:** - The border crossing point received the international status; - Incoming Russian leisure and shopping tourists facilitated the increase of encounters.	**Easier accessibility, increased interactions, and many personal ties across the border:** Mixed families, growing Russian community in Imatra, disappearing language barrier (Russian language education in Imatra, spread of English communication on both sides o the border) are factors that facilitate interactions.		
A: 'Twin City' Rhetoric	**"There is no 'twin city' of Imatra-Svetogorsk":** Despite the formal agreement, even the idea is little known among the population, especially on the Russian side.	**"It is more a 'neighbourhood cooperation' than twinning":** Despite joint events (festivals, sports competitions), there are not many instances of integrated practices between the two towns due to administrative barriers and asymmetries. Twinning depends on EU CBC funds.	**"The towns have good objective conditions and potentials for becoming a genuine twin city at some point":** Cycling is possible across the border, easier visa application facilitates daily contacts.	**"A true twin city, as an everyday experience":** Two cities became mutually familiar places with web of residents' bonds across the border.

Figure 19.9 Perception continuum model of town-twinning filled in for Imatra and Svetogorsk.

The model enables residents' survey and expert interviews to be synthesised and weighted. It also closely connects to residents' perceptions, thereby representing a bottom-up approach to assessing twinning. Finally, it offers the possibility of tracing the potentially volatile nature of town-twinning developments. Primarily, this continuum model is an analytical tool for research. However, its practical utilisation might enable municipal officials to monitor the volatility of each dimension of town-twinning and help adjust managerial policies to the needs of inter-municipal dialogue.

Analysis of the Imatra–Svetogorsk perception continuum demonstrates that inter-municipal collaboration has recently regressed. While most interviewees describe twin-town cooperation rather modestly, several (especially those active in CBC over the last 10–15 years) said that cooperation was more active in the past than now. A downgrading from stage 3 to 2 is present in three out of four criteria of town-twinning perception (i.e. in terms of all but general interaction). We conclude that the 'general interaction' criterion is important to having connected twin towns, without necessarily indicating full town-twinning: the ease and frequency of border-crossing are often dependent on other processes (e.g. currency rates and price differences). Despite losing some twinning robustness due to weakening governmental support for the twin-towns project, Imatra–Svetogorsk still enjoy quite considerable informal people-to-people cooperation. Such durable grass-roots bonding can be compared to a wound-up clock that keeps running after the watchmaker retired and seems likely to continue doing so.

Conclusions

Our surveys and in-depth interviews strongly reinforce the view that people-to-people interactions are far more important in determining success or failure of town-twinning than changeable political will – even and perhaps particularly at the local level.

The survey results suggest rather similar patterns of awareness among residents of Imatra and Svetogorsk of town-twinning generally, and their own town-twinning experiment in particular. It is indicative of the currently relatively weak image of Svetogorsk–Imatra as a twin city that only a fifth of the total sample and only one in three of those who know what 'twinning' means actually would apply it to their own cities. In terms of actual visitation, people from Svetogorsk seem the more active 'consumers' of space in the neighbouring city. Among Imatrans, there is a clear division between the Russian community, who find it easier to cross the border, and the Finnish community, for whom visiting Svetogorsk requires a visa and facing another culture and perhaps ignoring the perceived past or even recovering some prejudice rooted in the history of the border. Most informants on both sides cross the border several times monthly; however, some social groups cross several times weekly. Svetogorsk residents are most active here. Russians on both sides have more personal connections across the border, indicating that this ethnic mix is an important connecting factor. Other findings were that Imatra has a better image among Svetogorsk residents than Svetogorsk among Imatra residents; and that Imatran Russians have more critical opinions about Svetogorsk than Imatran Finns, due probably to higher expectations born of Finnish residency or close linkage.

The interview discussions indicated that town-twinning can become dormant and even start degrading if little attention is paid to its maintenance. The reasons for the erosion of twinning bonds include the rearrangement of political elites through elections or appointments (most influentially the mayoral change in Svetogorsk), increased competition for resources leading to a shift to more attractive CBC partners (beyond the local 'twin' scale), and a gap in CBC funding (between programming cycles, or after higher level political interference).

To understand residents' perspectives on town-twinning, we proposed a citizen-perception continuum model as an analytical tool. Filling in remarks from survey and interview respondents allows assessment of the degree of maturity of particular twinning examples. This model has potential for wider applicability, that is, to measure and compare integrative progress by various twin-town examples, and can help trace the drifting of town-twinning developments, which is important because, like all relationships, twinning has internal and external dynamics making it volatile, perhaps progressing, maybe regressing.

Placing Imatra–Svetogorsk on the perception continuum suggests that inter-municipal collaboration has recently regressed. The partnership has lost vibrancy due to many factors, particularly weakening governmental support. However, the synthesised results of the surveys and interviews indicate a relatively high performance of informal people-to-people cooperation. Nevertheless, wound-up clocks eventually run down. We believe that the most crucial dimension of twinning lies in how it impacts day-to-day life in the two communities. While grassroot networks can survive the lack of twin-town rhetoric, the absence of day-to-day manifestations of twinning in residents' lives is likely to lead the intercity relationship to premature withering.

Placing our work in the context of previous research, we should note that although some findings match those of an earlier survey on a similar subject carried out among Imatra and Svetogorsk residents in 2007 (Kaisto and Nartova 2008), we have some contradicting observations. In the 2007 survey, the majority of borderlanders defined town-twinning between Imatra and Svetogorsk as 'a mere political and administrative arrangement carried out between elites and with very limited impact on citizens' day-to-day life' (ibid., 106). Our own impression is very different. This may be partially explained by a shift in perceptions over the six years between the two surveys. Over this period, the dialogue between the two towns faced new challenges and opportunities: collaboration between the cities weakened due to changes of political leaders at the municipal level on both sides of the border, and especially due to Svetogorsk's decreased authority after the Russian administrative reform (see Chapter 18). As a result, the image of elitism earlier attributed to town-twinning has been replaced with that of grassroots connections. Thanks to permitting border-crossing by bicycle, a short trip to the neighbouring town has become more natural for Imatra and Svetogorsk residents. Also, the rise of incoming tourist flows to Imatra and the consequently created new jobs on the Finnish side helped reduce criticism voiced by Imatra residents earlier in 2007 regarding spending by their municipality on 'twinning' with Svetogorsk residents (Kaisto and Nartova 2008, 160).

Overall, although it takes time to engage residents in CBC activities or arouse interest in them to the point of cross-community dialogue, once involved, ordinary citizens constitute a reliable driving force for twinning. As other research indicates, knowing and sympathising with the 'others', treating them as 'selves' (Joenniemi 2015) are fundamental characteristics of twinning that are difficult for towns to achieve and not easy to lose.

Notes

1 The difference between speeches of municipal representatives and local news-papers is that the former refer to the public talks that ordinary citizens have a chance to hear (mostly at public gatherings), while the latter cover a broader range of twinning-related events including those closed to the public, like meetings of authorities (mayors, city councils, etc.) and official delegations (e.g. groups of athletes, pupils, social workers).
2 Residents' answers corresponded to results from a brief content analysis of local newspapers, where twinning is explained with either European or American examples.

References

Anishenko, A., and Sergunin, A. (2012) 'Twin cities: A new form of cross-border cooperation in the Baltic Sea Region?' *Baltic Region, 1,* 19–27.
Bucken-Knapp, G. (2001) 'Just a train-ride away, but still worlds apart: Prospects for the Øresund region as a binational city', *GeoJournal,* 54(1), 51-60.
Buursink, J. (2001) 'The binational reality of border-crossing cities', *GeoJournal, 54*(1), 7–19.
Dürrschmidt, J. (2006) 'So near yet so far: Blocked networks, global links and multiple exclusion in the German–Polish borderlands', *Global Networks, 6*(3), 245–263.
Ehlers, G. A. N. (2007) *The binational city Eurode: The social legitimacy of a border-crossing town,* Aachen, Shaker.
Ehlers, N. (2001) 'The utopia of the binational city', *GeoJournal, 54*(1), 21–32.
Figenschou, A. (2011), 'The Twin Cities Petchenga Rayon and Sør-Varanger Municipality'. Master thesis. The Faculty of Humanities. University of Oslo.
Schultz, H., Stokłosa, K. and Jajesniak–Quast, D. (2002) 'Twin towns on the border as laboratories of European integration', *FIT Diskussionspapiere,* (04). https://www.europa-uni.de/de/forschung/institut/institut_fit/publikationen/discussion_papers/2002/04-02-Schultz.pdf
Joenniemi, P. (2014) 'City-twinning as local foreign policy: The case of Kirkenes-Nickel', *CEURUS EU-Russia paper,* (15). http://ceurus.ut.ee/wp-content/uploads/2011/06/EU-Russian-paper-15_Joenniemi.pdf
Joenniemi, P. (2015) 'Others as selves, selves as others: Theorizing city-twinning', Manchester Twin Cities conference.
Joenniemi, P., and Sergunin, A. (2011) 'City-twinning in Northern Europe: Challenges and Opportunities', *Research Journal of International Studies, 22,* 120–131.
Kaisto, V., and Nartova, N. (2008) Imatra-Svetogorsk-kaksoiskaupunki. Asenne-barometri 2007, Lappeenranta.
Löfgren, O. (2008) 'Regionauts: The transformation of cross-border regions in Scandinavia', *European Urban and Regional Studies,* 15(3), 195–209.
Lundén, T., and Zalamans, D. (2001) 'Local co-operation, ethnic diversity and state territoriality - The case of Haparanda and Tornio on the Sweden–Finland border', *GeoJournal, 54*(1), 33–42.
Mikhailova, E. (2014) 'City twinning within the governance approach', Association of Borderlands Studies World Conference. Joensuu, Finland.

Németh, S., Fritsch, M., and Eskelinen, H. (2015) 'Cross-Border Cooperation and Interaction between Southeast Finland and Its Neighbouring Russian Regions of Leningrad Oblast and St. Petersburg', *Publications of the University of Eastern Finland*. Case Study Report, *Reports and Studies in Social Sciences and Business Studies No 7*. Joensuu. [online] Available at: http://epublications.uef.fi/pub/urn_isbn_978-952-61-2018-8/urn_isbn_978-952-61-2018-8.pdf [Accessed: 20/04/2017].

Trillo-Santamaría, J. M. (2014) 'Cross-border regions: the gap between the elite's projects and people's awareness. Reflections from the Galicia-North Portugal Euroregion', *Journal of Borderlands Studies*, *29*(2), 257–273.

20 Blagoveshchensk and Heihe

(Un)contested twin cities on the Sino-Russian border?[1]

Ekaterina Mikhailova, Chung-Tong Wu
and Ilya Chubarov

We now shift attention from the borders of Europe to those dividing Russia and China, and to the only cross-border continuously built-up area along that vast dividing line. Integrative possibilities might seem improbable here, particularly given the indifference (at best) of the two central governments. Yet, with both capitals very far away, and relations thus in the hands of regional and local authorities and local people, they seem alive and economically flourishing. Here, as in Chapter 21, flows are the principle inter-linkages of the two cities. However, the lack of mutual trust between national and regional officials born of historic burdens complicates cooperation in spheres beyond trade and tourism, and results in highly prescribed twin-city relations.

Introduction

There are few neighbouring settlements along the lengthy Sino-Russian border, but the cities of Blagoveshchensk (thereafter Blago) in Amur Oblast and Heihe in Heilongjiang Province are two notable ones. Sitting across the width[2] of the Amur River (Heilongjiang), Blago and Heihe's proximity point towards expectant border twin cities. This expectation is often dashed by local politicians, residents and international researchers, who point to the various asymmetries and contradictions between the two.

Most striking is the absence of a physical link. The advantages of one do not seem in dispute because, during the winter months when the river freezes over, a temporary pontoon bridge for vehicles has been built since 2011. During the warmer months, the only means of cross-border transportation is via ferry or hydrofoils (Figure 20.1). Perversely, it is during the months when traffic demand is highest that the transport link is physically most cumbersome since the goods have to be transported to the wharf, loaded onto boats, off-loaded on arrival and then loaded on trucks for their final journey – a triple process on both river banks. While a bridge is finally underway, it took nearly three decades of discussions and negotiations for its realization.

The built environment is also used to argue against applying the notion of a twin-city to Blago and Heihe. While previously Blago and Heihe were

Figure 20.1 Hydrofoil border-crossing over the Sino-Russian border from Heihe to Blagoveshchensk by Ekaterina Mikhailova (2017).

as contrasted as Europe and Asia (Wishnick 2000), the more recent juxtaposition is that of Heihe's modernity versus Blago's backwardness (Billé 2014). The obvious contrast is between each city's waterfront: Kucera (2009) described the Chinese side as a 'glitzy welcoming façade', while the Russian side was characterized as 'a barely-lighted promenade' with 'a Soviet-era World War II memorial that consists of a gunship with its barrels aimed across the river, towards China'. Since then Blago's riverfront has improved both in terms of backlighting and diversity of urban sculptures owing to extensive reconstructions in 2011 and 2016–17. However, Heihe's nightscape remains more dramatic and symbols of hostility like the Soviet gunship are still in place.

Even the use of the twin-city label on the two river banks has been asynchronous and asymmetrical. While the term made its debut in official political discourse in Heihe over a decade ago, the Russian side incorporated it only five years later. Despite perennial twinning efforts, no attempts to find mutually acceptable definition of the term 'twin city' or create a joint vision of how to capitalize on this particular kind of relationship have taken place. Underlining deliberate circumscription of twin-city relations between Blago and Heihe, Mikhailova and Wu (2017) characterize them as 'ersatz' twin cities. We argue that from the first, the call for a twin-city relationship was couched in strictly economic terms, with tourism branding and advertising as two key expressions. Systematic and conscious unwillingness to develop 'cultural empathy' and 'shared perception of the border' demonstrated by Blago and Heihe testifies to the same fact.

Cognizant of these divergent perspectives, we discuss the potential of placing Blago and Heihe within the 'twin cities extended family'. Stimulated by researchers who discussed cross-border economic exchanges in terms of trust and mistrust (Humphrey 2018), we posit our discussion of twinning in terms of a continuum of trust and mistrust. At one end would be complete mistrust which would indicate no attempts at twinning or even active opposition. At the other end would be complete trust indicating an enthusiastic embrace of twinning by both parties. Along this continuum would be varying degrees of mutual trust. A cautious level of trust could be expressed as cooperation in confined areas, like the economic activities that Blago and Heihe have chiefly focused on. We see the characteristics identified in this volume's introduction as dimensions to indicate the degree of trust between the two parties.

Starting from a brief historical account of Blago and Heihe, we focus on their contemporary ties by assessing the dynamism of flows between the cities, the evolution of challenges experienced and the aspirations expressed.

Urban development on the Amur River

The contrast between Blago and Heihe starts with their dissimilar urban development. While Blago is a historical city and long-standing administrative centre of the Russian Far East that celebrated its 160th anniversary in 2016, Heihe typically is projected as materialization of Chinese rapid urbanization: a village that tripled in size within three decades and became a full-fledged city. However, a closer look at historical maps and travelogues shows that this interpretation of urban development in the region is oversimplified.

Blago's history is well documented. Founded at the confluence of the Amur and the Zeya Rivers as a military outpost in 1856, the settlement was granted city status two years later by Emperor Alexander II, and renamed as Blagoveshchensk (Russian: 'the city of Annunciation') after the first Orthodox church founded there a few days before signing the Aigun Treaty of 1858 that established a border between the Russian and Chinese Empires.

Over its first 40–50 years, Blago grew 'almost at the American speed' due to development of its gold industry and trade, becoming, as some argue, 'the wealthiest city of Eastern Siberia' (Shipanovskaya 2016, 51). Many travellers paid compliments to Blago's orthogonal 'wide, straight, although not paved', streets, busy with Russian- and American-style carts, large rich shops, boulevard and numerous private gardens. Although much of the city consisted of log houses, its downtown had a number of exquisite red brick masons 'worthy of duplicating at Nevsky Prospect' (ibid.), the main avenue in Russia's then capital, St. Petersburg.

Life in late nineteenth-century Blago was expensive and vibrant. Social diversity is well in evidence with substantial and varied numbers of temples operating in the city by that time – nine Orthodox, one Catholic and one

Lutheran churches, one mosque and three sectarian prayer houses. By 1903, of its 70,000 inhabitants, every seventh was a sectarian and every 13th a foreign national. The majority were Chinese men, residing and working in Blago as servants, artisans or small traders selling everything from agricultural goods to 'contraband silk and opium' (ibid., 50).

The presence of Chinese (more precisely, Manchu) in Blago was mentioned in several historical memoirs. In the second part of the nineteenth century, Chinese traders (predominantly from Aigun) were visiting Blago monthly and staying for a nine-day city fair to trade (ibid., 49). Dyatlov pinpoints that, by the end of the nineteenth century, Blago's daily life and economic activities 'were unthinkable without the Chinese' (2002, 86). However, despite the vital necessity of their labour, Blago's Russian working people never perceived them 'as part of the urban community' due to their reluctance to assimilate (ibid., 86–87). Being a cheap and hard-working alternative to hiring Russian locals, the Chinese comprised the lion's share of seasonal workers in the gold industry, making them seem competitors in a limited labour market.

Life on the Amur's opposite bank is less well recorded. According to the Heihe Urban Planning Museum, the right river bank has been inhabited at least since the second half of the seventeenth century, first by Daurs and then Manchus. Some eighteenth- and nineteenth-century European maps depict *Sakhaliyan*, a Manchurian city situated at the present-day location of Heihe (e.g. Vandermaelen 1827), suggesting that Blago was not the first or only settlement at the confluence of the rivers.

In Manchurian language, *Sakhaliyan* means 'The Black River', the same as what *Heihe* means in Chinese. Curiously, one possible title considered for Blago was Chernorechensk (*Russian*: 'a city on the Black River') (Shipanovskaya 2016, 48). Had that title been chosen, the studied cities would have had identical names.

Blago's city plan of 1869 shows two villages across the river, 2 km apart – Big and Small Sakhaliyan (see www.wdl.org/en/item/21639/view/1/1/). More recent maps, those from the late nineteenth till the mid-twentieth centuries, have used both titles (Sakhaliyan and Heihe) interchangeably, one typically in brackets, to denote the settlement adjacent to Blago.

Blago's urban growth spilled over into both neighbouring Manchu villages: by 1900 they had about 130 'clay, bleached and unbleached one-story houses' (Shipanovskaya 2016, 50). In 1898, a Russian traveller described Sakhaliyan as a town that supplied Blago's bazaar with vegetables, watermelons, melons, poultry (ibid.) and baijiu, a Chinese strong distilled spirit. Besides Sakhaliyan dwellers were reselling gold mined in the Zey River at the Chinese domestic market.

The turn of the century brought serious disruption to Blago and Sakhaliyan's peaceful co-existence. Fear of the Boxer Rebellion, followed by attempts to capture a few river-going Russian ships on the Amur and shelling of Blago itself (Dyatlov 2002, 85), led first to Blagoveshchensk's massacre of about 4,000 Chinese residents in 1900 (ibid., 89) and then the seizure and destruction of both Sakhaliyan and Aigun by Russian troops (Mierzejewski 2012, 66). When hostilities ceased, the Chinese returned both to Sakhaliyan and to Blago: by 1907 their number had reached pre-conflict levels (Dyatlov 2002, 101). Events during the twentieth century did not fully remedy the fears and mistrust between the two sides at both local and national levels.

In the early twentieth century, the duty-free regime functioning within a 53-km zone on both sides of the Sino-Russian border along the Amur was over (Romanova 2006, 117–118). Blago's customs house was established in 1902; five years later customs started operating on the Chinese side too (http://video.amur.info/dom/2011/06/27/1639). Established in Aigun, the customs headquarters were transferred to Sakhaliyan in 1911. The rationale behind relocation was convenience of processing goods crossing from/to Blago. In 1913, Sakhaliyan became the seat of the military local governor (daotai) (Lukin 2013, 415). Chinese urban historians pinpoint the existence of the modern Heihe from that point, noting the preserved street grid.

In terms of size, regional significance and economic dynamism, Blago has doubtless been substantially more significant than its Chinese neighbouring settlement. On the other hand, presenting the opposite bank of the Amur as empty and unconnected is false. Multiple sources establish that Heihe can convincingly call itself a centenary settlement. Finally, communication and trade exchange between Blago and the settlement that eventually became known as Heihe has been developing since at least 1898.

Influence of domestic and foreign affairs on Blago and Heihe

The twentieth century was marked by Russia and China's long-lasting and bloody transition from empires to socialist states. Civil wars and foreign interventions occurred on both sides of the border, sometimes reducing cross-border contacts to zero. The first breach of bilateral relations occurred in 1917–18 when Beijing officially refused to cooperate with Soviet Russia. The other crisis in interstate communications, the Sino-Soviet split, started in the late 1950s, and included several border conflicts, leading to militarization of border regions and cities on both sides.

The reapproachment between the two states and their regions began only in the late 1980s. At that time, Blago and Heihe reinforced their role as a crucial social and economic contact zone along the Sino-Russian border. The year 1988 broke the ice for economic and cultural exchange between the cities. That year the port of Blago received the first portion of goods for trade and barter exchange from Heihe (Dolguileva 2012, 9), and sent a collection of paintings and sculptures made by local artists to exhibit across the river. A year later Blago received an exhibition of Chinese architectural projects from Heihe. The key role in arranging both cultural events were regional branches of the Soviet–Chinese Friendship Society (Lukin 2013, 450).

The USSR's demise, alongside economic reforms undertaken by both Russia and China, impacted greatly Blago and Heihe. For Blago, new economic reality meant the evaporation of Moscow financial subsidies (Billé 2014, 156), massive out-migration (over 250,000 have left Amur Oblast since 1989) and shutdown of multiple industrial enterprises. Heihe, to the contrary, experienced remarkable economic advance. The explosive demographic growth of Heihe's urban core (from 66,000 in 1982 to 124,901 in 2014)[3] coincided with Blago's stagnation, only recently replaced by modest repopulation (from 205,500 in 1989 to 224,400 in 2017).

China's unprecedented economic development stimulated the need for resources beyond the borders. Heihe sees itself as a conduit for investment across the border and beyond. This is evident in the growth-oriented slogan of the city – 'window to Russia, gateway to Europe and Asia, two countries one city, and fertile soil for investment'.

Both cities have initiated ambitious tourism plans aimed at attracting tourists from across the border. While Heihe is presently celebrating returning Russians as the dominant source of cross-border tourists after the recent downturn due to the Russian currency crisis, plans to expand tourism there include schemes to develop Heihe and its vicinity as a site for vehicle-testing in deep-winter conditions, a Russian village aimed at tourists from elsewhere in China, retirement villages for Russians alongside medical tourism and spas for elderly Russians in Wudalianchi, 250 km away from Heihe's urban core. Notably, all of these tourism strategies aim to reinforce Heihe as a destination; none involves cooperation with their Blago counterparts.

The lack of trust between regional officials is also manifested in Blago (officials often acting hyper-cautiously within nationally set remits). The go-it alone attitude and lack of aspiration to pool resources with Heihe are evident in the recent tourism development strategy of Amur Oblast, purporting to promote Blago as a centre of Sino-Russian trade and investment congresses and expositions. Achieving this goal involves grandiose comprehensive construction activities that encompass building a trans-border cableway as a smooth, fast and comfortable means of border-crossing, an Exhibition and Convention Centre, a Dinosaur Museum named 'A Cretaceous Period Park', an Aerospace Museum called 'A Space Odyssey', the Orthodox Cathedral, the Centre of Modern Art, the residential complex 'Little Holland' replicating the architecture of the Netherlands in the nineteenth century, the 'Venice' entertainment centre with a possibility to take a real gondola, a hotel complex, a park, a pedestrian street for selling arts and crafts, an aqua park and a sports complex with an indoor skating rink (Government of the Amur Oblast 2013). The list of intended construction work is exorbitantly long and seemingly random. Alongside the fact that, by June 2013, only two of the aforesaid building sites had actually got private investors, it calls into question the realization of this strategy anytime soon.

In the context of these projects, the absence of a bridge, the most obvious and efficient mode of connectivity across the Amur, is doubly bewildering.

This persistent inability to agree on such a project implies a lack of trust between Russia and China and strong national security concerns on the Russian side. While the Chinese have been pursuing the idea since 1988, it was only in 2015 that both sides finally agreed to commence construction. The facilitation of this long-standing issue can be traced to new domestic agendas on both sides. In 2014, as part of its 'turn to the East', Russia agreed with China to supply natural gas (Skalamera 2018) and the Chinese were aggressively pushing the Belt and Road Initiative first mooted by Xi Jian Ping in 2013. Since bridge building was never a technical nor a financial issue – the Chinese was reportedly willing to provide the full finance – it was chiefly a national political issue: neither Heihe or Blago having authority to construct anything across national borders.

Flows as principal inter-linkages of the cities

Trade was always on Blago and Heihe's agenda. As soon as the cities were independently declared 'open' by their respective national government from around the late 1980s, cross-border trade became the impetus for their interactions. As the relationships evolved, other economic ties blossomed to cater for the two cities' specific regional settings, influenced by their respective versions of post-socialist reforms.

Thanks to reciprocal liberalization of the border-crossing for tourist groups in 1999 and the unilateral Chinese declaration of Heihe's urban core as an area where Russians can visit visa-free in 2003, the two cities have developed several close links:

1   *Trade linkages*: trade has blossomed across the border. Blago/Amur Oblast exports resources, timber and agriculture products to Heihe and China. Heihe exports Chinese manufactured consumer goods. Trade is determined by the partners' economic conditions and in recent years, due to Russia's currency crisis, the volume and direction of trade has shifted. In 2016, Heihe's cross-border trade totalled US$580 million, of which export was 45% and import was 55%, representing a 14.8% drop from 2015 (Heihe City Statistical Bureau 2017). The downward trend continued during the first half of 2017 (ibid.). Official data on cargo traffic indicated a total of 2,000,000 tons between Blago and Heihe, but there was a decrease in Blago's imports and an increase in its exports to China, consisting chiefly of agricultural products (www.amur.info/news/2015/01/26/88688). For the Chinese, a significant shift in recent years is increasing food imports from Russia due to domestic concerns about food safety and Chinese consumers' perception that Russian food is of high quality due to lack of additives. Significantly, Heihe has also become more reliant on Amur Oblast for part of its electricity supply, especially to support its export-processing zone. Despite an 11% drop in 2015, electric power is still Heihe's largest import commodity

(Zou and Li 2017, 27). With the expected completion of the Power of Siberia gas pipeline from Russia, resulting from bi-lateral agreements, this dependency will increase. The gas pipeline will run via a tunnel underneath the river from Blago to Heihe terminating in Shanghai.

2   *Consumption linkages*: during 2000–2012, benefitting from a strong Russian currency, Heihe has become a shopping haven for residents of Blago, Amur Oblast and beyond. Blago, on the other hand, became the base for shuttle traders and the source of consumers for Heihe. Precise data is lacking, but reportedly in the summer months annually before 2013, some 2,000 Russians used to visit Heihe daily to take advantage of much cheaper consumer goods. For Heihe, this flow of Russian tourists has produced significant growth of wholesale and retail businesses, as well as hotels and restaurants. There is also anecdotal evidence of Russians moving to Heihe to find employment, Russian pensioners settling in Heihe for economic reasons and Russians visiting Heihe for medical and dental care (Dolguileva 2012).

In Amur Oblast, Chinese tourists have dominated the flow of international incoming tourists since the late 1980s. Till 2005, Blago's main tourist attraction for Chinese was gambling. After Russia's ban on slot machines, the tourist flow from China halved to about 10,000 per year. By 2010, Chinese tourists reached 21,200 people, equalling 24.2% of organized tourism with the Amur government touting the cooperation programs as a crucial instrument for revitalizing the touristic flow from China.

Although the flow of Chinese to Blago was more modest through the 2000s than that for Russians into Heihe, changes have also occurred in Blago. Since command of Chinese language enhances job opportunities for Blago citizens engaged as shuttle traders and related employment (Dolguileva 2012, 16), the city witnessed sharply rising demand for Chinese language courses culminating in the establishment, with Chinese government funding, of a Confucius Institute in the Blago State Pedagogical University, Russia's third (http://portamur.ru/~Ehryc). Blago residents who are frequent visitors across the border also developed a taste for Chinese cuisine inducing the opening of many relatively authentic Chinese restaurants in Blago (Ryzhova and Ioffe 2009, 355–356).

While the recent fall of the Russian currency has dampened the flows of people and goods, quantities remain largely the same. Unavoidably, a stronger yuan has not only produced a slow-down among hotel and catering businesses in Heihe, but also stimulated Chinese customers to increasingly visit shops in Blago. As Heihe's retail sector fell into the doldrums, Blago's retailers found new consumers in Chinese tourists. According to data provided by Oblast officials in a recent fieldtrip to the cities, the steepest increase in Chinese visitors occurred in 2014 and 2015 when the flow grew 28% and 25% per year, respectively, and equalled 52,600 and 70,000 tourist arrivals. In 2016, around 80,000

Chinese visited Blago using the visa-free tourist-group arrangement accounting for 88.5% of the total flow of tourist groups. Although a 15% reduction of Russians visiting Heihe was reported in both 2014 and 2015 (Heihe City Statistical Bureau 2016), 2017 data on cross-border tourism indicate a small upturn in 2016 and 2017 (Qiu 2018).

3   *Investment linkages*: unsurprisingly, cross-border investments vary vastly between the two cities. Glazyrina, Faleichik and Faleichik have shown through an investment cluster analysis that regions of the Russian Far East bordering dynamically growing China are in no better position that other Russian border regions (2012, 59). In 2010, Chinese foreign direct investments in Amur Oblast totalled US$ 15,800,000 representing a 3.4% share of regional total investment capital and 2.6% of China's investments in Russia. The main sectors of investment are timber and construction industries. In general, Russian foreign direct investment in China is very modest, in 2011 reaching only US$ 31,000,000. Through land leases, in 2009, Heihe has already invested in 53,360 hectares of agriculture land in Amur. In 2015, reports emerged that Chinese investment in agriculture in Amur has expanded to 77,000 ha due to the falling rouble (Mikhailova and Wu 2017, 520) and expanded to 86,667 ha by 2017 (Xie 2017). Chinese investors in Blago/Amur Oblast are however hampered by uncooperative Russian bureaucrats regarding business and labour permits. This has become a source of many complaints from Chinese investors.

It is evident over the last two decades that Blago and Heihe people have a direct role in the unfolding cross-border space. Consumers from Blago and Heihe have expanded the economy, especially the retail and service sectors, altered its economic structure and impacted the investment and employment structure of both cities. Unsurprisingly, the respective city governments responded to these changes in their own separate ways. Residents from both cities have engaged in negotiating and transcending in various ways the political obstacles related to the border (Ryzhova and Ioffe 2009). While consumption linkages are the most visible, the other two are equally important but can become the source of conflicts. The Chinese complain about Russian bureaucratic inefficiencies and inflexibility, hampering investment and businesses. Some Russians became increasingly concerned about what they perceived to be widespread flaunting of visa conditions among Chinese employed in construction, catering, agriculture and forestry sectors in Blago and Amur Oblast (Ryzhova and Ioffe 2009).

Conclusions

For Blago and Heihe, the road to twin-city formation is constrained by their respective centralized governments, regional politics, historic burden, bureaucratic self-interests and local jalousies. These have produced highly prescribed twin-city relations, both illustrating the complexities of international twinning and specific local limitations. Despite the widespread idea that ties between international twin cities tend to be more cooperative and constructive, relationships between Blago and Heihe are not immune to intercity competition and tensions, typically associated with intra-national twins. Even though there is tacit acceptance of the twin-city idea by both Blago and Heihe through participation in trade fairs and joint tourism promotion events, both sides still promote projects independent of or even competitive with each other. One possible reason might be the inspirational example of Hong Kong–Shenzhen that motivated the Blago–Heihe twinning effort. There too primary attention has been paid to economic integration, while institutional and social dimensions of integration demonstrate little progress and cities prefer 'co-opetition' to full-heated cooperation (Shen 2014 and Chapter 9 of this volume).

Being products of two post-socialist societies with vastly different momentums of reforms, Blago and Heihe learnt to interact notwithstanding lack of support from national capitals and within a restrictive legal footing of cooperation. Their twinning experience expands our understanding of the dimensions of the twin-city characteristics identified in this chapter's introductory chapter. Over the last two decades, the shifting impacts of Russian and Chinese consumer demands have been dictated by national political agendas and national economic fortune. Under these conditions, the two cities became economically interdependent for consumer goods, agriculture products, food and energy even though these interdependencies have evolved within the national and regional context beyond the control of either city.

The unequal relationship of Blago and Heihe has numerous manifestations starting with the 'asymmetry of travel regulations' (Ryzhova and Ioffe 2009, 353), namely no visa facilitation for individual Chinese visitors to Blago, and culminating in divergent objectives pursued by local officials promoting intercity cooperation. Under China's driving national paradigm of economic growth above all else, local officials in Heihe, fearing being seen as ineffectual, have been pushing hard on the twin-city idea. To protect their careers and perhaps anticipating promotion, they focus on initiatives to stimulate economic development for Heihe, twin-city formation being one. The border identity of some Amur Oblast residents relies on ideas of 'fortress'/'national outpost' (Okunev and Tislenko 2017, 598). No wonder regional civil servants conduct themselves in ways proving their loyalty to the Russian state by making no concessions to the Chinese and maintaining the status of an equal partner at whatever joint activity. Hence, Heihe's eagerness to foster twinning is at best seen by some Amur Oblast officials as being too pushy, ignoring ideas they may have about the relationship. Worse still, Heihe becomes perceived as an upstart trying to show off to a neighbour that used to have much stronger political and economic prowess.

On-going formal negotiations both within the various levels of each state and at various levels across the border are essential to any cross-border initiative supervised by the highly centralized Russian and Chinese states. The frequency and complexity of problems arising between and within twin cities generally, and between Blago and Heihe specifically, often require informal networking. Chinese entrepreneurs in Blago are well known for their talents for personal contacts and informal connections (Chinese *guanxi*) generally attributed to Oriental culture. Ryzhova and Ioffe bluntly admit that the success of Chinese construction firms in Blago 'depends on their ability to endear themselves to the municipal government' (2009, 356).

Modern societal and technological advances make informal negotiations increasingly important and broaden the range of negotiating parties. Mass media, NGOs, business networks and residents increasingly participate in intercity negotiations along with state and local authorities. However, joint negotiations between the cities are hard to realize and moderate. The study on cooperation between regional newspapers of Blago and Heihe shows that they fail to provide a shared dialogue platform, due to discrepancies in valuing information and different interpretations of the mass-media mission. While the Amur Oblast daily newspaper is community concerned and seeks to debate such burning questions as illegal import to Russia and cross-border criminal activities, the Heile daily is full of slogans and exaggerations, typical of state-owned periodicals (Mikhailova 2017, 12–13).

A history of border conflicts and territorial disputes sowed mistrust between the two parties. Mass media on both sides of the border do not tend to foster a platform for dialogue. This history of mistrust is often used as a convenient excuse to blame the other side for lack of progress or the existence of perceived or real obstacles to investments or cooperation. At the local and regional levels, some Chinese blame cumbersome investment and labour import regulations on unhelpful bureaucracy, motivated partly by discriminatory attitudes towards Chinese, while some Russians blame the Chinese for contaminating leased agricultural land with pesticides.

Furthermore, mistrust is occasionally stirred by asymmetries in the economic fortunes and consequent political clout of the two countries. National security concerns are conveniently touted to deter closer contacts except in limited spheres. Local and regional officials in Blago were not reassured by the persistent, and to them, overly enthusiastic Heihe officials, who pushed for the construction of a bridge and promoted the concept of twinning. The tension between the drive for openness and the call for caution, pragmatism and economic realities led local/regional officials of both cities to focus on trade and tourism or economic exchanges, where a degree of trust can be established for mutual benefit. Their cautious and hesitant approach towards twinning resulted in an outcome falling well short of complete mutual trust or full embrace of the twin-city concept. However, it musters sufficient trust to accept a realistic twinning, leaving much to constant negotiations, testing of relationships and gradual movements towards cooperation in practical issues.

Notes

1 This study was carried out within the project MK-2007.2017.6 financed by the Russian Presidential Council for Support of Young Scientists.
2 After summer 2013, when the water level near the cities increased by 8.2 m due to the Amur River flooding, both sides carried out substantial bank protection work. Thereby, the distance between the river banks has shrunk. Recent measurement at one online map service shows that Blago's central square is separated from the Big Heihe Island by just 570 m of open water.
3 The population of Heihe refers to the non-agriculture population of Aihui District, which includes the urban core of Heihe City located across the river from Blago. As a prefecture-level city, Heihe controls the territory of 54,390 km^2 with the total population of 1,227,585 (Heihe City Statistical Bureau 2016).

References

Billé, F. (2014) 'Surface Modernities: Open-Air Markets, Containment and Verticality in Two Border Towns of Russia and China', *Journal of Economic Sociology* 15 (2), 154–172.

Dolguileva, A. (2012) 'Perception of home in the process of migration and transmigration: the experience of Russian teachers in Heihe, China', Master thesis. Massey University, Manawatu, New Zealand.

Dyatlov, V. (2002), 'Blagoveshchenskaya "Utopiya": iz istorii materializatsii fobiy' [Blago's 'Utopia': the history of phobia materialization], *Vestnik Evrazii* 4, 84–103.

Glazyrina, I., Faleichik, A., and Faleichik, L. (2012), 'Cross-Border Cooperation in the Light of Investment Processes: More Minuses Than Pluses so Far', *Problems of Economic Transition* 55 (6), 43–62.

Government of the Amur oblast (2013) 'Kontseptsiia razvitiia goroda Blagoveshchenska kak mezhdunarodnogo rossiisko-kitaiskogo tsentra kongressno-vystavochnoi deiatel'nosti – investitzionnyi proekt "Zolotaia milia" [Strategy of Developing Blagoveshchensk as a Centre of International Russian-Chinese Exhibitions and Congresses – The Investment Project "Golden Mile"].

Heihe City Statistical Bureau (2018), 'Foreign trade of Heihe 2017' (in Chinese) 2017年全市对外贸易运行情况.

Heihe City Statistical Bureau (2016), '2015 Heihe Social Economic Development Report', available: www.heihe.gov.cn/info/1402/58549.htm (accessed 22/03/2018).

Humphrey, C. H. (ed.) (2018) *Trust and Mistrust in the Economies of the China-Russia Borderlands*, Amsterdam: Amsterdam University Press. DOI: 10.5117/9789089649829

Kucera, J. (2009), 'Where Russia Meets China: Don't Call Them Twin-Cities', available: www.slate.com/articles/news_and_politics/dispatches/features/2009/where_russia_meets_china/dont_call_them_twin_cities.html (accessed 01/03/2018).

Lukin, A., ed. (2013), *Rossiya i Kitay: chetyre veka vzaimodeystviya* [Russia and China: four-century relationship], Publishing House 'Ves' Mir', Moscow.

Mierzejewski, D. (2012), 'Reading Years of Humiliation. Sino-Russian Border and China's National Identity', *Sensus Historiae* 8, 59–70.

Mikhailova, E. (2017), 'Collaborative Problem-Solving in the Cross-Border Context: Learning from Paired Local Communities along the Russian Border', *Journal of Borderlands Studies* DOI: 10.1080/08865655.2016.1195702.

Mikhailova, E. and Wu, C. T. (2017), 'Ersatz Twin City Formation? The Case of Blagoveshchensk and Heihe', *Journal of Borderlands Studies* 32 (4), 513–533.

Okunev, I. and Tislenko, M. (2017), 'Geopolitical Positioning of Twin Cities: A Case Study of Narva/Ivangorod, Valga/Valka, and Blagoveshchensk/Heihe', *Teorija in Praksa* 54 (3/4), 592–702.

Qiu, Q.-L. (2018), 'Both tourists and income have increased in 2017 for Heihe' (In Chinese) 我市2017年旅游人数收入双增, *Heihe Daily*, available: www.heihe.gov.cn/info/1103/87399.htm (accessed 15/03/2018).

Romanova, G. (2006) 'Porto-Franco and Protectionism', *Russian and Pacific RIM* 3, 110–122.

Ryzhova, N. and Ioffe, G. (2009), 'Trans-Border Exchange between Russia and China: The Case of Blagoveshchensk and Heihe', *Eurasian Geography and Economics* 50, 348–364.

Shipanovskaya, L. (2016) 'Blagoveshchensk v vospominaniyakh sovremennikov: po materialam knigi "Istoriya Blagoveshchenska 1856–1917" [Blago in memoirs of his contemporaries: based on the book 'History of Blago 1856–1917]', *Slovo* 13, 47–52.

Shen, J. (2014), 'Not Quite a Twin City: Cross-Boundary Integration in Hong Kong and Shenzhen', *Habitat International* 42, 138–146.

Skalamera, M. (2018), 'Explaining the 2014 Sino-Russian Gas Breakthrough: The Primacy of Domestic Politics', *Europe-Asia Studies* 70, 90–107.

Vandermaelen, P. (1827) 'Partie, l'Empire Chinois. Asie 34', in: *David Rumsey Historical Map Collection*, available: www.davidrumsey.com/luna/servlet/ detail/RUMSEY~8~1~25100~980076:Partie,-l-Empire-Chinois--Asie-34- (accessed 30/03/2018).

Wishnick, E. (2000), 'Russia in Asia and Asians in Russia', *SAIS Review* 20 (1), 87–100.

Xie, B.-L. (2017). 2018 Heihe Government Report (in Chinese) 黑河市政府工作报告 (2018年). Heihe.

Zou, J. and Li, C. (2017) 'Analysis of the Current Situation of Heihe Port's Trade with Russia', *World Rural Observations* 9(3), 25–28.

21 Flows making places, borders making flows

The rise of Mae Sot–Myawaddy hub in the Thai–Burmese borderland

Indrė Balčaitė

We now shift to the Thai–Myanmar border and the rise of Mae Sot and Myawaddy from small settlements on either side of the border into a thriving and increasingly co-joined trading hub. This first involved Mae Sot from the 1960s and then Myawaddy from the 1990s. Key factors were the changing security and economic politics of the two central governments, both internally and externally (involving trading) directed. However, what crucially shaped development in the area were the responses of the growing army of borderlanders, impelled by their own agendas regarding the opportunities the border was creating in terms of 'legal' and 'illegal' trade and migration. Central governments created the framework, but were hardly in control of the way these thriving communities developed.

This chapter traces the development of a cross-border hub between Thailand and Myanmar.[1] The rise of these border twin cities from obscurity to busy regional centre provides a case study in borderland socio-spatial relations as an ongoing interaction of networks, place, territory and scale (Jessop, Brenner and Jones 2008). Mae Sot is a border town in northwest Thailand, almost 500 km away from Bangkok but under 10 km from Myawaddy on the opposite side of the Moei River marking the Myanmar–Thailand boundary. The two cities are increasingly merging, but, until the late 1990s, there was no solid bridge between them (Figure 21.1). Mae Sot was a remote outpost before the 1960s and Myawaddy a small underdeveloped settlement into the 1990s. Mae Sot first developed as a transit point for smuggling to and from Burma, but growing legal trade in the late 1980s twinned it with Myawaddy, 430 kilometres from Yangon (Rangoon). I locate the emergence of this cross-border hub within the growing human and commodity flows through the surrounding borderland. Growing volumes of trade, migrations of soldiers, traders and people seeking refuge or work in Thailand together produced the growth of Mae Sot and Myawaddy. The trigger that initiated and rerouted these flows was national policy changes in the two states. Flows defying borders make places, influencing their centrality in hierarchal spatial orders, but are themselves shaped by the territorial implications of national policies.

Figure 21.1 The view of Myawaddy from the Thailand–Myanmar Friendship Bridge by Indrė Balčaitė (2016).

Rigorous quantitative analysis of both towns' population growth, volumes of trade and migration over time would strengthen my argument. However, statistical data is patchy, incomparable or altogether unavailable regarding flows designated as 'illegal'. Population numbers for Mae Sot district are unavailable before 1993 and do not capture Mae Sot's vast undocumented or semi-documented Burmese population. Myawaddy data is even scarcer, with only 1983 and 2014 censuses available. Trade volumes specifically through the Mae Sot–Myawaddy border crossing are hard to find for earlier years. Daily human traffic through the border is mostly informal and not rigorously monitored.

Hence, my study[2] is exploratory and based on eye-witness accounts from borderland inhabitants and foreign visitors. The argument rests on interviews and less formal conversations conducted in 2012–16.[3] My interlocutors were predominantly Karen migrants and Thai and Burmese business people who themselves or whose families lived in this borderland between the 1960s and the 2010s or traversed it between the mid-1980s and the early 2010s. I have changed interlocutors' names to protect their identities. I complement their accounts on the period before 1990 with the reports of foreign journalists (Sitte 1979; Boucaud and Boucaud 1985; Lintner 1999; M. T. Smith 1999).

The chapter has three parts. The first briefly outlines the theoretical framework based on literatures of world cities and borderlands. It builds a case for a borderland-centred perspective to observe the flows 'feeding' the two cities and connects it to the Indochinese context. The other two sections apply this framework to Mae Sot and Myawaddy, demonstrating how human and commodity flows converged on these places. The second part traces Mae Sot's initial growth (1960s–80s) due to Burmese socialist policies that produced intense smuggling from Thailand through insurgent-held territories. In the long run, the policy changes reversed not only commodity but also migration flows. The third section covers the period since the late 1980s, the transforming political topography of the Thailand–Myanmar borderland and the resulting rerouting of people and goods through Mae Sot–Myawaddy, an ever-expanding cross-border hub.

Flows making places and borders making flows

Arjun Appadurai (1996, 199) has argued that today's 'displaced, de-territorialized, and transient populations … are engaged in the construction of locality as a structure of feeling'. However, it is not just a location's image that emerges in such fluid landscapes. Flows of people and commodities literally build hubs. Cities constitute nodes of networks of economic, social, demographic and information flows (Smith and Timberlake 1995, 85–86). Immigration is one such city-building flow (Sanderson et al. 2015). Moreover, changes in flows passing through cities affect the networks of relations upon which those cities' centrality or marginality rests (Andrucki and Dickinson 2015, 208; Doel and Hubbard 2002, 361). A location's relative centrality thus changes over time. It also depends on the scale chosen. Although the argument originally emerged from world-cities literature, it holds when applied to smaller locales. They, too, are powered by flows: towns and even villages are changing and interconnected (Hedberg and do Carmo 2012, 2; Soi and Nugent 2017).

However, fluidity faces obstruction. Earlier claims of de-territorialisation proved unfounded: fluidity was countered by continuing spatial bordering (van Houtum, Kramsch and Zierhofer 2005). The policies nation-states implement are based on the premise of territoriality (Terlouw 2012); borders are thus sites of constant 'othering' (van Schendel 2005, 46). Meanwhile, differentials resulting from discrepant policy regimes on each side of the border become resources that borderlanders can exploit, thereby initiating human and commodity flows. Both unauthorised cross-border trade and undocumented migration work against the spatial inequalities that state territoriality produces (van Schendel 2005, 55). In my case, increasingly diverging levels of economic development between Myanmar and Thailand from the 1960s onwards produced growing cross-border flows of commodities and eventually people. In the late 1980s, the Thailand–Myanmar rapprochement and efforts to control the borderland, dislodging the wedge of insurgent forces, gradually reconfigured the cross-border flows. To use van Schendel's (2005, 51–52) terms again, a much denser funnel shape replaced the capillary border-crossing mode.

Practices of spatialisation become apparent at the margins (Shields 1992, 3–5). The South-East Asian borderlands thus reveal the relative lure of regional centres. The rugged Karen frontier has been shifting between Thai and Burmese lowland centres of power through the centuries. Under British colonial administration, Yangon was an industrial centre, surrounded by rice-exporting areas, while Bangkok was capital to a buffer state wedged between British and French colonial possessions. The balance began to shift after Burma's troubled independence in 1948, eventually reversing the flows through the Thai–Burmese borderland. Being part of the network connecting Bangkok and Yangon, the volumes, contents and direction of the flows that produced the Mae Sot–Myawaddy hub provide a litmus test of the power balance between the two centres across time.

Insurgent economy and the making of Mae Sot's transit hub

Mae Sot's history is tangled with that of the Karen insurgency against the central Burmese government across the border (1949–2012). However, it was the Burmese government's post-independence policies that triggered the human and commodity flows centring on the Thai–Burma borderland. Multiple rebellions and turmoil in the country eventually triggered the 1962 military takeover. Subsequent economic policies, labelled the 'Burmese Way to Socialism', aimed to withdraw from the international economy (Taylor 2008, 375), but the Yangon government did not control the borders it declared closed. Clandestine human and commodity flows through the insurgent-held borderlands kept the country connected to regional capitalist networks until legal trade re-emerged in the late 1980s. By then, regional migration patterns had also reversed. This was the initial period of Mae Sot's growth.

Between 1962 and the early 1990s, most of the Thai–Burmese border was controlled by insurgent armies on the Burmese side: Mon, Karen, Karenni, Shan, Pa-o and Communist Party of Burma (CPB) rebels. A 640 kilometre segment of the total 2,000 kilometre border was under the control of the Karen National Union (KNU) (Smith 1999, 283) fighting against the central government in the name of the Karen ethnic minority in Burma. KNU's control of the remote borderland made it a key player when General Ne Win, the leader of 1962 coup, announced nationalisation of production, distribution, import and export of commodities (Charney 2009, 123–24; Lintner 1999, 178–79). The economic reforms undertaken by the Burma Socialist Programme Party's (BSPP) government (1962–88) produced chronic deficit and the rise of smuggling to satisfy basic Burmese consumer needs. Smith (1999, 283) notes:

> Many key sectors of the economy collapsed ... and the hitherto backward and neglected hill tracts along the Thai border assumed an unexpected geographic and economic importance ... The first new Karen customs gate was opened ... at Phalu, south of Myawaddy, in the 6th brigade[4] area in 1964. The following year Bo Mya [KNU's leader] opened another gate at Kawmoorah (Wangkha[5]) in the 7th brigade area to the north.

There were certainly earlier cross-border exchanges. The border was hardly enforced before 1962 and people moved freely in the borderland. Mae Sot was a remote outpost without good roads to other Thai towns, including the provincial centre of Tak, where the highway from Bangkok finished. Hence, the town relied on Myawaddy and its connection to Mawlamyine (Moulmein) port in Burma for supplies. That reversed after BSPP's decision to close the country (Zaw Aung 2010, 23–24), which led to surplus production, on the one hand, and great demand for consumer products, on the other. Borderlanders' perspectives help to grasp the huge change induced by simultaneous Thai industrialisation and declining Burmese manufacturing. Channarong, a Sino-Thai businessman in Mae Sot whose father had settled to trade at the border, said that before the 1960s Mae Sot youth would study at university in Mawlamyine as English used to be the medium of instruction, and education was considered to be of better quality than in Thailand. At the time, imports like seasoning, tennis balls, cakes and pears would still reach Mae Sot from Myawaddy, he said. However, by 1970, commodity flows were reversing. Channarong's father stopped importing goods from Burma to Bangkok and started sending clothing from Bangkok to Yangon via Mae Sot and Mawlamyine.

This shift turned the borderland into a trade hub. Flows of goods kept growing. With the help of black-market traders and their carriers, cattle, agricultural products, gemstones, gold and teak (from KNU-controlled forests) flowed to the border and manufactured goods returned from Thailand. With Myawaddy under central government control and the border officially closed, Kaw Moo Rah–Wang Kha (approximately 20 km north of Mae Sot) became the border's busiest crossing point, followed by Pha Lu in the south. Mae Sot was a transit node between Bangkok and these border-trade gates. Sitte (1979, 31–38) recorded an average of 500 carriers checking in daily at Kaw Moo Rah with loads of 40–50 kg. Smith (1999, 282–83) reported that up to 1,000 cattle a day could be driven across the Moei River to Wang Kha by the 1970s. Cross-border trade via rebel-held areas reputedly generated between one-half and one-third of legal imports (World Bank 1976, 8) and approximately 40% of Burma's General National Product in early 1988 (Smith 1999, 25).

Inside Burma, these commodity flows were considered 'illegal'; thus, transportation routes were dispersed. To avoid arrest for contacts with the rebels, goods were transported on foot through the borderland (Falla 1991, 344) before switching to the mainstream infrastructure in government-controlled areas: trucks, railways or boats (Boucaud and Boucaud 1985, 67; Sitte 1979, 33–34). Many older Karen men from around Hpa-an (Pa-an, capital of Kayin State adjacent to Thailand) were carriers in their youth. Some worked alone, only bringing supplies from Thailand to their native villages, whereas others formed groups. U Kaw Thu worked in a group of ten in the mid-1970s before having all his wares confiscated and quitting the trade, whereas U Khaing traded clothing alone in the 1980s. From central Kayin State, they usually went via To Kaw Ko village in Kawkareik Township that hosted a KNU base and crossed the border at Kaw Moo Rah–Wang Kha.

Extensive networks developed, reaching well inside the country, supplying Yangon and other cities. Ukrit, a long-term Burmese resident of Mae Sot, had previously run a small factory in Mawlamyine, making luggage items from smuggled Thai materials and selling them in Yangon. After losing a tailoring workshop to nationalisation, Ma Lay's grandfather ran a clothing shop in central Yangon from the 1970s to the 1990s, procuring supplies from black-market operations. In Ukrit's words, prior to 1988, the 'black market was everywhere'.

The policy changes in Yangon had implications on both sides of the border, triggering both commodity flows and movement of people. Constant demand meant steady traffic. Thai entrepreneurs moved to the border to supply the black-market traders in Burma. Five Mae Sot businessmen, whose families had engaged in cross-border trade for two or even three generations, told me that their families moved to the border in the 1960s and 1970s. Four interlocutors specifically cited Mae Sot's proximity to Yangon as a reason. Three interviewees were Sino-Thai merchants who themselves or whose fathers came to the border for business opportunities. They moved from Bangkok to the malarial and still-lawless frontier, separated by mountain ranges both from Central Thailand and from inner Burma. Such movement occurred all along the border (see Sitte 1979, 18; South 2003, 155). Some traders arrived even from inside Burma. The grandfathers of two other Mae Sot businessmen interviewed were Muslims of Indian or Rohingya origin, who had faced discrimination under Burmese military rule. One had his passport revoked and another lost land and a factory in Yangon to nationalisation. Ukrit remembered that even the deposed Burmese Prime Minister U Nu retreated with his supporters to the Thai border in 1968, becoming insurgent militias and joining the cross-border trade.

Human flows followed cattle trails to Thailand but not all related to trade. As inhabitants of the eastern frontier and co-ethnics of Thailand's indigenous Karen, Burma's Karen were among the first to settle in the Thai borderland. As the economy deteriorated and civil war intensified, they moved to Thailand's shadow labour market and refugee camps. Inhabitants of central Kayin State (Hpa-an area) used to seek paid work in Yangon or neighbouring Mon State. In the mid-1980s, some still considered whether to head to the lowlands west or east of the Dawna Range (Hayami 2011, 21), that is, Lower Burma or Thailand. But the tide had already turned: growing numbers of locals started climbing the mountains to reach Thailand. Those leaving before 1990 usually trekked for days, avoiding the Burmese army (Tatmadaw), but over-nighting in Karen villages or KNU bases. Those who knew any black-market traders had an advantage. Thus, already in the late 1980s when Daw Mya and her husband toiled on a Thai-owned farm near Mae Sot, all their fellow workers were Karen from Burma. By the mid-1990s, going to Mae Sot had got so popular in her village in Hpa-an Township that three people per household could be there, Daw Hla estimated. Ukrit remembered that the first Thai police raid in 1993 rounded up over 8,000 Burmese in Mae Sot. At the time, the Mae Sot municipality reportedly had 23,512 inhabitants (ระบบสถิติทางการทะเบียน 2017).

According to Channarong, Mae Sot had been altogether rural in the 1960s. Yet inflows of goods and people grew the town. It became part of the spider's web of trails connecting Yangon with Bangkok through Kaw Moo Rah–Wang Kha and Karen villages. It thrived on the alliance among the KNU, Thai merchants and Thai military authorities. Smith (1999, 283) writes that smuggling turned Mae Sot 'into a bustling new market town'. Sitte (1979, 17) also describes it as having 'sprung off the ground like a gold-rush town within the shortest time' due to smuggling. He was impressed by brick buildings, many shops and company branches, new cars, trucks being loaded in many compounds, motorbikes and scooters (Sitte 1979, 18). However, my interlocutors resident in the borderland for a long period argued that Mae Sot was still very small compared to the present. Somchai's father's warehouse in Mae Sot, on the edge of town before the 1980s, now finds itself in Mae Sot's expanded town centre.

According to Somchai as well as Saw Eh Tu who grew up in Pha Lu, Mae Sot was smaller than the Kaw Moo Rah–Wang Kha hub where many Thai traders lived and business transactions occurred. In the mid-1970s, André and Louis Boucaud (1985, 65) found Karen, Indian and Chinese merchants, a monastery, a church, a mosque, a school and a basic hospital there along with KNU barracks. The smuggling that created Kaw Moo Rah–Wang Kha also uplifted its satellite Mae Sot. By 1970, a new road through the mountains all the way to the Moei riverbank (Lintner 1999, 222–23) drew the once remote outpost into Bangkok's orbit. A new period in cross-border trade was approaching where Myawaddy would play a much bigger role.

The rise of the Mae Sot–Myawaddy gateway

If Mae Sot's rise was initially intertwined with smuggling through KNU-held border checkpoints, Myawaddy's role in modern history was that of a government military stronghold. Until the early 1980s when the first KNU border bases fell to the Tatmadaw, Myawaddy and Tachileik, well to the north, were the only points along the Thai–Burmese border under central government control (Smith 1999, 283; South 2003, 125). It was the Thai–Burmese rapprochement in the late 1980s and efforts to strengthen border control that transformed Myawaddy's role in the border economy. The Kaw Moo Rah–Wang Kha hub turned to ruins in the 1990s, whereas Mae Sot was poised to grow further with Myawaddy as its twin town. The new hub formed around the official border crossing. The dispersed human and commodity flows through the Burmese borderland gradually centralised, converging on the Yangon–Hpa-an–Kawkareik–Myawaddy road and changing Myawaddy beyond recognition.

The smuggling routes had developed during the period of low-intensity civil conflict when the Burmese army was bogged down in fighting the CPB (Smith 1999, 283). From the mid-1980s, government forces turned against the KNU in earnest. Repeated army attacks on KNU bases disrupted trade, making thousands flee into Thailand. Pha Lu was captured in 1989. In 1994, a split within KNU produced the Democratic Karen Buddhist Army (DKBA) that allied itself with the government. In 1995, two important KNU bases – Ma Ner Plaw and Kaw Moo Rah – fell. Merchants left early due to lack of security. Somchai's father relocated to Mae Sot in the 1980s after a blast destroyed his shop in Wang Kha. By the late 1990s, government forces or their allies controlled most of the border.

The Tatmadaw offensive may have failed without a Thai foreign-policy overhaul. Burmese currency (kyat) demonetisations in 1985 and 1987, aiming to collapse the black market, impoverished many especially in inner Burma with no access to alternative currency. With the economy tumbling, nation-wide demonstrations occurred in 1988. A bloody crackdown produced another opposition exodus to the border. Despite international sanctions, the Thai government, hungry for Burma's resources, cut large-scale extraction deals with the Burmese military junta. Concessions awarded in contested areas bypassed the insurgents, the Thai military's former allies, but gave the Burmese government much needed hard currency. Mae Sot–Myawaddy became the poster-child for Thai Prime Minister General Chatichai's new foreign policy calling to turn 'battlegrounds into marketplaces' (Lang 2002, 143–44).

An alternative route for both goods and people was emerging. When U Khaing started working as a carrier importing clothing from Thailand in 1984/85, the black-market trade was about to decline. Due to Tatmadaw's advances, the road to Myawaddy was slowly opening. U Pu remembered an army truck going from Hpa-an to Myawaddy once a month carrying personnel and supplies in his youth (1970s). For villagers, there was a truck every three months. The Burmese army already controlled its route, but fighting made it unsafe. In the 1980s, there was one car a month with departures not guaranteed. Saw San who went in 1989 recalled one car every 15 days, but their trip to the border took five days (longer than most walking routes), as they had to wait till it was safe to go. Nan Nay, first travelling to Thailand around 1993, boarded a pick-up truck from Hpa-an to Kawkareik (mid-way town) and then paid the Burmese army for a lift in a car delivering supplies to a military base in Myawaddy. Later, trips became more regular. In the early 1990s, there was a thrice-a-month service between Hpa-an and Myawaddy, later weekly, though government soldiers had to accompany each car. By the late 1990s, the DKBA controlled the route, and traders could count on daily departures. An overloaded car could carry goods worth 10,000 USD, U Khaing said, eventually putting carriers travelling on foot out of business. He quit the trade in 1991/92.

Co-operation between Thai and Myanmar state authorities (often with the involvement of the militia such as DKBA) produced growing official trade. In 1989, Myanmar's total exports to Thailand were worth 2.156 billion Thai baht (83.89 million USD at contemporary exchange rate) and Thailand's total exports to Myanmar of 638 million baht (24.8 million USD). Whereas the former fluctuated, the latter continued growing steadily, topping 14 billion baht by 1998 (Maung Aung Myoe 2002, 28) – an increase of 22 times. Myawaddy became a gateway for this growing flow, receiving 130 billion baht worth of Thai exports a month by 1998 (Ibid.). By 2012, Thai exports worth 1.47 billion baht crossed through Mae Sot–Myawaddy, more than doubling just two years later (Tak Chamber of Commerce information supplied in an interview).

Human flows also gradually redirected. With the weakened KNU now unable to guarantee safety, Kayin State inhabitants moved into government-controlled zones or to Thailand. Cross-border human flows continued growing, especially after the 1997 Asian financial crisis, but now crossed Myawaddy. By the 2000s, almost all migrants from central Kayin State were hiring a broker to arrange travel to Thailand. With the KNU dislodged and the 'buffalo track' routes neglected, Karen migrants came into direct contact with state agents along the Hpa-an–Myawaddy route, where Burmese officials collected bribes to allow travel to the border.[6] On the Thai side, new undocumented migrants no longer stayed in the borderlands but travelled to Central Thailand. Their Thai brokers usually co-operated with the Thai police to ensure passage. By the 2010s, one-third of the inhabitants of the village in central Kayin State where I conducted fieldwork were in Thailand, mostly in Bangkok. An undocumented migration industry had developed.

The passing flows and long-neglected infrastructure development brought buzz to Myawaddy. Before the 1990s, it was a small militarised settlement, isolated from the rest of the country by the Dawna Range. The old road cutting across was narrow, curvy and dangerous. Earlier undocumented Burmese migrants avoided Myawaddy by travelling through KNU areas. In the mid-1980s, Myawaddy was tiny, said Daw Mu whose family lived there due to her father's civil service assignment. Electricity was available just a few hours daily and only the hospital, police station and the military base were connected to the grid. Everybody else relied on generators even into the 1990s, said U Khaing. According to Saw Eh Tu, Myawaddy was smaller than Kaw Moo Rah and Wang Kha and had fewer cars than Pha Lu in the 1980s. The 1983 Burma population census recorded 6,389 inhabitants and only 23 solid houses (out of 1,292) in the urban areas of Myawaddy Township (Burma Immigration and Manpower Dept. Census Division 1986, 21, 9).

As Channarong put it, 'everything changed because the policy changed'. The 2014 census found 113,155 inhabitants in Myawaddy Township (MIMU 2015, Table A-3). Mae Sot town municipality had 37,998 inhabitants and the whole district (including rural areas) 87,060 inhabitants in 2016 (ระบบสถิติ ทางการทะเบียน 2017). It may seem that Myawaddy has already outgrown Mae

Sot. However, as Phil Thornton (2006, 74) noted a decade ago, nobody knows exactly but, given the undocumented Burmese workforce, the population of Mae Sot district could be closer to 250,000. Myawaddy, too, continues expanding along the new highway to Yangon completed in 2015. Since 1997, when the first sturdy bridge over the River Moei connected them,[7] the settlements have been growing ever closer. Having relied on Mae Sot for supplies for decades, Myawaddy's centre is right beside the border bridge and the busy highway is its axis. Mae Sot's centre, initially a satellite of Kaw Moo Rah–Wang Kha further north, is further from the border. However, the area in-between is urbanising fast, a trend expected to strengthen further with the development of special economic zones on both sides of the border. Human movement continues shaping the borderland by establishing new centres such as Lay Kay Kaw, a settlement of former refugees and internally displaced people in Kayin State.

Conclusions

Flows make and shrink places by centring or de-centring them. They literally build cities, the intersections of movement. The heavier the flows, the more central the hub becomes. Whether it is a hut hamlet, a town or a global city, a settlement's growth is largely powered by flows. In 1960–2000, flows passing through the network connecting Yangon and Bangkok reconfigured the Thailand–Myanmar borderland twice. Growing volumes of goods transported through KNU-controlled Kaw Moo Rah developed Mae Sot into a transit hub between the 1960s and the 1980s. Flows re-centred the borderland when they abandoned Kaw Moo Rah–Wang Kha hub amidst war. In the 1990s, they converged on the Myawaddy–Mae Sot checkpoint, creating the new busiest crossing on the Thailand–Myanmar border.

On the other hand, bordering that creates spatial differentiation initiates and reroutes or reverses flows. Burmese socialist policies derailed the economy and eventually reversed the regional human and commodity flows. The widening economic asymmetry with neighbouring Thailand turned into a lucrative asset exploited by the borderlanders as well as arrivals from Bangkok and Yangon. The deficit in Burma reversed and intensified the pre-existing flows. Once able to manufacture goods for export, Burma became a supplier of raw materials by 1970 and labour by the 1980s. Yangon attracted fewer people, which meant gains for Mae Sot and later Bangkok.

In the late 1980s and the early 1990s, another policy shift reconfigured the borderland again, rerouting and growing the flows. The Thai and Burmese governments started co-operating to dislodge the KNU, assume direct control of their border and tap the profitable flows. With the KNU border gates out of business, official trade growing and infrastructure constantly improving, flows converged on Mae Sot–Myawaddy. The spatial change that took place demonstrates that national policies shape flows even when governments do not directly control them. The officially shut Burmese border was in fact very porous in the 1960s–80s, but the designation of the flows as 'illegal' implied a dispersed rather than centralised route.

Notes

1 In 1989, the military regime changed the official country name from 'Burma' to 'Myanmar' along with many place names. I use 'Myanmar' to refer to the post-1989 state but 'Burma' in pre-1989 or trans-historical context. I adopt the Burmese versions of place names ('Kayin State', 'Yangon'), but call the ethnic group in question 'Karen' rather than 'Kayin'. I add the Anglicised names in parentheses at first mention.
2 First presented at the 2016 Annual Meeting of American Geographers.
3 My PhD and subsequent fieldwork in Myanmar and Thailand (September 2012–July 2013, April 2014, October–November 2016) were part-funded by a SOAS Field-work Award, SOAS Centre of South East Asian Studies (CSEAS) Small Grant and Santander Mobility Grant.
4 The KNU administrative units are 'brigades' and 'districts', whereas the Burmese government uses 'states', 'districts' and 'townships'.
5 Kaw Moo Rah was the KNU base on the Burmese side and the settlement on the Thai side was called Wang Kha (in Karen).
6 Unlike the KNU, the state authorities would not allow people to leave Burma undocumented, whereas emigrating legally implied very complicated procedure under BSPP rule, often leaving no right to return.
7 A second border bridge and road are scheduled to open in 2019.

Bibliography

Andrucki, Max J., and Jen Dickinson. 2015. 'Rethinking Centers and Margins in Geography: Bodies, Life Course, and the Performance of Transnational Space'. *Annals of the Association of American Geographers* 105 (1): 203–18.

Appadurai, Arjun. 1996. *Modernity at Large: Cultural Dimensions of Globalization.* Public Worlds Series. Minneapolis; London: University of Minnesota Press.

Boucaud, André, and Louis Boucaud. 1985. *Birmanie: Sur La Piste Des Seigneurs de La Guerre.* Paris: L'Harmattan.

Burma Immigration and Manpower Dept. Census Division. 1986. *Burma 1983 Population Census.* [Rangoon]: Socialist Republic of the Union of Burma, Ministry of Home and Religious Affairs, Immigration and Manpower Dept.

Charney, Michael W. 2009. *A History of Modern Burma.* 1st edition. Cambridge, New York: Cambridge University Press.

Doel, Marcus, and Phil Hubbard. 2002. 'Taking World Cities Literally: Marketing the City in a Global Space of Flows'. *City* 6 (3): 351–68.

Falla, Jonathan. 1991. *True Love and Bartholomew: Rebels on the Burmese Border.* Cambridge: Cambridge University Press.

Hayami, Yoko. 2011. 'Burmese Migrants to Thailand: Vignettes from the Border as In-Between Space'. *Kyoto University Center for Southeast Asian Studies Newsletter No. 63*, Spring 2011.

Hedberg, Charlotta, and Renato Miguel do Carmo. 2012. 'Translocal Ruralism: Mobility and Connectivity in European Rural Spaces'. In *Translocal Ruralism Mobility and Connectivity in European Rural Spaces*, edited by Charlotta Hedberg and Renato Miguel do Carmo, 1–9. Dordrecht: Springer.

Houtum, Henk van, Olivier Kramsch, and Wolfgang Zierhofer. 2005. 'B/Ordering Space'. In *B/Ordering Space*, edited by Henk van Houtum, Olivier Kramsch, and Wolfgang Zierhofer. Border Regions Series. Aldershot: Ashgate.

Jessop, Bob, Neil Brenner, and Martin Jones. 2008. 'Theorizing Sociospatial Relations'. *Environment and Planning D: Society and Space* 26 (3): 389–401.

Lang, Hazel J. 2002. *Fear and Sanctuary: Burmese Refugees in Thailand*. Studies on Southeast Asia, no. 32. Ithaca, NY: Southeast Asia Program Publications, Southeast Asia Program, Cornell University.

Lintner, Bertil. 1999. *Burma in Revolt: Opium and Insurgency since 1948*. 2nd edition. Chang Mai: Silkworm Books.

Maung Aung Myoe. 2002. *Neither Friend nor Foe: Myanmar's Relations with Thailand since 1988. A View from Yangon*. IDSS Monograph. Singapore: Institute of Defence and Strategic Studies.

MIMU. 2015. 'The 2014 Myanmar Population and Housing Census (Kayin State)'. Myanmar Information Management Unit. 6 May 2015. www.themimu.info/sites/themimu.info/files/documents/BaselineData_Census_Kayin_with_Pcode_MIMU_05Jun2015_Eng.xlsx.

Sanderson, Matthew R, Ben Derudder, Michael Timberlake, and Frank Witlox. 2015. 'Are World Cities Also World Immigrant Cities? An International, Cross-City Analysis of Global Centrality and Immigration'. *International Journal of Comparative Sociology* 56 (3–4): 173–97.

Schendel, Willem van. 2005. 'Spaces of Engagement: How Borderlands, Illicit Flows, and Territorial States Interlock'. In *Illicit Flows and Criminal Things: States, Borders, and the Other Side of Globalization*, edited by Willem van Schendel and Itty Abraham, 38–68. Tracking Globalization. Bloomington and Indianapolis: Indiana University Press.

Shields, Rob. 1992. *Places on the Margin: Alternative Geographies of Modernity*. International Library of Sociology. London, New York: Routledge.

Sitte, Fritz. 1979. *Rebellenstaat Im Burma-Dschungel*. Graz: Styria.

Smith, David A., and Michael Timberlake. 1995. 'Cities in Global Matrices: Toward Mapping the World-System's City System'. In *World Cities in a World-System*, edited by Paul L. Knox and Peter J. Taylor, 79–97. Cambridge: Cambridge University Press.

Smith, Martin T. 1999. *Burma: Insurgency and the Politics of Ethnicity*. Rev. and updated edition. London: Zed Books.

Soi, Isabella, and Paul Nugent. 2017. 'Peripheral Urbanism in Africa: Border Towns and Twin Towns in Africa'. *Journal of Borderlands Studies* 32 (4): 535–56.

South, Ashley. 2003. *Mon Nationalism and Civil War in Burma: The Golden Sheldrake*. London: Routledge.

Taylor, Robert H. 2008. *The State in Myanmar*. New edition. London: C. Hurst.

Terlouw, Kees. 2012. 'Border Surfers and Euroregions: Unplanned Cross-Border Behaviour and Planned Territorial Structures of Cross-Border Governance'. *Planning Practice & Research* 27 (3): 351–66.

Thornton, Phil. 2006. *Restless Souls: Rebels, Refugees, Medics and Misfits on the Thai-Burma Border*. Bangkok: Asia Books.

World Bank. 1976. *Burma - Development in Burma: Issues and Prospects*. 1024. Washington, D.C: The World Bank. http://documents.worldbank.org/curated/en/1976/07/1561280/burma-development-burma-issues-prospects.

Zaw Aung. 2010. *Burmese Labor Rights Protection in Mae Sot*. Bangkok: Chulalongkorn University Center for Social Development Studies.

ระบบสถิติทางการทะเบียน [Official Statistics Registration Systems]. 2017. 'จำนวนประชากร [Population Numbers]'. 2017. http://stat.dopa.go.th/new_stat/webPage/statByYear.php.

22 So close, so far

National identity and political legitimacy in UAE–Oman border cities

Marc Valeri

In this final chapter, we turn attention to borders in the Middle East, more particularly that between Oman and the UAE, and the border towns of al-Buraymi and al-Ayn. Here policies pursued by the two central governments, driven by their own priorities over legitimacy, security, socio-economic engineering and the consequent desire for a hard border were again central, but in far more successfully controlling ways. Thus, what had been a single thriving oasis community increasingly became two cities on either side of the border, with former business and family relationships increasingly disrupted.

Oman–United Arab Emirates (UAE) border, Thursday 5 May 2016 early morning. As has been the case for years on weekends and holidays, endless queues of cars from Oman wait to cross the border into the UAE and enjoy Dubai attractions and entertainment their country does not offer. The Omani border city of al-Buraymi, across the contiguous UAE city of al-Ayn, is grid locked.

Many border cities are contiguous urban areas, either 'dependent on the border for [their] existence' or coming 'into existence because of the border' (Buursink 2001, 7–8). As noted in this book's Introduction, they have usually either developed on either side of long-established borders, or were originally single cities that have been split by a subsequently drawn border, often with forcible population transfer. In al-Ayn/al-Buraymi, urban settlement long predates the border. They previously existed as a unique oasis (known as al-Buraymi) comprising nine small villages until the late 1950s. At that point, on British initiative, the Sultan of Oman and the Emir of Abu Dhabi signed agreements for reciprocal recognition of territorial jurisdiction. Six villages were deemed to pay tribute to Abu Dhabi's ruler and the remaining three, including al-Buraymi itself, historically the oasis' largest village, were recognised as loyal to Oman. When the sultanate of Oman and the federation of the UAE became independent from the British (1970 and 1971), this political division of the oasis was confirmed, but without any physical borderline or border control materialising except on maps. Individuals crossed and recrossed regularly, the two towns effectively working like one unit. Only in 2002, did the UAE start installing a fence along its Omani border, including through the al-Ayn/al-Buraymi conurbation.

The political and economic trajectories of al-Ayn and al-Buraymi, whose populations reached 519,000 and 104,000, respectively, in 2016, have been adversarial, epitomising the two countries' evolving relationship. Their common past and 'notion of being one community with common interests and belonging together' (ibid., 17) have been present among their inhabitants since independence, with close cross-border contacts maintained through family/marriage, and business. However, the top-down political priorities of both national leaderships have contributed to substantially re-defining border perceptions in these cities, and the local populations' lives and actions. In particular, drawing an international demarcation across the oasis has started a new dynamic resulting from increasingly divergent trajectories of social and economic development. Given these conflicting evolutions and the absence of like-mindedness and 'intention to construct and maintain the common space or reach common goals' (Mikhailova 2015, 438), al-Ayn and al-Buraymi partly differ from typically co-operative international twin cities in Buursink's understanding (ibid., 15) and as described in the book's Introduction.

While most studies on this oasis have concentrated on the pre-independence era, particularly the 1950s dispute over its control (Schofield 2011; Morton 2013), this chapter focuses instead on its evolution since the early 1970s. Studying this border pair, sharing history but growing apart since the early twenty-first century, provides insight into how political sovereignty and legitimacy are constructed in post-colonial states and the link between building national identity and the physical demarcation from the (br)other. Since the early 1970s and Oman and the UAE's independence, the building of both a state apparatus and nation has been central to the two rulers' political projects to assert their legitimacy and control over their respective territories. Their mutual relationship has been particularly crucial, given the tribal and ethnic proximity between inhabitants spanning the new boundary, but also the two countries' shared modern history.

The chapter has two parts, showing how the opposing trajectories followed by these border cities post-independence parallel the way the UAE and Oman have grown separately. Part 1 highlights the slow mutual recognition between the two newly independent countries from the 1970s to the early 2000s. While Abu Dhabi and Muscat's pursued similar and simultaneous state-building projects, al-Ayn and al-Buraymi began diverging due to the roles those projects gave them. Part 2 explores how the question of separating from the (br)other re-emerged in the early 2000s as both countries faced domestic challenges calling into question the socio-political order established in the 1970s. Growing suspicions and mistrust in both capitals meant that al-Ayn and al-Buraymi increasingly turned their backs on each other.

Post-independence nation-building projects and borderland management

Former President of the UAE, Sheikh Zayed Al Nahyan (r.1971–2004), and Oman's Sultan Qaboos (r.1970–) accessed power in similar circumstances, overthrowing incumbent rulers (Zayed's brother and Qaboos's father) helped by British advisers. Their rooms to manoeuvre regarding the British became minimal due to the imperial policy developed since the mid-nineteenth century. The Gulf had become a place 'where time had stood still ... frozen into the requirements of the *pax Britannica*' (Zahlan 1989, 13–14) and where political actors were isolated from the outside world, tributary to Britain alone. However, this started changing during the Arab nationalist heyday in the late 1950s: the Gulf's southern shore was no longer isolated from political convulsions. In particular, Southern Oman's Dhofar region saw an uprising against the authority of Sultan of Oman, Said bin Taymur (r.1932–70), from the 1960s. Gradually, this morphed into the Marxist–Leninist Popular Front for the Liberation of the Occupied Arabian Gulf. In July 1970, with the uprising about to spread to northern Oman, the British orchestrated the Sultan's overthrow by his son Qaboos. This coup enjoyed the Shah of Iran's support alongside other Gulf rulers, notably Abu Dhabi's Zayed Al Nahyan, worried by potential revolutionary contagion from Dhofar (Takriti 2013, 153). Unsurprisingly, Zayed provided substantial financial help to Oman's new ruler during the Dhofar war, including providing a garrison for Sohar in northern Oman in 1974 to free Qaboos's troops for fighting the war in the South (ibid., 285–6), which officially ended in 1975.

Oman–UAE relations after the 1970s: 'Twin palm trees on the same land?'

This war proved decisive for both rulers, destroying the most credible political alternative to their national leadership and helping them assert control over the newly independent states by defining a new order wherein they alone held the keys. Both implemented ambitious national unification policies, which became pivotal to regime ideologies. These policies have been based on oil rent, a godsend enabling widespread redistribution games reserved for nationals only within the framework of an allocation state. The public sphere became an inexhaustible tank of jobs at all levels of skill, on territories characterised for several decades by emigration. The new civil servants, employed in the national civil and military services, were all less inclined to rise against their regimes, depending on them for their daily lives. Similarly, both regimes began rewriting identity frames of reference around the ruler's person, identified in the new historiography with the contemporary welfare states and thus the countries themselves. This political work on history has aimed at 'naturalising' the special pantheon (contemporary regime, welfare state, and the institution and persons of the rulers, portrayed as 'fathers of the nation'), underpinning both regimes' legitimacies. However, bilateral relationships between the rulers remained cool for a long period, despite (or, rather, *because of*) the intimacy of respective national-building projects and the two countries' shared modern history.

While Muscat recognised the new UAE state in 1971, it refused to appoint an ambassador in Abu Dhabi because relations between the two countries were so close historically that it was seen as unnecessary. In a 1970s inter-view, Sultan Qaboos explained:

> Brotherly relations between the two peoples are flourishing, for this has surpassed the diplomatic and routine and complex procedures. By this I mean: we are two brothers, you [the UAE] in your house and we in ours, but the two houses are next to each other. Do you really need a mediator? In my opinion, what affects you affects me automatically
>
> (Kechichian 1995, 80)

Even Dhofari revolutionaries viewed the UAE and Oman 'as one national entity', believing 'in the "Natural Oman" idea, the unification of the Trucial States with Muscat and Oman' (Takriti 2013, 165 and 300).

Mutual relations deteriorated after the UAE's 1974 Jeddah agreement with Saudi Arabia over border demarcation. While Saudi Arabia dropped its claim on al-Buraymi oasis, Oman considered that the two other countries had reached agreement over part of its territory. Armed clashes happened in December 1977 at the border, with many Omanis serving in the UAE army refusing to fight against Omani troops (al-Sayegh 2002, 132). After over a year's negotiation, agreement was reached in April 1979 to fix 10% of the total Oman–UAE border. Oman's Sultan realised that too much inflexibility endangered economic and social co-operation with the UAE, vital for Oman.

The two rulers came closer personally during the 1980s; producing en-hanced economic and security relations, and exchanging ambassadors in 1991 (Allen and Rigsbee 2000, 195). In April 1992, Oman and UAE nationals were granted permission to cross the border with just identity cards. In May 1992, a joint UAE–Oman commission began examining the border issue. In 1993, an Omani Foreign Affairs Ministry official declared 'Oman and the Emirates were twin palm-trees on the same land' (personal interview, Muscat, 31 May 2003), while Dubai's daily *al-Khaleej* headlined the start of border negotiations: 'One nation, two states' (al-Sayegh 2002, 134). Government newspapers and official statements increasingly portrayed mutual relations in brotherly terms. The two rulers ratified the final delim-itation agreement in June 2002. In July, Sheikh Zayed expressed 'his deep satisfaction over the firm and distinguished fraternal relations ... which have further enhanced affinities of brotherhood, kinship and common his-tory binding the peoples of both countries' (*Gulf News* 21 July 2002).

For al-Ayn and al-Buraymi, their trajectories, geographically peripheral in their new states and intimately linked to each other, started diverging inexorably from the 1970s. While, as noted, UAE and Oman rulers were driven by similar but inward-looking political priorities in the decades following independence, the parts al-Ayn and al-Buraymi played in these political constructions contrast dramatically. Elites in al-Ayn have re-tained crucial influence in Abu Dhabi, unlike al-Buraymi's populations within Oman's regime. Furthermore, both states' highly centralised and

authoritarian decision-making allowed local authorities no latitude for institutional cross-border co-operation initiatives.

al-Ayn and al-Buraymi: contrasting roles in the nation-building project

The al-Ayn/al-Buraymi oasis has been highly coveted since the eighteenth
century, due both to its abundant agricultural resources in a very infer-
tile hinterland and its strategic location in the Eastern Arabian Peninsula,
almost equidistant from Abu Dhabi, Dubai (on the Persian Gulf) and Sohar
(on the Gulf of Oman). In 1915, it was described as 'verdant and fruitful.
The soil, though thin, is fertile, and streams of running water abound on
every side' (Lorimer 1986, 263). Unlike its hinterland, populated by nomadic
tribes, oasis society long comprised sedentary town-dwellers, engaged in
agriculture, and semi-nomad groups settling in the oasis during summer
and living nomadically during the winter, eventually becoming sedentary
cultivators in the twentieth century's second half. Total population was estimated at around 5,500 in 1915 and 25,000 in 1960 (Table 22.1). Communications remained extremely difficult until the 1960s.

Until the early twentieth century, political control over the main tribal
confederations within the oasis oscillated between Abu Dhabi, Muscat and
Saudi Arabia. In the early nineteenth century, Abu Dhabi's ruler, Shakhbut
Al Nahyan, built a fort there, reflecting the old personal connection between

Table 22.1 Populations of the wilaya (province) of al-Buraymi and al-Ayn urban area

	1975	1980	1993	2003	2005	2010	2016
No. (%) of Omanis in al-Buraymi			28,988 (60%)	33,455 (49.2%)	34,615 (45.7%)	37,964 (60.1%)	46,950 (45.1%)
No. (%) of expatriates in al-Buraymi			19,299 (40%)	34,508 (50.8%)	41,102 (54.3%)	25,195 (39.9%)	57,123 (54.9%)
Total population of al-Buraymi			48,287	67,963	75,717	63,159	104,073
No. (%) of UAE nationals in al-Ayn						103,897 (27.6%)	131,963 (25.4%)
No. (%) of expatriates in al-Ayn						273,032 (72.4%)	387,309 (74.6%)
Total population of al-Ayn	50,704	102,329	202,968		284,040	376,929	519,272

Sources: Oman's National Centre for Statistics and Information (www.ncsi.gov.om); Abu Dhabi's Statistics Centre (www.scad.ae/).

the ruling family and the oasis. His grandson Zayed bin Khalifa (r.1855–1909) helped the Imam of Oman take the oasis from the Saudis in 1869 (Gervais 2011, 77), consolidating his control over local tribes by buying lands there (Lorimer 1986, 261–4).

In 1922 and 1923, the British imposed treaties, whereby local rulers would grant no oil concession except to somebody British government approved. Prospecting commenced in 1948, but the Iraq Petroleum Company (IPC) soon realised that Oman's Sultan had little control over the relevant territories, despite bribing local tribes. In 1949, Saudi Arabia's King Abd al-Aziz (r.1932–53) awarded a concession to the Arabian American Oil Company (Aramco) south of al-Buraymi; the company started exploration, highlighting a new British–American (IPC vs Aramco) tussle over oil-bearing areas. Aware that their area of influence was threatened by this US–Saudi Arabian oil alliance, the British raised a military force in Abu Dhabi, the Trucial Oman Scouts. With rising tension over oil concessions, several tribal groups (especially the Bani Qitab, Naim and Bani Kaab) settled in the contested areas and exploited the struggle for influence between Abu Dhabi, Muscat and Riyadh to retain their autonomy. The Saudis occupied al-Buraymi in August 1952, appointing a governor who secured allegiance from many oasis tribal leaders (Gervais 2011, 82–84). The Saudis and British signed a temporary agreement in October 1952 to appeal to outside arbitration. However, in October 1955, the British, 'convinced they were going to lose' (Takriti 2013, 22), stopped the discussions unilaterally; the Trucial Oman Scouts rapidly recaptured al-Buraymi oasis. In 1959 and 1960, Oman and Abu Dhabi rulers signed agreements under British auspices to delineate their respective authority-areas, mainly to facilitate awarding concessions to British companies in areas where boundaries had previously been non-existent.

Future UAE President Zayed Al Nahyan himself was born in the oasis in 1918 and represented Abu Dhabi's ruler there 1946–66; this personal connection helps explain al-Ayn's special role in the new UAE after independence. By the mid-1960s, the villages under Zayed's authority already had a hospital and Jordanian-staffed school – long before such facilities became available in Oman's part of the oasis and Abu Dhabi itself (Davidson 2009, 46–47). Since 1971, the ruler's representative in al-Ayn has been Zayed's cousin's son, Tahnoon Al Nahyan. A prominent businessman, he has held top decision-making positions at emirate level. Zayed's son and current UAE ruler Khalifa bin Zayed (r.2004–), who was his father's representative in al-Ayn (1966–70), has maintained his principal family home there (Davidson 2006, 52).

This political influence of al-Ayn connections does not stop at Abu Dhabi's ruling family. Beyond their own tribe, Zayed and his son Khalifa heavily rely on tribes native to al-Ayn to fill sensitive civil and military positions – the latter being considered the most reliable partners to strengthen Al Nahyan's grip on national power. This is the case for the Dhawahir tribe (in the army, in the Interior and in ruling family members' personal offices) and the Bani Qitab (the army and the police).

Since independence, Zayed Al Nahyan has used the oil rent to develop local infrastructure and increase influence over the borderlands. UAE' first university (UAE University) opened in al-Ayn in 1976. By the late 1970s, a four-lane highway connected al-Ayn to the capital and, in 1994, an international airport opened. The year 1981 saw the UAE's first dairy farm in al-Ayn, becoming one of the largest in the Middle East. The area also acquired the Middle East's first Coca-Cola bottling plant and a very large date producer (Al Foah). Two private universities arrived in the 2000s.

Al-Ayn has been trying to develop as a tourism and recreation hub too. As early as 1971, al-Ayn national museum, the country's first, was opened. Three major shopping malls emerged by the late 1990s. In 2011, the local cultural sites (including prehistoric tombs and irrigation systems dating from the Iron Age) were registered on UNESCO's World Heritage List.

In 2009, an ambitious *Plan Al-Ain 2030* identified aviation and defence as key engines of city development, alongside existing agribusinesses. The Nibras Al Ain Aerospace Park started in 2010, intended as an industrial platform for manufacturing and maintenance facilities and training and research institutions. Home of Etihad Airways' pilot training centre, it is supposed to create 10,000 jobs by 2030, by when the Plan expects that the city population will reach one million.

Obviously, al-Buraymi has not held the same political role in Oman under Sultan Qaboos. The disparity of treatment emerged quickly. Until the early 1990s, the Sultanate was divided into two governorates and nine regions, including the 'Jaw and al-Buraymi' covering the UAE borderland. In 1991, Oman was re-divided into five new regions and three new governorates. These new administrative entities were designed around regional capitals destined as conveyor belts between Muscat and their respective hinterlands. The former Jaw and al-Buraymi region was merged into the Dhahira region. Ibri, instead of al-Buraymi, became the new entity's capital. The result was that al-Buraymi became just one town among others in northern Oman, remaining semi-rural until the early 2000s, broadly neglected by the post-1970 centralised modernisation process. Beyond its fort and a souk (market place), tourist resources have remained notably underdeveloped in al-Buraymi. Two private higher education institutions (Buraymi University College and University of Buraymi) opened, but not until 2003 and 2010. Until the 2000s, few measures emerged aimed at avoiding the increasing extraversion of borderland regions like al-Buraymi towards the UAE. Central government's only noticeable action was the prevalence in cabinets of personalities from Northern Omani tribes, said to receive preferential treatment aimed at diverting them from UAE enticements (Peterson 2004, 10).

The contrast between the two cities after three decades of contiguous but separate development was summarised by Christopher Davidson:

> Throughout Zayed's reign as ruler of Abu Dhabi, an enormous amount of the emirate's wealth was channelled into Al-Ayn, transforming it into a modern commercial hub, a booming agricultural centre, and of course into the 'garden of the Gulf,' with many miles of beautiful parks and landscaped areas. In complete contrast, one can walk just a few hundred yards outside of Al-Ayn into the Omani-controlled and comparatively undeveloped town of Buraimi and perhaps imagine what Al-Ayn would have looked like today if it had not received such munificence.
>
> (2006, 51–52)

Unsurprisingly, the UAE's symbolic, cultural and economic influence on northern Oman has strengthened since the 1970s. The UAE, which is Oman's first economic partner, has been a geographical sales outlet for Omani exporting companies, a supply source for petty illicit dealing and private trade, and a place where Omanis, especially from the north, go at weekends: for fun or commercial deals. More particularly, Northern Oman's lack of employment opportunities, aside from local branches of government and public sector administrations, explains why the UAE's job market has remained attractive to many low-skilled Omanis. Omani nationals, many from border regions, have been highly represented in the UAE Armed Forces until the early twenty-first century. Kechichian estimates that over 30,000 Omanis already lived and worked in the UAE in the mid-1970s, 85% in the military (1995, 78). This consistently increased until the early 2000s. A northern Omani civil servant highlights the underlying political issues:

> In [some towns], 80% of the population is dependent on the UAE; people are working on construction projects in [the UAE]. Local sheikhs are related to Sharjah and Dubai families, by marriage, family agreements or business. [The towns are] completely turned towards the UAE.
>
> (personal interview, Muscat, 14 February 2005)

As noted elsewhere in this book, borders create differentials in prices, revenues, regulations and thus opportunities for locals. al-Buraymi has been a dormitory town for al-Ayn since the 1990s. By the mid-2000s, around 40,000 al-Buraymi residents commuted daily to work in al-Ayn (*Gulf News* 14 June 2008). Nationally, rents and salaries have been much lower in Oman than the UAE; the border situation reflects this. House rents have been 3–5 times lower and salaries 2–3 times lower, in al-Buraymi than in al-Ayn. Thus, many expatriates, particularly from the Indian subcontinent, Egypt and the Philippines, lived in al-Buraymi but worked in al-Ayn. Meanwhile, many al-Ayn inhabitants have gone to al-Buraymi for cheaper everyday goods.

Similarly, regulatory differences around hiring foreign domestic workers and issuing work visas (both allegedly much quicker in the UAE) mean that recruitment agencies in al-Ayn and al-Buraymi have created a human-trafficking economy, bringing housemaids across the border from the UAE into Oman, instead of hiring them directly from say the Philippines (https://wikileaks.org/plusd/cables/07MUSCAT206_a.html).

The two cities' contrasting trajectories is best illustrated by their demographic evolutions. The total oasis population was around 25,000 in 1960, probably equally distributed across the border. By 1993, and Oman's first census, Al-Ayn's urban population had quadrupled compared to 1975 and was four times higher than al-Buraymi's. The gap increased during the following two decades, al-Buraymi's population doubling during 1993–2016, while al-Ayn's expanded over two-and-a-half times. al-Ayn is now five times bigger than al-Buraymi.

For long, this extraversion of its borderlands towards the UAE was probably not problematic for Oman's government, who considered that the many Omanis employed in the UAE helped alleviate the financial burden of developing the sultanate's peripheral regions. However, the early 2000s saw a dramatic turn in UAE–Oman relations, in turn impacting the relationship between al-Ayn and al-Buraymi.

The UAE and Oman turn their backs on each other

At the very time, the two countries reached agreement delineating the border and observers were considering that 'both sides concluded that their destinies were inevitably and permanently tied' (al-Sayegh 2002, 136), the issue of the symbolic and physical separation from the (br)other returned to prominence. In 2002, the UAE announced that it was installing a fence along the UAE–Oman border. In spring 2003, Oman introduced an exit toll on crossing the UAE border. By 2004, a 12-foot barbed-wire border fence divided the al-Ayn/al-Buraymi oasis, and the UAE severely tightened border control on vehicles and pedestrians. On the Omani side, al-Buraymi became a no man's land, with checkpoints established on roads 40 km inland from al-Buraymi to Muscat and Ibri. Residents of al-Buraymi who were not nationals of a Gulf Cooperation Council (GCC) country were restricted to this 40-km radius (*Gulf News* 29 June 2004).

Abu Dhabi justified materialising its separation from Oman, over three decades after independence, by claiming that Oman was doing insufficient to fight illegal immigrants from Pakistan and Afghanistan passing through the Gulf of Oman and Sultanate to find jobs in the UAE. In fact, the UAE and Oman jointly launched a three-day co-ordinated crackdown in July 2004 in and around al-Buraymi, arresting 1,000 illegal immigrants (*Gulf News* 28 July 2004). However, the border's militarisation and cooling relations reflect more structural political and economic challenges these two young regimes faced.

New challenges impacting relations with the brother

By the early twenty-first century, both countries started facing domestic challenges calling into question the socio-political order established in the 1970s. In Oman, the regime's *état de grâce* was ending. Given Oman's limited hydrocarbon resources compared to its neighbours, a long-term economic program (*Oman 2020: Vision for Oman's Economy*) emerged in 1995, promoting economic diversification and nationalisation for private-sector jobs. However, these policies had limited results, not preventing growing inequalities, endemic unemployment and poverty, especially outside the capital. This situation was already reviving frustration and mutual prejudice, re-polarising society. This was occurring both socio-economically, through clientelism and favouritism, and symbolically via competitive bidding in declarations of Omanity along ethno-linguistic and tribal lines. These dynamics suggested that growing sectors of society, particular among the young, were becoming reluctant to support a system from which they felt politically and economically excluded.

The UAE was not facing such challenges. However, with Abu Dhabi by far the largest contributor to the UAE federal budget, and its oil reserves representing 95% of the UAE's total, questions surrounding the approaching end of Zayed Al Nahyan's reign were becoming crucial. As Davidson explains, the UAE regime's legitimacy was intimately linked to the person of Zayed Al Nahyan, 'father of the nation', embodying the UAE welfare state and financial redistribution to the citizens and among the different emirates (2005, 73).

Among other questions concerning the authorities was the nationals/expatriates imbalance, and the alleged impact of demographics on Emirati identity, culture and national security. In April 2008, Dubai's Police Commander-in-Chief warned a conference on national identity that the Emiratis, a minority in their own country, could induce regime collapse. Thus, questions around the succession to Zayed bin Sultan, Abu Dhabi's political and economic pre-eminence within the federation, and UAE national identity, were all interrelated, especially when dealing with the Omani neighbour and brother.

2003 and 2004 also witnessed the political ascension of Mohammed, Zayed's third son. His appointment as Abu Dhabi's Deputy Crown Prince was followed by his becoming Deputy Supreme Commander of the UAE Armed Forces in January 2005, after his father died. Mohammed quickly started replacing Omanis serving in the federal security forces, perceiving their presence becoming problematic if UAE–Oman tensions increased. The impact of this policy change was felt immediately in Oman: many low-skilled and unskilled Omanis, living in the UAE and sending remittances to their relatives at home, returned to compete for jobs with home Omanis in an already saturated job market.

Physical demarcation with the (br)other

In February 2005, Sultan Qaboos' first annual country-wide tour since Shaykh Zayed's death took place in the provinces bordering the UAE. This provided the setting for long-awaited development projects in housing, road infrastructures and social services, marking a drastic change in Omani government policies towards the border regions. Due to Northern Oman's growing economic extraversion towards UAE development, the Sultan worried about these regions breaking away in favour of Saudi Arabia in the 1950s, and nowadays looking enviously across the border. Oman's Seventh Five-Year Plan (2006–10) illustrated this re-evaluation of the border region. Additional to 108 million Omani Rials (OMR) (US$281 million) specifically allocated in the annual budget, the Plan put aside a further OMR 50 million (US$130 million) for public investment (transport infrastructure, schools, etc.) alongside private-sector tax concessions to encourage industrial and tourism projects in these provinces.

In September 2006, in a new sudden unexpected change by the UAE authorities, non-GCC nationals holding Omani residence visas, suspected of working and/or living in the UAE, could no longer enter the UAE without purchasing a 100 UAE Dirham (AED) entry-visa. This measure, initially implemented at some border checkpoints outside the cities, did not affect expatriates with UAE residency living in al-Buraymi, who could move freely (*Gulf News* 29 September 2006). It was also decided that only GCC nationals could pass through the city centre's border-point (al-Madeef), while non-GCC nationals (even holding UAE residence) must use another border point north of the cities. Again, the Omani authorities considered that these decisions were taken without prior notice and badly affected al-Buraymi's economic situation. The Sultan of Oman issued a royal decree in October 2006 establishing a new al-Buraymi governorate comprising three border provinces. Several ministers rushed to al-Buraymi to announce new infrastructure investments. Unsurprisingly, al-Buraymi residents received these short-term belated decisions sceptically, perceiving them as too late to offset their city's Kafkaesque situation: with checkpoints now 40 km inland since 2004, expatriates holding Omani tourist visas cannot enter al-Buraymi without having their visas cancelled (*Times of Oman* 8 February 2016).

In summer 2008, the border-point for non-GCC nationals (al-Hili) temporarily closed, requiring them to use a new one, 20 km south (Khatam al-Shakla). As described by a local businessman, overnight al-Buraymi shops were 'no longer local' (*The National* 14 July 2008) for al-Ayn residents, the vast majority of al-Buraymi shops' customers, who suddenly had to drive 40 km for daily shopping. Moreover, a new UAE visa rule was implemented. Until July 2008, people whose UAE visit visa had expired could make short trips to neighbouring countries like Oman and Qatar to renew it. The new rules meant that this was impermissible for all but 33 exempt nationalities. Visitors from elsewhere had to return to their home countries for at least a month before getting another visa. Thousands were stuck outside the country (many in al-Buraymi) after attempting visa runs, caught unawares by the new law (*The National* 6 June 2008). These various UAE policies meant that the number of expatriates based in al-Buraymi dropped significantly after around 2005 (Table 22.1).

The Arab spring: security first

Tensions peaked in late 2010 when Omani security officials claimed to have uncovered a spy ring affiliated to the UAE security service targeting the Omani regime, and had 20 Omani officials arrested (*The Financial Times* 30 January 2011). The UAE denied the allegations, but other Gulf monarchies considered the crisis serious enough to require mediation by the Emir of Kuwait in March 2011.

This revelation came a few weeks before the Sultanate experienced its most widespread popular protests since the Dhofar war ended. These were part of the Arab Spring. Major popular mobilisations occurred in the northern town of Sohar, where two protesters died (Valeri 2015). Demonstrations and rallies also produced clashes with the security forces in al-Buraymi. After riot police cleared these peaceful gatherings, a creeping militarisation of the territory occurred with Sohar and al-Buraymi transformed into fortified cities in April and May 2011: with the army's deployment in town, and police controls and check-points on roads to the UAE steadily increasing since.

Throughout this troubled period, and more generally since independence, there have never been contacts and relations at local level for solving problems, let alone attempts at establishing local channels of institutional co-operation. This reflects the authoritarian and hyper-centralised decision-making processes of both regimes, where all initiatives come from central government. Far from being a decentralising measure, the creation of a governorate in al-Buraymi in 2006, where the governor is sultan-appointed and acts under the Interior Ministry's strict supervision, was clearly intended to re-assert central control on the border region. When visiting al-Buraymi in May 2008, Oman's US Ambassador had opportunity to experience the governor's total absence of latitude:

> Of interest is what was not discussed … In submitting the request for the appointment, the Ministry of Foreign Affairs agreed to schedule it on condition that the Ambassador not raise any substantive issues with the governor. The meeting itself was devoid of genuine substance, an indication that the sheikh was following strict [Ministry of Foreign Affairs] orders to keep discussion to the weather and the immaculate condition of Buraimi's streets.
>
> (https://wikileaks.org/plusd/cables/08MUSCAT359_a.html)

In October 2011, the sultan established municipal councils in all governorates. However, this attempt to show attentiveness to popular aspirations for participation in decision-making had limited impact. Members are elected by universal suffrage for four-year renewable terms, but sit alongside ex-officio members representing ministries, and are chaired by sultan-appointed governors. They have only advisory powers, providing opinions and recommendations only about developing municipal services. Across the border, the ruler's representative in al-Ayn belongs to Al Nahyan's ruling family and Abu Dhabi's inner circles of power, suggesting that no intention of cross-border co-operation has ever been present among local authorities.

In October 2013, in a rare co-ordinated move, the two countries further hampered mobility by introducing a new border-crossing tax for expatriates. All non-GCC nationals would pay AED35 for UAE entry and, if lacking Omani residence visas, OMR5 to enter al-Buraymi from Oman. This applies to expatriates with UAE residency permits living in al-Buraymi and working in al-Ayn, and expatriate students living in al-Buraymi and attending school in al-Ayn; Abu Dhabi had previously issued unlimited visas. In October 2015, the UAE tightened rules further by requiring all citizens from countries like India, Pakistan and the Philippines – regardless of professional status or even visa status in Oman – to apply for a visa to travel to the UAE (*Times of Oman* 12 October 2015).

Underpinning these new regulations were obviously both capitals' security concerns. Following the Arab Spring, Muscat and Abu Dhabi presume the other a cause of political destabilisation: for the UAE, a porous border supposedly produces more smuggling and illegal-migrant entry; for Oman, proliferating Islamist activism and/or spying networks. Since 2011, Omani officials and the national media have repeatedly accused protesters of being foreign influenced to discredit them and their demands. Illustrative is the sentencing of a Salafi Parliament Member in 2014 to three-years custody charged with illegal gathering and undermining the state's status and prestige. In 2014, it was estimated that 200 Omanis, mainly native from northern Oman, including al-Buraymi, had joined Sunni military groups in Syria. Omani intelligence and internal security unambiguously referred to possible support for Salafi networks in Oman from UAE individuals. Alongside these security concerns, these sealing measures aim to increase competition between the two cities, forcing them to turn their backs on each other and develop independently and at the other's expense.

Conclusion

Since the 1970s, these border cities have followed divergent, often opposing, trajectories; reflecting how the UAE and Oman have diverged. Despite shared history and the undisputable feeling of belonging together long presiding over relations between inhabitants of the two cities, both regimes have turn their backs on each other, for legitimacy and security purposes: translating in al-Ayn/al-Buraymi in a complete absence of local cross-border co-operation and growing asymmetries – in size, wealth, attractiveness and political influence on national arenas. The fence across the conurbation in the 2000s and accompanying materialisation of separation from the br(other) have played key roles in increasing this. Even more than allowing the UAE (and al-Ayn) to project its power over Oman (and al-Buraymi), physical demarcation from the neighbour, and the accompanying devices and measure, must be understood as part of the legitimation strategies used by Abu Dhabi's ruling family for the last decade. For a small state like the UAE, branding itself as the epitomy of a globalised country, the fence with Oman has been critical in justifying its existence as a nation – after the demise of the 'father of the nation' Sheikh Zayed; and as a federal state, distinct from its Omani brother. This question has been crucial for the Omani regime too, considering the structural economic challenges the Sultanate has faced from the early twenty-first century, the increasing dependency of many northern provinces on the UAE and the ensuing political consequences, of which the Arab Spring protests have been just one illustration. No wonder these issues have crystallised in the peripheral border cities of al-Ayn and al-Buraymi, where questions of political allegiance and belonging to the national community have been critical since the mid-twentieth century.

References

Allen, C., and Rigsbee, L. (2000) *Oman under Qaboos: From Coup to Constitution, 1970–1996*, London: Frank Cass.

Al-Sayegh, F. (2002) 'The UAE and Oman: Opportunities and Challenges in the Twentieth-First Century', *Middle East Policy*, 9/3, 124–37.

Buursink, J. (2001) 'The Binational Reality of Border-Crossing Cities', *GeoJournal*, 54/1, 7–19.

Davidson, C. M. (2009) *Abu Dhabi. Oil and Beyond*, London: Hurst.

Davidson, C. M. (2006) 'After Shaikh Zayed: The Politics of Succession in Abu Dhabi and the UAE', *Middle East Policy*, 13/1, 42–60.

Davidson, C. M. (2005), *The United Arab Emirates. A Study in Survival*, Boulder, CO: Lynne Rienner.

Gervais, V. (2011), 'Du pétrole à l'armée. Les stratégies de construction de l'état aux Emirats Arabes Unis', *Etudes de l'IRSEM* 8.

Kechichian, J. (1995) *Oman and the World: the Emergence of an Independent Foreign Policy*, Santa Monica: Rand.

Lorimer, J. G. (ed.) (1986) *Gazetteer of the Persian Gulf, Oman, and Central Arabia*, Gerrards Cross: Archives Editions (1st ed. 1915).

Mikhailova, E. V. (2015) 'Border Tourism on the Russian-Chinese Border', *Journal of Siberian Federal University. Humanities and Social Sciences*, 8, 437–51.

Morton, M. Q. (2013) *Buraimi. The Struggle for Power, Influence and Oil in Arabia*, London: I. B. Tauris.

Peterson, J. E. (2005) 'The Emergence of Post-Traditional Oman', *Durham Middle East Papers* 78. Durham Middle East papers no. 78 (Sir William Luce Fellowship Paper no. 5). University of Durham, Institute for Middle Eastern and Islamic Studies, Durham. http://dro.dur.ac.uk/92/1/Peterson.pdf (accessed March 3, 2018)

Schofield, R. (2011) 'The Crystallisation of a Complex Territorial Dispute: Britain and the Saudi-Abu Dhabi Borderland, 1966–71', *Journal of Arabian Studies*, 1(1), 27–51.

Takriti, A. (2013) *Monsoon Revolution. Republicans, Sultans, and Empires in Oman, 1965–1976*, Oxford: Oxford University Press.

Valeri, M. (2015) 'The Suhar Paradox. Social and Political Mobilisations in the Sultanate of Oman since 2011,' *Arabian Humanities* 4 [Online]. http://journals. openedition.org/cy/2828 ; DOI : 10.4000/cy.2828

Zahlan, R. S. (1989) *The Making of the Modern Gulf States: Kuwait, Bahrain, Qatar, the United Arab Emirates and Oman*, London: Unwyn Hyman.

Conclusion

Ekaterina Mikhailova and John Garrard

In June 2015, scholars from 10 countries and 3 continents gathered in Manchester for a conference on 'Twin Cities in Past and Present'. From the outset of our debates, a joint book on the various manifestations of twinning around the globe seemed to be the logical and necessary next step. The idea had a number of enthusiastic (and patient) contributors, thanks to whom the long journey of preparing a coherent edited volume has been completed. The team of authors and consequently regions they work on expanded over time, allowing us to fill important gaps. Alongside well-established work exploring European and North American twin cities, we have drawn attention to twins in South America, the Middle East, Africa and Asia. While some twins and areas have been the focus of well-established attention and debate, others have had little or no notice (certainly in English) as examples of twin cities. These include: Wervik–Wervicq Sud, French and Belgian Comines; Indian twin cities; Islamabad–Rawalpindi; Leticia–Tabatinga–Santa Rosa; Manchester–Salford; Newcastle–Gateshead; Hong Kong–Shenzhen and Blagoveshchensk–Heihe. And nothing has previously gathered all these twins and regions into one book.

Equally important, by suggesting five characteristics shared by both intranational and international twins, the book's introduction has created a possible framework for the cross-national and interdisciplinary analysis of what we argue to be the twin city family. With full attention to each case's unique features, contributors have unveiled enduring patterns among twin cities of dominance–subordination, tensions between inwardness and openness, ongoing formal and informal negotiations as well as interdependencies and persistence. Considering the book as an integral publication with 54 cases in 32 countries examined within one theoretical framework, we hope we have promoted the comparative approach to twin-city study. Every third chapter in the book focuses on two and more examples, thereby correlating with the recent call for overcoming the empirical slant in the twin-city literature (Joenniemi and Jańczak 2017, 423–4).

The book testifies both to the diversity of twin cities and the variety of contemporary twin-city research. Depending on our contributors' very mixed academic backgrounds and interests, some chapters have analysed economic life (Chapters 9, 12 and 20); others emphasise the human dimension of twin-city relationships (Chapters 19 and 21) or their architectural evolution (Chapters 13 and 15). Several chapters have used twin cities as laboratories for studying global events, trends and challenges (Chapter 8 on human rights and gender-transport accessibility in Pakistan, or Chapter 14 on the transformative influence that accession to the EU has had on Central European states). Many chapters dealing with cross-border twins underlined the divergent interests of national and local governments, and the problems and opportunities their mismatched agendas can bring for borderlanders (Chapters 16, 20 and 22). Some chapters point out that urban history provides a useful lens through which to understand how and why twin cities originate and develop (Chapters 1–6).

Studying historical examples of twin cities is one methodological contribution our authors have made. Three chapters exemplify twin cities that have ceased to exist: the frequently used example of Budapest (Chapter 4) alongside the far less well-known Nigerian instances of Lagos–Ikeja (Chapter 5), Kukawa and Maiduguri (Chapter 6). Although showcasing scenarios of dissolving twin cities (due to the destruction of one city, merger initiated at national level, followed by profound urban redevelopment, or simple metropolisation), these three chapters also testify to the persistence of twin cities – each of them revealing traces of former twins perceptible today, several decades after twin-city pair has disappeared (see Figure 4.2 with the Budapest transport corporation logo consisting of two parts 'Buda' and 'Pest' divided with a city crest as an example).

The other methodological novelty our book presents is emphasis on the dynamic character of twin-city relationships and their volatility. Chapter 19 offers a citizen-perception continuum model that allows for the drifting of town-twinning developments 'perhaps progressing, maybe regressing'. Chapter 20 suggests using a continuum of trust and mistrust as an indicative tool for understanding and predicting the changing balance between embracing opposing twinning. Four chapters highlight evolutionary stages in twin-city relationships, pointing to how far their ups and downs (Chapters 3, 15, 20 and 22) are conditional on external factors – political and economic (in)stability in the country and the world, technological advances, administrative reforms, not to mention wars and natural disasters.

Furthermore, the book clarifies and supplements the range of instrumental attitudes to twinning (Mikhailova and Wu 2017). Some chapters present examples where cities themselves behave entrepreneurially, willingly exploiting the twin-city brand for place promotion, hoping to jointly improve their economic fortunes (Chapters 1, 3 and 9). Two focus on twin-city initiatives launched by central governments, either to improve bilateral relations (Chapter 17) or resolve domestic challenges like rapid urban development in India (Chapter 7). Both show that governmental support for adopting a twin-city model enhances the status of the intercity relationship and sometimes materialises in financial, legal and institutional assistance to selected cities.

Two more chapters contextualise twinning within colonial/post-colonial history. Chapter 6 vividly shows how the British colonial government, wishing to avoid intermixing with local population, preferred establishing a European reservation next to an African settlement of Maiduguri, thus creating duplicated twin cities. Interestingly, this chapter (in its treatment of West and East Towns in Kukawa) also suggests similar urges for social stratification and security among elites (Sheikhs and their servants) within Nigerian feudal society. Chapter 9 explores how twin cities have been employed for colony-absorption purposes by the Chinese government, and how such staged twinning prevents full-hearted integration.

While some twin cities have evolved largely without supervision from parent states/regions (Chapters 1–3, 10, 15 and 21), others (predominantly cross-border examples) could exploit opportunities granted by state and regional authorities. Notable examples of governmental measures aimed at twin cities' prosperity mentioned in the book include maquiladoras on the US–Mexican border (Chapter 11), special economic zones usually with simplified visa regime in Chinese cities on the border with Russia (Chapter 20) and the targeted financing of project activities in twin cities on EU's internal (Chapters 13, 14 and 16) and external borders (Chapters 14, 16, 18 and 19). Although these measures produced varied results, they always successfully mobilised cross-border networks and local economies by expanding the client base of local enterprises. Thereby they created new jobs or sustained ones already existent, and increased the number and scale of transactions. Since these supportive practices appeared somewhat randomly among our twins, some transfer of policy initiatives and experience across borders might be in order. This and the fact that twin cities continue emerging suggests the need for more detailed analysis of accumulated experience in supervising twin-city development and for further experiments in this sphere.

We presume that future twin-city research will be more comprehensive and contextualised. On the one hand, there is a deficit in analysing both official and unofficial practices and discourses, top-down and bottom-up initiatives, urban identities and rituals. Such all-round approaches require applying mixed methods (particularly interviewing, historical, statistical and cartographic analysis) and provide the best way towards rendering a more objective picture. However, as mentioned in the introduction, twin cities do not exist in isolation; they must be contextualised within full accounts of ongoing urbanisation and globalisation, alongside technological and social advances in the urban environment adequate to the needs of modern society.

It is also clear that twin-city research, particularly that concerned with border cities, will need to explore the consequences of the ongoing and multi-headed migration crisis: within the Middle East and from that region into and across Europe; from Myanmar into Bangladesh; from and across Mexico into the USA. This is already impacting massively the internal politics, and external relations, of the many countries affected or who feel themselves to be. It is already impacting their borders, President Trump's 'wall' being just one manifestation, and will certainly profoundly affect border cities in many of the areas this book has been concerned to explore.

Finally, another promising avenue for research lies in comparing twin cities with twinned cities. Currently, two bodies of literature briefly cross-reference each other warning the reader that their research unit is different (Zelinsky 1991, 3–4). However, as the homonymy of the terms suggests, they share extensive commonalities and potentialities. Pondering this direction has already started with articles drawing on both twinned cities and twin cities (Joenniemi and Jańczak 2017; Langenohl 2017; Mikhailova and Wu 2017). This may well become more common.

References

Joenniemi, P. and Jańczak, J. (2017) 'Theorizing Town Twinning—Towards a Global Perspective', *Journal of Borderlands Studies*, 32 (4), 423–428.

Langenohl, A. (2017) 'The Merits of Reciprocity: Small-Town Twinning in the Wake of the Second World War', *Journal of Borderlands Studies*, 32 (4), 557–576.

Mikhailova, E. and Wu, C. T. (2017), 'Ersatz Twin City Formation? The Case of Blagoveshchensk and Heihe', *Journal of Borderlands Studies*, 32 (4), 513–533.

Zelinsky, W. (1991) 'The Twinning of the World: Sister Cities in Geographic and Historical Perspective', *Annals of the Association of American Geographers*, 81 (1), 1–31.

Index